[redacted]

MAGILL'S SURVEY OF SCIENCE

MAGILL'S SURVEY OF SCIENCE

PHYSICAL SCIENCE SERIES

Volume 6
2353-2796

The Standard Model—X-Ray Determination of Molecular Structure

Edited by
FRANK N. MAGILL

Consulting Editor
THOMAS A. TOMBRELLO
California Institute of Technology

SALEM PRESS
Pasadena, California   Englewood Cliffs, New Jersey

∞ The paper used in these volumes conforms to the American National Standard for Permanence of Paper for Printed Library Materials, Z39.48-1984.

Library of Congress Cataloging-in-Publication Data
Magill's survey of science. Physical science series/edited by Frank N. Magill, consulting editor, Thomas A. Tombrello.

p. cm.

Includes bibliographical references and index.

1. Physical sciences. 2. Computer science. I. Magill, Frank Northen, 1907- .

Q158.5.M34 1992 91-32962
500.2—dc20 CIP
ISBN 0-89356-618-7 (set)
ISBN 0-89356-649-7 (volume 6)

PRINTED IN THE UNITED STATES OF AMERICA

CONTENTS

MAGILL'S SURVEY OF SCIENCE

THE STANDARD MODEL

Type of physical science: Elementary particle (high-energy) physics
Field of study: Systematics

The standard model is an elegant and concise theory of the fundamental constituents of matter and the forces that govern them. It is the outgrowth of a number of crucial experiments starting in the late 1930's, the development of accelerators of increasing energy, and a parallel development of the theoretical synthesis of experimental results.

Principal terms

BARYONS: a subset of a class of subatomic particles called hadrons, baryons are combinations of three quarks; the lightest baryon is the proton; the neutron is also a baryon

COLOR CHARGE: an attribute, analogous to electrical charge, which is carried by quarks and gluons; fundamental particles carry one kind of electrical charge, which can be positive or negative; there are three kinds of color charge

FERMIONS: a class of particles, including leptons and quarks, that obey the Pauli exclusion principle

GAUGE BOSONS: the particles that mediate the forces; the electromagnetic force is mediated by the photon, the weak force by the Z^0, W^+, and W^-, and the strong force by the gluons

HADRONS: subatomic particles that are made of quarks

LEPTONS: the fundamental building blocks of matter, along with quarks; the leptons include the electron, the muon, the tau, and their associated neutrinos

MESONS: a subset of a class of subatomic particles called hadrons; combinations of quarks and antiquarks

QUARKS: six fundamental building blocks of matter, along with leptons: the up, down, charmed, strange, top, and bottom quarks

STRONG FORCE: the force of attraction that binds the protons and neutrons together in a nucleus because of the strong force between the quarks making up the neutron and proton

WEAK FORCE: the force responsible for such processes as radioactivity, the decay of the neutron, and the processes that fuel the sun

Overview

The theory known as the "standard model" specifies which subatomic particles are truly elementary, or fundamental, and it describes the forces that are mediated between these particles. After decades of experimental work, studying the ways in which subatomic particles scatter off one another, the conclusions are the following:

The truly elementary particles can be classified into two general groups—leptons and quarks, and gauge bosons. The leptons and quarks are the particles of which matter is made, and the gauge bosons are responsible for the forces. Leptons and quarks belong to a class of particles called fermions. Fermions obey the Pauli exclusion principle of quantum mechanics, which states that no two fermions can exist in the same quantum-mechanical state. This principle accounts for the chemical properties of the elements and is the cornerstone of the periodic table of the elements. Leptons and quarks are further divided into three families. Each family consists of two leptons and two quarks.

The first family of leptons and quarks consists of two leptons, the electron and the neutrino associated with the electron, and two quarks, the up quark and the down quark. The existence of the electron was known as early as 1897. The neutrino is assumed to be a massless, electrically neutral particle that interacts extremely rarely with matter. According to special relativity, since neutrinos have zero mass, they always move at the speed of light. Neutrinos are elusive and copious. Theoretical estimates place the number of neutrinos per cubic centimeter at between one hundred and one thousand. Many of these neutrinos come from nuclear reactions taking place in the sun. Neutrinos are able to pass through matter so readily that the number of solar neutrinos passing through Earth during the day is very nearly equal to the number of solar neutrinos passing through Earth during the night, even though those neutrinos first have to punch through the entire Earth. Neutrinos have no noticeable effect on Earth, but they appear to be essential in the makeup of matter.

The up quark and down quark have peculiar properties, one of which is their electrical charge. Except for quarks, all electrically charged particles carry an integer multiple of the absolute value of the electrical charge on the electron, e. The sign of the charge is either positive or negative. Quarks, on the other hand, carry a fractional charge. The charge of the up quark is $+2e/3$ and the charge of the down quark is $-e/3$. Another peculiar property of quarks is that they do not appear to exist as free particles. A quark is always bound with one or more quarks to form a subatomic particle. For example, a proton is composed of two up quarks and one down quark, giving it a total electrical charge of $+2e/3 + 2e/3 - e/3 = +e$, or the familiar result that the magnitudes of the proton and electron charge are the same. A neutron is composed of one up quark and two down quarks, giving a total electrical charge of $+ 2e/3 - e/3 - e/3 = 0$; that is, the neutron is electrically neutral.

The first family of quarks and leptons accounts for normal matter. The other two families account for the subatomic particles found in cosmic rays or those produced in accelerators.

The second family of fermions consists of the muon and its associated neutrino, the muon neutrino, and the charmed quark (charge $= +2e/3$) and the strange quark (charge $= -e/3$). The third family of fermions consists of the tau lepton and its associated neutrino, and the top quark (charge $= +2e/3$) and the bottom quark (charge $= -e/3$). The electron, muon, and tau have the same electrical charge. Like the electron neutrino, the muon neutrino and the tau neutrino are assumed to be electrically neutral and massless, but they are distinct from each other in their interactions with the other fundamental particles.

Four forces govern the way leptons and quarks interact: gravitation, electromagnetism, and the weak and the strong forces. Particle interactions can involve the collision of two particles, which causes the simple transfer of energy and momentum between the two particles. Another possibility is that particles can be created as the result of the collision. An interaction can affect a particle in isolation, causing it to decay spontaneously into other particles.

Gravity has not been studied on the scale of elementary particles. Because of the extremely small masses of the elementary particles, the effects of the other three forces ovewhelm gravity. The force of electromagnetism is something that human beings confront in almost all aspects of everyday existence. There are obvious examples such as electric lights and motors, but there are also subtle effects. For example, electromagnetism explains why water can exist as a liquid or a solid (ice) or a gas (vapor). The weak force is responsible for radioactivity and is at the heart of the physical processes that fuel the sun. The strong force is responsible for the existence of nuclei. The nucleus of an atom is made up of protons and neutrons. It is well known that electrical charges of like sign repel each other and electrical charges of unlike sign attract each other. The neutron is electrically neutral, but protons carry an electrical charge. The strong force explains why nuclei can exist. At small distances, the attractive strong force overcomes the repulsive electromagnetic forces between the protons. If the strengths of the four forces are compared at a distance that is comparable to the size of a proton (10^{-13} centimeters), the results show that relative to the strong force, electromagnetism is 10^2 times weaker, the weak force is 10^{13} times weaker, and gravity is 10^{38} times weaker.

Within the standard model, quarks can interact with other quarks by means of the strong, electromagnetic, or weak force. The leptons can interact with each other or with quarks by means of the electromagnetic (only if the leptons are charged) or the weak force.

According to quantum mechanics and relativity, the force between particles is mediated by means of "force particles," also known as gauge bosons, which carry energy and momentum. The electromagnetic interaction between two particles carrying an electrical charge takes place when one particle emits a photon and the other particle absorbs the photon, or vice versa. The quantum of light is the gauge boson of electromagnetism. The theory of electromagnetism that assimilates James Clerk Maxwell's equations, quantum mechanics, and relativity is called quantum electrodynamics (QED), which is the most successful physical theory and has made predictions of incredible accuracy. This accuracy was made possible by a method for doing QED calculations developed by Richard P. Feynman. For his work in QED, Feynman, along with Shin'ichirō Tomonaga and Julian Seymour Schwinger, won the 1965 Nobel Prize in Physics. QED calculations incorporate the idea that electrically charged particles interact by exchanging photons.

Over the years, theorists attempted to model the theory of the weak interactions after QED. The result, during the 1960's, was the GSW theory (named for its inventors, Sheldon L. Glashow, Abdus Salam, and Steven Weinberg, who won the 1979 Nobel Prize in Physics), which unifies the weak and electromagnetic interactions into a single force, the electroweak force. According to the GSW theory, the bosons responsible for the electroweak force are the photon and three other bosons, called Z^0, W^+, and W^-. The photon has zero mass, but the other bosons were predicted to have masses about one hundred times the size of the proton mass. Just as particles carrying an electrical charge can emit or absorb photons, particles carrying a "weak charge" can emit or absorb the Z^0, W^+, or W^-.

The theory of the strong force is also modeled after QED. The result is the theory known as quantum chromodynamics (QCD). QED, the GSW theory, and QCD are known as gauge theories. Analogous to the electrical charge or the weak charge, there is another kind of charge, which is whimsically called the color charge. Unlike the electrical charge, of which there is only one kind (which appears as positive or negative), there are three kinds of color charge that are arbitrarily called red, blue, and green (which can also appear as red and antired, blue and antiblue, and green and antigreen). Only quarks carry a color charge. Because of their color charge, quarks can interact by exchanging gluons, of which there are eight types. Like the photon, gluons are massless. Unlike the photon, which does not carry an electrical charge, gluons carry a color charge.

The six leptons and six quarks are the fundamental constituents of matter. In addition, each of the leptons and quarks has a corresponding antiparticle. Each quark also comes in one of three colors.

The standard model does not limit the number of families, but experiments performed in 1989 seem to indicate that there are only three families of particles. Many experiments in elementary particle physics have been performed, and all of their results are consistent with the predictions of the standard model.

Applications

The standard model has the essential feature of a good theory: It makes definite predictions that suggest experiments to test the model. It provides a framework within which experimental and theoretical progress is made. The next real advancement in the knowledge of elementary processes very likely will be made when deviations from the standard model are discovered.

The standard model makes a definite statement about which particles are fundamental, which, in turn, suggests experiments to test this assumption. Determining whether a particle is fundamental is equivalent to asking whether a particle behaves as if it is made of more elementary constituents. To answer such a question requires experimentation, which involves probing at scales of distance that are exceedingly small. This requires the use of a very special type of microscope.

An optical microscope uses visible light to look at tiny objects. The wavelength of visible light is about 400 nanometers (a nanometer is one-billionth of a meter). Objects that are about the size of the wavelength of visible light cannot be studied with an optical microscope, since the resolution is insufficient. To understand this in more familiar terms, consider a tank of water in which water waves are generated. An object placed in the tank will disturb the waves in a noticeable way if the object is larger than the wavelength, the distance from crest to crest, of the water waves. If the object is very small compared to this wavelength, it will have no measurable effect on the water waves. To examine an object that is smaller than the wavelength of visible light requires the use of an electron microscope, in which electrons are accelerated and scatter off of the object, revealing its details, in much the same way that visible light scatters off of an object being studied in an optical microscope. According to quantum mechanics, particles behave like waves. The wavelength associated with a particle is related inversely to the momentum of the particle. The momentum of a particle is equal to its mass times its velocity, so increasing the momentum of a particle will decrease its wavelength. The microscopes used to probe the subatomic world are particle accelerators. The scales of distance that can be probed with current technologies are 10^{-16} centimeters, which is about one-thousandth of the diameter of a proton (10^{-13} centimeters), which in turn is 100,000 times smaller than the diameter of an atom (10^{-8} centimeters). In the future, higher-energy accelerators will allow probing at even smaller distances. The experimental evidence accumulated to date indicates that leptons and quarks have no substructure. Future experiments may very well overturn this conclusion.

The standard model predicts the way in which particles decay. A good example is the decay of a muon, which lives for about 2.2 microseconds (a microsecond is one-millionth of a second) and then decays into an electron, a muon neutrino, and an electron antineutrino. The standard model tells how to calculate the muon lifetime and also explains why the muon does not decay by other modes that would otherwise be allowed by the conservation of energy or electrical charge (in other words, the muon does not decay into an electron plus a photon).

The standard model explains the pattern of more than one hundred known subatomic particles. There exists a class of subatomic particles, called baryons, that consist of three quarks. The proton and the neutron are two examples of baryons. Many other examples of baryons exist in nature. The proton is the lightest baryon, and all other baryons eventually decay, ending up as protons. Baryons also belong to the fermion family. A quark-antiquark combination is called a meson, examples of which include the pi-meson and the K-meson. Mesons and baryons together form a class of particles called hadrons, which can interact with each other by means of a force that at short distances is greater than the electrical force. This force is the residual strong force between the quarks that form hadrons.

The GSW theory of electroweak interactions has a structure that was motivated by the remarkable success of QED. Similarly, the formulation of QCD was guided by the electroweak theory and QED. Theorists are attempting to unify the four forces of nature, in much the same way that electromagnetism and the weak forces were unified. The standard model will undoubtedly be an important stepping-stone in the achievement of that goal.

Context

The lightest lepton, the electron, was discovered in 1897 by the English physicist Sir Joseph John Thomson, who was awarded the 1906 Nobel Prize in Physics for his work. The neutrino was postulated in 1930, by the physicist Wolfgang Pauli, to account for the data on the decays of radioactive nuclei. The interactions of neutrinos with matter were first observed in 1955 by physicists Clyde L. Cowan and Frederick Reines at the Savannah River reactor. The muon was discovered in 1937 by Seth Neddermeyer and Carl David Anderson. That the neutrinos associated with the electron and muon were distinct was the result of an experiment carried out at the Brookhaven National Laboratory on Long Island, New York, in 1962, for which the experimenters Leon M. Lederman, Melvin Schwartz, and Jack Steinberger won the 1988 Nobel Prize in Physics. The tau lepton was discovered in 1975 at the Stanford Linear Accelerator Center (SLAC) in California.

The existence of quarks was postulated, independently, by California Institute of Technology theorists Murray Gell-Mann (winner of the 1969 Nobel Prize in Physics) and George Zweig to account for the plethora of subatomic particles being discovered in the late 1960's in much the same way that the model of atoms made up of electrons, protons, and neutrons accounted for the periodic table of the elements. At first, three types of quarks were hypothesized, and these were eventually called up, down, and strange quarks. Experiments that looked for isolated quarks gave null results. Quarks were viewed as a convenient mathematical description with little basis in physical reality until a number of crucial experiments were performed at the SLAC near San Francisco in the late 1960's and early 1970's. Electrons were scattered off of protons and neutrons. When analyzed, the results showed that the proton and the neutron were made of constituents that had the properties of quarks. The experiment team leaders, two Americans, Jerome I. Friedman and Henry W. Kendell, and a Canadian, Richard E. Taylor, were awarded the 1990 Nobel Prize in Physics for this work.

The strange quark was postulated as part of the original quark model to account for the subatomic particles that were known at that time. In 1974, a new type of subatomic particle was discovered almost simultaneously at SLAC by Burton Richter and coworkers and at Brookhaven National Laboratory by Samuel C. C. Ting. Richter and Ting shared the 1976 Nobel Prize in Physics for this work. This new particle could be explained only by postulating the existence of a fourth quark, which was called the charmed quark. The beauty quark was discovered in 1978 at Fermilab, near Chicago, by Lederman and coworkers.

In 1983, the Z^0, W^+, and W^- bosons were discovered at the European Laboratory for Particle Physics (CERN) in Geneva, Switzerland, resulting in Carlo Rubbia and Simon van der Meer being awarded the 1984 Nobel Prize in Physics. Many other crucial experiments were performed that lent credence to the standard model.

Bibliography

Calder, Nigel. *The Key to the Universe.* New York: The Viking Press, 1977. This book includes a treatment of the important building blocks of the standard model, along with a description of the experimental tools.

Carrigan, Richard A., and W. Peter Trower, eds. *Particle Physics in the Cosmos.* New York: W. H. Freeman, 1989. A companion volume to *Particles and Forces at the Heart of Matter.*

____________. *Particles and Forces at the Heart of Matter.* New York: W. H. Freeman, 1990. This book, a collection of *Scientific American* articles, is a companion volume to *Particle Physics in the Cosmos.* The articles discuss the development of the theory of elementary particles, tools, crucial experiments, and the connection with the knowledge of the cosmos.

Close, Frank E. *The Cosmic Onion.* New York: American Institute of Physics, 1986. A comprehensive and clearly written account of the standard model that is intended for the informed layperson. Close, a practicing theoretical physicist, is known for his ability to present abstract ideas to a general audience.

Close, Frank E., Michael Marten, and Christine Sutton. *The Particle Explosion.* New York: Oxford University Press, 1987. A very well-written and illustrated text.

Crease, Robert P., and Charles C. Mann. *The Second Creation.* New York: Macmillan, 1986. This is a fascinating and vivid account of the history of modern physics, leading to the evolution of the standard model. The authors bring this history to life by stressing the human side of this development.

Davis, Paul, ed. *The New Physics.* Cambridge, England: Cambridge University Press, 1989. A collection of well-written articles on the latest developments in physics, including the standard model.

Glashow, Sheldon L., with Ben Bova. *Interactions.* New York: Warner Books, 1988. Glashow (the "G" of the GSW theory) has written an entertaining, readable, and personal account of his contributions and those of others to the building of the standard model.

Lederman, Leon, and David Schramm. *From Quarks to the Cosmos.* New York: Scientific American Library, 1989. The authors are excellent expositors of science for the general public.

Weinberg, Stephen. *The Discovery of Subatomic Particles.* New York: Scientific American Library, 1983. A historical account that focuses on the period of particle physics before 1950.

Alex R. Dzierba

Cross-References

Baryons, 195; Detectors on High-Energy Accelerators, 643; The Exclusion Principle, 860; Feynmann Diagrams, 868; Gauge Theories, 942; Grand Unification Theories and Supersymmetry, 985; Leptons and the Weak Interaction, 1242; Mesons, 1374; Quantum Electrodynamics, 1965; Quarks and the Strong Interaction, 2006; Special Relativity, 2333; The Unification of the Weak and Electromagnetic Interactions, 2635.

STAR FORMATION

Type of physical science: Astronomy and astrophysics
Field of study: Stars

Star formation occurs as the result of gas and dust collapsing, which is caused by gravity. Heat builds up until temperatures of 10 million Kelvins initiate hydrogen burning. Many examples of star formation are seen in nebulas within the Milky Way.

Principal terms

ASSOCIATION: a loose cluster of stars whose position, spectra, or motions suggest that they have a common origin

BOK GLOBULES: small units of dark clouds that are nearly spherical and that contain both dust and interstellar molecules

EMISSION NEBULA: a glowing gaseous nebula whose light comes from fluorescence caused by a nearby star

HELMHOLTZ-KELVIN CONTRACTION: a theory that the sun and all stars' energy was created by the star contracting and slowly gaining heat

HERBIG-HARO OBJECTS: clumps of glowing gas that are in their final stages of pre-main-sequence contraction

HI REGION: a region of neutral hydrogen in interstellar space

HII REGION: a region of ionized hydrogen in interstellar space

O AND B STARS: very hot young stars with enormous amounts of energy; because of their rapid burning of fuel, they live for a short time

PROTOSTAR: the embryonic stage of a young star that is in the process of formation

T TAURI STARS: very young stars vigorously ejecting gas prior to burning hydrogen and becoming a main sequence star

Overview

Near the end of a star's life, it casts much of its matter out into space by either a slow gentle process, in which the outer layers are gradually expelled, or the violent explosion of supernova. Either way, the interstellar gases in the galaxy become enriched with heavy elements from the dying star's interior, where they have been created. New stars that form from this enriched material have heavy elements for planet formation. All elements, other than hydrogen and helium, were cast off by earlier generations of stars. Stars can create different elements in different amounts. Carbon, oxygen, silicon, and iron are readily produced within the interiors of stars. All elements heavier than iron are produced in supernova explosions.

Many cold clouds of gas and dust are scattered throughout the Milky Way galaxy, which are termed "giant molecular clouds." In some cases, they appear as dark regions silhouetted against glowing background nebulosity. In other cases, they appear as dark areas that obscure background stars. These dark objects are also called

Barnard objects, after their discoverer, Edward E. Barnard, near the end of the nineteenth century.

Smaller dark clouds are termed Bok globules, named for Bart J. Bok, who called attention to them in the 1940's. More than two hundred of these Bok globules are known in the Milky Way. These are spherical dark nebulas that are typically 1 to 2 lightyears across and have a mass ranging from 20 to 200 solar masses. Spectroscopic analysis of these globules finds that they have a cosmic abundance of elements. This is 75 percent hydrogen, 23 percent helium, and 2 percent heavier elements, formed from supernova or other star death. Infrared observations have demonstrated that the internal temperature of globules are 5 to 15 Kelvins. Larger globules are warmer, on the order of 20 Kelvins. At these very cold temperatures, the dense regions inside the globule contract gravitationally, and these globules coalesce gradually into lumps called protostars.

At first the protostar is a cool blob of gas and dust several times larger than the solar system. The low pressure that exists inside the protostar is incapable of supporting the cool gas that surrounds it. The protostar will begin to collapse under the influence of gravity. As the atoms within the protostar become more crowded, more collisions take place between the atoms. These collisions increase the temperature of the protostar. Gravitational energy is thus converted to thermal energy. The gases that make up the protostar begin glowing as they heat up. After only a few thousand years of gravitational contraction, the surface temperature of the protostar reaches 2,000 to 3,000 Kelvins. As this cloud collapses, all solid grains or ice trapped in the cloud vaporize by 2,000 Kelvins. The protostar that is in the state of collapsing is very large. Because it is so large, therefore, it has a low surface temperature; it has a great luminosity. After a thousand years of contraction, a protostar of 1 solar mass would be twenty times larger and one hundred times more luminous than Earth's sun. Much of this radiation is visible only as infrared light. When this protostar begins to shine at visible wavelengths, it is luminous, red, and cool. Thus, it is located in the upper right of the Hertzsprung-Russell diagram, which plots stars by their temperature and luminosity.

A protostar of 5 or more solar masses will become hotter without much change in overall luminosity. This is caused by the effect of decreasing surface area, which is counterbalanced by an increase in surface temperature. The evolutionary track of massive protostars is to transverse the Hertzsprung-Russell diagram from right to left, as they age toward the main sequence stage; that is, they get hotter while keeping the same luminosity. Protostars of 100 solar masses, which are very rare, rapidly develop such extremely high temperatures that radiation pressure becomes the dominant force supporting the star against gravitational collapse. In less massive protostars, the increase in surface temperature is less rapid than massive protostars. This does not compensate fully for the shrinking surface area, resulting in a luminosity decrease. Protostars less than 0.08 solar mass never manage to develop the necessary pressure and temperatures to initiate hydrogen burning. These become planetlike objects. If Jupiter had been ten times more massive, it would have had enough material to begin nuclear fusion. The solar system then would have had a tiny red dwarf star along with the sun.

The more massive the protostar, the faster it evolves. A protostar of 15 solar masses takes ten thousand years to evolve to the main sequence, that point when a protostar's shrinking builds heat to 10 million Kelvins, which is when nuclear fusion begins. For a 1-solar-mass protostar, the journey to the main sequence requires a million years. Nuclear fusion involves hydrogen combining to make helium and release energy. This thermonuclear process releases tremendous amounts of energy. The outpouring of energy creates conditions inside a protostar that finally halt gravitational collapse. Hydrostatic and thermal equilibrium are established, and a star is created.

Astronomers are unlikely to observe the birth of a star in visible wavelengths because the surrounding globule or interstellar clouds hide the protostar. A cocoon nebula, a brilliant nebula embedded in clouds of interstellar grains that absorb visible light, surrounds this protostar. Heated to a few hundred Kelvins, it reradiates infrared radiation; both infrared and radio waves can penetrate these dust clouds.

The most massive stars are the first stars to form. The more massive the protostar, the sooner it develops the necessary central pressure and temperature to initiate fusion reactions. The most massive main sequence stars are of spectral class O and B, which are luminous and hot. Surface temperatures of O and B stars range from 15,000 to 35,000 Kelvins. Much of their energy is radiated in ultraviolet light. These energetic ultraviolet photons radiated from these newborn massive stars ionize the surrounding hydrogen gas. Most of the hydrogen gas in the galaxy is neutral hydrogen, that is, uncharged, with temperatures at or below 60 Kelvins. Astronomers refer to these areas as HI regions. The ionization energy has a dramatic effect on the globule or dark nebula in which the cluster of stars is forming. The hydrogen glows in the red portion of the spectrum at 6,563 angstroms. Nebulosities around these newborn stars shine red and are called emission nebula.

The intense ultraviolet radiation of O and B stars can produce atomic excitation of HI regions. If the energy is of very high frequency and very short wavelength and strikes a neutral atom, it may cause the expulsion of an electron, termed photoionization. The free electron may then combine with a free proton. It would likely orbit in a high-energy orbit. As it descends to a lower-energy orbit, it gives off light. The Orion nebula is an example of this.

Because the star-forming clouds are predominantly ionized hydrogen, they are termed HII regions. The few O and B stars at the core of the nebula producing ionizing radiation are called an OB association. With a density of one hydrogen atom per cubic centimeter, one O star may produce complete ionization for a radius of 600 light-years. In HII regions, the ionizing ultraviolet radiation from nearby stars is plentiful, and the electrons are ejected at the time of ionization with sufficiently high speeds to produce temperatures in interstellar gas of 10,000 Kelvins. Less than 10 percent of interstellar gas in the Milky Way is ionized. A newborn HII region buried deep within such a dark cloud would be detectable with radio rather than visible wavelengths. As new stars continue to develop, the HII region grows and eventually becomes visible at optical wavelengths. From the first O and B stars that develop in this dark cloud, a cavity is carved out by stellar winds and ultraviolet radiation. A shock wave forms where the outer edge of the expanding HII region encounters this giant molecular cloud. The shock wave compresses hydrogen gas, stimulating a new round of star birth.

Any mechanism that can compress these interstellar clouds can trigger star formation. Other mechanisms include the expanding shell of a supernova remnant. As this encounters an interstellar cloud, it triggers star formation. The supernova explosions produce a considerable mass of heavy elements. The tremendous energy from the shock wave, plus the ionization energy from ultraviolet light, would help form stars. It is estimated that 5 to 10 million years after the explosion, an OB association would form. There is strong evidence that the sun was once a member of one of these stellar associations created by a supernova. The shock wave of the explosion caused the collapse of the solar nebula some 5 billion years ago. A collision between two cosmic clouds of gas and dust would produce a compression at the interface between the clouds. These clouds move up to 10 kilometers per second relative to each other and must collide.

Applications

There is ample evidence in the skies that leads astronomers to believe that they have a good understanding of star formation. One of the best-studied areas of star formation is in Orion. The Orion nebula is a spectacular region of turbulent gas and dust heated by the radiation of hot young stars that have recently condensed from the matter of the cloud. The diameter of the Orion nebula is twenty thousand times that of the solar system and contains 10,000 solar masses. In the heart of the Orion nebula is the Trapezium star cluster, which constitutes the four brightest components of an expanding cluster containing hundreds of fainter stars. These O and B stars make the Orion nebula an HII region. The Orion nebula glows from fluorescence from these stars. Radio observations show a large cloud of neutral hydrogen with a diameter of 320 light-years and 50,000 to 100,000 solar masses. The Orion nebula is a small spot of ionized hydrogen within a larger complex.

The Orion aggregate is an enormous interstellar boiling pot, made of mostly neutral hydrogen. Some of it is condensed apparently into relatively young stars, which ionize some of the gas and cause it to glow. No globules are seen within the Orion nebula, though within the complex are several strong infrared sources which are likely protostar formations. Within this complex are other well-known nebulas, such as the Horsehead. Evidence of a violent past for this complex exists in Barnard's loop, which may be an ancient supernova remnant. At least three O and B stars seem to have been ejected from the Orion aggregate 2.5 million years ago to scattered parts of the sky.

There are many T Tauri stars found near the Orion nebula. These are young stars that are vigorously ejecting gas just before reaching the main sequence. It is theorized that the fusion reactions are preceded by stars propelling their outermost layers back into space. Infant stars in the T Tauri stage can lose 0.4 solar mass before entering the main sequence. The T Tauri stage is the gap between protostars and young stars. A T-association has an abundance of T Tauri-type stars. They occur in groupings, some near O and B associations, others by themselves, but always in regions of the sky where cosmic dust is plentiful. A T-association is formed around the Orion nebula and its surroundings, especially in regions of dark nebulosity.

It is theorized that star formation is presently occuring in the Orion complex. Plates obtained in 1947 and 1954 have shown conspicuous progressive changes in several of the small nebulous knots in the Orion cloud. These are Herbig-Haro objects, which are in the final stages of pre-main-sequence contraction. They are found near T Tauri stars. From year to year, these clumps of glowing gases change slightly in size, shape, and brightness. Even after stars arrive on the main sequence, there is sporadic activity.

The Rosette nebula in the constellation Monoceros is found near the Orion complex, which is a glowing wreath of gas surrounding a dark cluster of young stars. Photographs show many dark tendrils, or globules, of gas, which are most likely stars in formation. These globules are collapsing under gravity to become self-luminous stars eventually.

Other emission nebula show similar evidence of star formation. The Lagoon nebula in Sagittarius contains many T Tauri stars and globules. The North America nebula in the constellation Cygnus contains 12 hydrogen atoms per cubic centimeter. By earthly standards, this would be a nearly perfect vacuum; however, in space, this gas is thick. The North America nebula contains enough gas to create more than one hundred suns.

These nebulas have lanes of obscuring material that obviously overlie the brighter structure of the nebula, and numerous dark spots against the bright background. Some of these spots have a windblown, turbulent appearance, whereas the globules have a rounder appearance. The smallest of these globules has a diameter of 0.1 light-year and a hydrogen concentration 100,000 times that of interstellar space. A star may be formed by the process of collapse and fragmentation of the original interstellar cloud. This may lead to multiple stars or a star cluster.

Context

The process of star formation follows the well-established path of gravitational contraction. This theory is similar to the idea formulated in the 1800's: The Helmholtz-Kelvin contraction suggested that the sun was powered by a slow contraction. The gravitational collapse is turned into thermal energy. Though it is now known that the sun is powered by nuclear reactions, as observed, the contraction was theorized to create stars.

Observations with radio telescopes have detected regions called giant molecular clouds. These clouds have a mass of 10,000 to 1 million solar masses with a diameter of 60 to 240 light-years. Inside one of these clouds, the density is 200 hydrogen molecules (2 hydrogen atoms joined) per cubic centimeter. Approximately five thousand such clouds are located in the Milky Way. These giant molecular clouds and globules contribute materially to the total mass of the galaxy. Halfway between the sun and the center of the galaxy is a ring of three thousand molecular clouds with 2 billion solar masses. The estimated twenty-five thousand globules have a mass of 2 to 3 million solar masses. Large, dark nebula that have a radii of more than 12 light-years and a mass of 2,000 solar masses may be slated for fragmented collapse into a cluster of protostars. Some of the larger nebula are observed in the act of breaking into many units. In the Southern Coalsack (a dark nebula found in the constellation Crux), fragmentation is already in progress. The Coalsack is a nearly spherical ball of cosmic dust that is a conglomerate of smaller units, some very dense and headed for star formation.

Star formation is an ongoing process in a spiral galaxy. The spiral form of the galaxy is maintained by density waves. A high-pressure shock wave accompanies each density wave that maintains the spiral form. If clouds of gas and dust of more than average density are present in the interstellar medium, they may be compressed by the shock wave five to ten times the original density. Star formation begins in the densest region of these clouds. Massive stars, the first born, emit ultraviolet radiation, which ionizes the surrounding hydrogen, forming an HII region. This is evident as protostars and newborn stars are along the inner edge of the spiral arm, whereas O and B stars are brilliant and slightly older, marking the center line of the galaxy.

Astronomers now have a good general understanding of stellar formation. By observing various stages of star birth with radio, infrared, and visible light, it is well known that stars form from gravitational collapse. Star formation will be most evident in spiral arms.

Bibliography

Bok, Bart J., and Priscilla F. Bok. *The Milky Way.* 5th ed. Cambridge, Mass.: Harvard University Press, 1981. Bart and Priscilla Bok have been among the great researchers of the Milky Way galaxy. Bart Bok developed theories of Bok globules, which are a cornerstone of star formation. Describes the galaxy in detail, with intense treatment of the nucleus and of the spiral structure.

Burnham, Robert Jr. *Burnham's Celestial Handbook.* 3 vols. New York: Dover, 1978. This is the standard reference book for deep-sky objects. In a survey of each of the eighty-eight constellations, Burnham describes how to find deep-sky objects, how they will appear, and historical and modern theories as to their nature. Much of the numerical data on deep-sky objects was derived from Burnham.

Ferris, Timothy. *Galaxies.* San Francisco: Sierra Club Books, 1980. This coffee-table book has wonderful, large observatory photographs of nebulas and galaxies. The text briefly covers star formation, especially in the Orion complex. Very clear on the structure of galaxies and galaxy evolution.

Hartmann, William K., and Ron Miller. *Cycles of Fire.* New York: Workman, 1987.

This is a spectacularly illustrated book that helps the reader visualize difficult concepts, such as protostar development. The text is brief, clear, and concise. Many of the illustrations are original paintings by Hartmann.

Kaufmann, William J. *Universe*. New York: W. H. Freeman, 1985. This freshman-level textbook covers all aspects of astronomy with some mathematics. The chapters on planetary formation, the sun, and stellar formation are particularly useful. The mathematics and nuclear reactions are well presented. Illustrations clarify more difficult concepts.

Raymo, Chet. *365 Starry Nights*. Englewood Cliffs, N.J.: Prentice-Hall, 1982. Introduces the reader to the night sky on each night of the year. A number of essays cover star formation in nontechnical terms. The reader is left with a clear understanding of astronomical theories.

David R. Teske

Cross-References

The Hertzsprung-Russell Diagram, 1050; Infrared Astronomy, 1118; Interstellar Clouds and the Interstellar Medium, 1167; Nuclear Synthesis in Stars, 1621; Planet Formation, 1828; Protostars and Brown Dwarfs, 1951; Radio Astronomy, 2050; Supernovas, 2434.

STATISTICAL MECHANICS

Type of physical science: Classical physics
Field of study: Statistical mechanics

Statistical mechanics is the study of the physical properties of systems containing very large numbers of atoms or molecules. The macroscopic properties of these systems are regarded as averages over the microscopic properties of their constituents, and their behavior is therefore explained in statistical terms.

Principal terms

DISTRIBUTION FUNCTION: gives the probable number of particles having positions and velocities (or more generally, energies) within arbitrarily small ranges

ENTROPY: a measure of the energy of a system that is unavailable for work

EQUILIBRIUM: the state in which the entropy is a maximum

IRREVERSIBLE PROCESS: a process for which the total entropy increases

MAXWELL-BOLTZMANN DISTRIBUTION: the form adopted by the distribution function when the system reaches equilibrium

MOLECULAR CHAOS: a fundamental assumption that asserts the absence of correlations between the particles of the system

STATISTICAL WEIGHT: the number of microstates encompassing a given macroscopic state

TRANSPORT PROCESSES: processes that take place when a fluid (gas or liquid) is in a nonuniform state and some physical property—such as mass, density, or thermal energy—is transported, by collisions, through the fluid

Overview

A macroscopic system, such as a gas, possesses certain quantitative properties, such as pressure, temperature, and volume. Since these describe the bulk properties of the system, they are known as macroscopic variables. The relationships between these variables are then given by a number of definite laws. The perfect gas law, for example, states that the pressure of a dilute gas multiplied by its volume is proportional to its temperature; the first law of thermodynamics states that the increase in energy of a system is equal to the heat supplied to the system, plus the work done on the system; the second law of thermodynamics states that the entropy of an isolated system always increases and reaches its maximum value in the state of equilibrium; and so on.

According to the kinetic theory of matter, such macroscopic systems are composed of extremely large numbers of atoms and molecules in rapid motion that undergo repeated collisions. The behavior of these constituents of macroscopic bodies is also determined by certain laws: Newton's laws in classical physics and the laws of quantum mechanics in quantum physics. These laws express the relationships between certain microscopic variables, such as position and velocity.

The question then arises: How are the macroscopic variables of bulk matter related to the microscopic variables of its constituents? To put it another way, how are the perfect gas laws and the laws of thermodynamics to be explained in terms of Newton's laws or the laws of quantum theory? Given the huge number of atoms or molecules in a macroscopic system, such as a gas (on the order of 10^{23}), it would clearly be impossible to track every single particle, solve the relevant equations for the motions of all the particles, and hence determine all the properties of the system. Statistical mechanics solves this problem by considering the macroscopic properties of the system to be averages over its microscopic ones. Thus, the laws obeyed by the macroscopic variables are regarded as being probabilistic, or statistical, in character.

Consider the pressure exerted by the air in a balloon, for example. This is a macroscopic property, which one can measure directly and easily. Yet, the pressure is caused by the air molecules hitting the inside of the elastic skin of the balloon and transferring momentum to it. The pressure one feels is actually the average over these momentum transfers, per unit area. One does not feel the individual collisions because the number of molecules hitting the skin per second is so large. Likewise, the temperature of a macroscopic body is related to the average kinetic energy of its constituent particles. It makes no sense to talk of the temperature of an individual atom or molecule; temperature is a concept that emerges as a collective property of many atoms or molecules. Thus, pressure, temperature, and the other macroscopic variables are average properties of the distribution of motions of collections of many particles.

Thus, rather than try to determine the individual motions of the particles comprising a gas or other substance, statistical mechanics invokes mathematical statistics in order to obtain the "probability distribution" of particle motions. This is represented by a function known as the "distribution function," which gives the probable number of particles that have positions and velocities within certain arbitrary, extremely small ranges. The actual number of particles in these ranges will fluctuate about the most probable value, as the positions and velocities of individual particles change because of collisions. These random fluctuations can be observed in the phenomenon known as "Brownian motion," where very small macroscopic bodies, such as grains of pollen suspended in a liquid, can be seen to undergo small irregular movements resulting from the impact of individual molecules of the liquid.

As the molecules of the gas collide with one another and the walls of the container, their positions and velocities will be altered and hence the distribution function will change. It can be shown, mathematically, that if there are no external forces acting, this function tends toward a certain value, known as the Maxwell-Boltzmann distribution function, which is independent of both position and time. When this value is reached, there exists an overall balance between the numbers of particles gaining and losing a given change of velocity resulting from collisions. The gas is then said to be in a state of "thermodynamic equilibrium." Its macroscopic properties are constant, and the values of the pressure, temperature, and density are the same throughout its volume. Once the Maxwell-Boltzmann distribution function has been obtained, the equilibrium properties of the gas can be calculated. Hence, the thermodynamic properties of systems in equilibrium can be understood in microscopic terms.

It is observed that macroscopic systems, if left to themselves, pass irreversibly from a state of nonequilibrium to one of equilibrium. Heat will flow from a hot body to a cold one, for example, until the temperatures of the two become equal. This irreversible behavior is understood in terms of the concept of "entropy," which can be regarded as a measure of the quantity of energy unavailable for work. According to the second law of thermodynamics, the entropy of an isolated system always increases until it reaches its maximum value at equilibrium, when the distribution function takes on the Maxwell-Boltzmann form. It can be shown that this is the most probable distribution in the following sense: A system in a given macroscopic state can be in any one of a huge number of microscopic states and the "coarse-grained" description offered by the macroscopic variables cannot distinguish between these states. It cannot distinguish between the different microscopic states obtained by simply interchanging the energies of two particles, for example. The number of microscopic states corresponding to the state of equilibrium is overwhelmingly large compared with all other possible microscopic states, and hence this particular macroscopic state is more probable than any other. The probability of appreciable fluctuations away from this state is correspondingly extremely small.

More specifically, the entropy is proportional to the logarithm of the number of microstates composing a given macrostate, as specified by the set of macroscopic variables, where it is assumed that all microstates compatible with this specification of the state of the system are equally likely to occur. The number of microstates of which a given macrostate is composed is known as the "statistical weight" of that macrostate. As this number increases, so does the entropy, until it reaches a maximum. This will occur when the number of microstates corresponding to a given macrostate is at maximum, that is, at equilibrium. Thus, irreversibility can be regarded in terms of a move toward more and more probable states until the most probable macroscopic state of all—the state of equilibrium—is reached.

As a simple illustration of this result, consider a glass jar in which a layer of pepper sits on a layer of salt of equal depth. This initial configuration corresponds to a highly improbable distribution. If the jar is shaken vigorously, a gray mixture will result. If the shaking is continued, it is extremely unlikely that the original configuration will ever return. There are so many more arrangements of the salt and pepper particles that result in a gray mixture than there are that give the original configuration, that the gray state is overwhelmingly more probable. Thus, shaking the jar represents an irreversible process, in which the system evolves from less probable states to those that are vastly more probable.

In evaluating the statistical weight of a certain macrostate, due consideration must be given to the number of different arrangements of the particles that are regarded as possible. In classical statistical mechanics, an exchange of energies of two indistinguishable particles is taken to give rise to a different arrangement, whereas in quantum statistical mechanics it is not, leading to a reduction in the statistical weight. Two forms of quantum statistical mechanics arise as a result: Fermi-Dirac statistics, according to which no more than one particle can occupy a given quantum state (this is the famous Pauli exclusion principle), and Bose-Einstein statistics, which apply to particles that tend to "condense" into the same quantum state. Fermi-Dirac statistics explain a wide range of phenomena, including the nature of the periodic table of elements, the properties of atomic nuclei, the formation of white dwarf stars, and the conduction properties of metals; Bose-Einstein statistics apply to electromagnetic radiation, considered in terms of photons, and explain the superfluid behavior of liquid helium 4.

Applications

Consider the example of a column of gas, at thermal equilibrium, extending to a great height. This can be viewed as a model of the atmosphere; it is highly idealized, since the atmosphere is not at equilibrium—it gets colder as the height increases. The following question can be asked: How does the density of this idealized atmosphere change with height? Since the temperature is constant, the number of molecules per unit volume, or particle density, is proportional to the pressure. The pressure decreases with increasing altitude and therefore so must the particle density. On this basis, it is easy to show that the density decreases exponentially with height. Furthermore, the rate of decrease is proportional to the mass of the molecules: the density of heavier gases decreases more rapidly with increasing altitude than the density of lighter gases. Although this is not strictly observed in the earth's atmosphere, because it is not at equilibrium and there are winds and atmospheric disturbances that mix the gases, there is a tendency for lighter gases, such as hydrogen, to become predominant at very high altitudes.

This exponential variation of density is an example of Boltzmann's law, which states that the probability of finding molecules in a given spatial arrangement varies exponentially with the negative of the potential energy of that arrangement, divided by the absolute temperature, and multiplied by a constant term (known as Boltzmann's constant). This result explains a wide range of phenomena, such as evaporation and thermionic emission in cathode-ray tubes. More generally, the Boltzmann distribution gives the probability that a system in equilibrium at a given temperature is in a state with a certain energy as varying exponentially with the negative of the energy of that state, divided by the product of Boltzmann's constant and the temperature. This is one of the fundamental results of equilibrium statistical mechanics.

With regard to the approach to equilibrium, it is fairly straightforward to give more or less qualitative explanations of irreversible behavior. Take the example of a gas contained in a box. At equilibrium, the density will be uniform and there will be almost exactly half the total number of molecules in each half of the box. There will be fluctuations, where one half of the box contains slightly more molecules than the other half, but these will be small compared with the total number of molecules. The larger the fluctuation, the more improbable it is, since it will correspond to a macrostate for which there are fewer microstates than the number composing the equilibrium state. It is therefore extremely improbable that the gas should suddenly spontaneously change from its equilibrium state to one in which all the molecules are contained in one half of the box—so improbable, in fact, that the gas would have to be observed for a very long time, longer than the universe has been in existence, before this spontaneous change would be seen. Obviously, the gas can be specially prepared to be in this latter state: by compressing it into half of the box and inserting a partition, for example. This effects a change in the macroscopic variables of the system (by reducing the volume), and the molecules will take up the most probable configuration compatible with this new macroscopic state. When the partition is removed, however, the gas will very rapidly expand to fill the box. With the macroscopic variables returned to their original values, the configuration with all the molecules in one half of the box is now extremely improbable, and the gas will quickly pass through a sequence of ever more probable states until equilibrium is reached once more.

Nevertheless, a complete microscopic understanding of irreversibility must take into account the detailed dynamics of the processes involved and, in particular, the nature of the collisions between the particles. The simplest example is that of a dilute gas, where only binary collisions occur between the molecules. As these collisions redistribute the velocities of the molecules, the distribution function changes with time. The change in this function is given by the Boltzmann transport equation, which governs the three fundamental transport processes of diffusion, viscosity, and thermal conduction. It is based on a statistical assumption known as the assumption of "molecular chaos," which refers to the absence of any correlations—either in position or velocity—between the two particles involved in the collision.

This is a plausible assumption to make in the case of a dilute gas because the particles are traveling freely most of the time, and it is very unlikely that a recollision will occur between the same two particles. Hence, particle correlations are negligible. For a dense gas, or liquid, however, a large proportion of the molecules will be undergoing collisions at any given time, and a number of these collisions will be recollisions, so the positions and velocities of the molecules may be correlated. In these cases, molecular chaos can no longer be assumed and the Boltzmann equation no longer holds. Although attempts have been made to generalize this equation by including a correlation function, most of the results that have been obtained have been restricted to certain special and highly idealized cases. The statistical mechanics of fluids is, in general, extremely complex and relies heavily on computer simulations.

Context

The development of statistical mechanics can be traced back to the work of James Clerk Maxwell and Ludwig Boltzmann. Maxwell demonstrated that the effect of collisions was to produce a statistical distribution of particle velocities in which all velocities would occur with a known probability. This is given by the velocity distribution function, first introduced in 1859. In a memoir published in 1866, Maxwell showed that the fundamental linear transport processes in a gas, involving diffusion, heat conduction, and viscosity, could all be conceptualized as special cases of a generalized transport process, in which a physical quantity is carried by molecular motion and transferred from one molecule to another by collisions.

The Boltzmann transport equation, obtained by Boltzmann in 1872, is a special case of Maxwell's result, in which the physical quantity of interest is the velocity distribution function. The first definitive solution of this equation was obtained in 1917 by D. Enskog, although a similar method of solution for Maxwell's original results was obtained one year earlier by Sydney Chapman. The Chapman-Enskog method, as it is now known, establishes a connection between the microscopic behavior of particles and the hydrodynamic behavior of dilute gases. Much of modern research in this field concentrates on obtaining solutions to generalized forms of the Boltzmann equation for nondilute fluids.

In 1872, in an application of his own equation, Boltzmann discussed the behavior of the entropy function in irreversible processes by considering the way in which the velocity distribution changes with time because of interparticle collisions. By considering the details of collision processes, he was able to deduce that a certain quantity, known as the "H-function," must always decrease or remain constant. This function was then shown to be proportional to the negative of the entropy in the equilibrium state, thus providing a molecular interpretation of the second law of thermodynamics and, consequently, of irreversible phenomena.

Nevertheless, criticism of this result, in terms of the impossibility of explaining irreversibility solely in terms of time-reversible Newtonian mechanics, forced Boltzmann to acknowledge the statistical nature of the second law of thermodynamics. In 1877, he considered the distribution of particles over all possible energy states and was able to show not only that the equilibrium distribution (as given by the Maxwell-Boltzmann distribution function) was overwhelmingly the most probable one but also that the second law could be understood as the statement that systems tend to evolve from less probable states to more probable ones. Boltzmann's combinatorial approach was subsequently used by Max Planck to explain the distribution of energy in electromagnetic radiation at thermal equilibrium. The division of the energy into discrete "quanta" involved in this explanation led to the development of quantum theory.

The consideration of collections, or "ensembles," of particles was further developed and generalized by Josiah Willard Gibbs, who coined the term "statistical mechanics" in 1901. Boltzmann's approach of 1877 effectively introduced the "microcanonical" ensemble, involving systems of constant energy. Gibbs went on to discuss the "canonical" ensemble, in which the temperature is kept constant but the energy fluctuates, with these fluctuations being, in general, extremely small, and the "grand canonical" ensemble, in which the temperature is kept constant but the number of particles in the system is not.

Despite the power of this approach and the clear exposition given by Paul and Tatiana Ehrenfest, who demonstrated the connection between Boltzmann's work on the H-function and the combinatorial approach involving the microcanonical ensemble, a number of fundamental questions remain. These questions have centered on the assumption of molecular chaos, invoked by Boltzmann to explain irreversible phenomena. Certain modern approaches to irreversibility avoid probability arguments and attempt to derive irreversibility on the basis of a consideration of the dynamic instability of systems of many particles. The central idea here is that two such systems, whose initial states are very similar—in fact, arbitrarily so—may evolve with time into states that are very different. This idea is the basis of the theory of "chaos," which has been developed to explain a wide range of phenomena, from atmospheric turbulence to the beating of the human heart. Although much attention has been given to this kind of approach, a number of fundamental problems remain. In general, it can be said that although the problems of equilibrium statistical mechanics are essentially mathematical in nature, those of the nonequilibrium theory are ones of principle, involving the very foundations of the subject.

Bibliography

Brush, Stephen G. *The Kind of Motion We Call Heat: A History of the Kinetic Theory of Gases in the Nineteenth Century*. 2 vols. New York: Elsevier, 1976. A comprehensive account of the history of kinetic theory and statistical mechanics up to the beginning of the twentieth century. The discussions of Maxwell and Boltzmann's work is particularly illuminating.

Cohen, E. G. D. "The Kinetic Theory of Fluids: An Introduction." *Physics Today* 37 (January, 1984): 64-73. A useful survey of attempts to explain the hydrodynamic behavior of dense fluids in statistical mechanical terms. The contrast with dilute fluids is very well drawn and the usefulness of computer simulations is emphasized.

Feynmann, Richard P., Robert B. Leighton, and Matthew Sands. *The Feynmann Lectures on Physics*. Vol. 1. Reading, Mass.: Addison-Wesley, 1963. This classic set of lectures contains a useful introduction to the principles of statistical mechanics, with an emphasis on applications. There is, however, little discussion of irreversibility from a statistical mechanical as opposed to a thermodynamical point of view.

Mandl, Franz. *Statistical Physics*. New York: John Wiley & Sons, 1971. This is an excellent exposition of the fundamental principles of statistical physics and includes a number of interesting and useful examples of their application. The discussion of quantum statistical mechanics is particularly clear.

Prigogine, Ilya. *From Being to Becoming: Time and Complexity in the Physical Sciences*. San Francisco: W. H. Freeman, 1980. Written by a Nobel Prize winner, this is an introductory account of a novel approach to nonequilibrium statistical mechanics that emphasizes the role of fluctuations and the open nature of self-organizing systems.

Steven R. D. French

Cross-References

Black Body Radiation, 231; Brownian Motion, 266; Entropy, 800; The Behavior of Gases, 935; The Maxwell-Boltzmann Distribution, 1360; Quantum Statistical Mechanics, 1993; Laws of Thermodynamics, 2486.

STATISTICS

Type of physical science: Mathematical methods
Field of study: Probability and statistics

Statistics is the field concerned with methods, based on mathematical theory and probability, that allow scientists to summarize many observations concisely. It also tests hypotheses about conditions within or between populations or universes from which these data were drawn.

Principal terms

ACCURATE: referring to how close to the real or actual value an estimator is; an unbiased estimator gives an accurate estimate

DATA: the individual real number or count of information that one is measuring for use in statistical calculations

HYPOTHESIS: what it is that one wishes to test; always written as an either-or choice; the null hyphothesis (H naught or Ho) is the first one stated, the alternative hypothesis (Ha) is the second one listed

POPULATION or UNIVERSE: the entire group or area for which descriptive and inferential statistical statements are made based on samples (n) from within these areas

PRECISION: how close together in value repeated estimates from the same population or universe actually are

SAMPLE: the total number of individual observations or pieces of data ($=n$) drawn from the population or universe on which the estimates of parameters for that population or universe are based

Overview

As a plural noun, "statistics" refers to numbers, such as the number of individuals living in a certain city. When used as a singular noun, statistics refers to the methods used to summarize, analyze, and draw conclusions from numerical data. These methods make sense out of the apparent chaos of many individual observations. This article discusses the singular definition of statistics.

Some consider the field of statistics to be both an art and a science, similar to the practice of medicine. As in medicine, the statistician often must draw conclusions from a set of data or diagnostic symptoms that are often incomplete, vague, or represent "too small a sample." Statistical methods can be used by all who deal with numerical data. Statistical theory and techniques are the same for all fields of study.

Statistical activity can be divided into two areas: theoretical or mathematical statistics, and applied or practical statistics. Theoretical statistics deals with the mathematical foundations of methodology. This study of abstract mathematics provides the tools used to attack practical problems. Theoretical or abstract statistics is often an end in itself.

To the applied or practical statistician, statistics is a means to an end. It is a kit of tools or the technology of the scientific method. Solving problems in statistics involves three steps: definition of the problem, collection of the data, and analysis of the data. How the data are collected and the units of measure used for these data determine how these observations are analyzed and summarized by the statistician.

Five sampling designs are available to use to collect data from a population. Random sampling is the most common type used to collect data. It assumes that all individuals within a population or universe have an equal chance of being selected in a sample. If one hundred iron rods were present and ten were to be selected to measure their tensile strength, each rod should have a one-in-ten chance of being selected for testing. Stratified random sampling is the procedure used when the population being sampled is not uniform. The population is then separated into strata, which have similar characteristics, and a portion of the total sample to be taken from the entire population is randomly taken from within each strata. If strata cannot be established and then stratified, random sampling cannot be used. If the strata are properly drawn, stratified random sampling will provide a more precise estimate of the population parameters than will simple random sampling with the same sampling effort. Perhaps the iron rods of the example come from two different foundries, one of which produces better iron than the other. The rods would be separated, based on which foundry they were produced by, into two strata. Stratified random sampling would provide a better estimate of the overall tensile strength of the average iron rod for all the iron rods than, perhaps, would simple random sampling.

Systematic sampling is another way one might collect information on the tensile strength of the iron rods. Assume all one hundred rods are in a pile; every tenth rod would be selected that was picked up as the one that would be used to determine tensile strength. Although easier to do, systematic sampling does not meet the mathematical assumptions needed to provide unbiased descriptive statistics and should not be used in sampling. Cluster sampling requires that many samples are taken once the initial area or sample site is located. This type of sampling, used when getting to where the data are to be collected, such as on a mountain top, is very difficult. Once on top of the mountain, many samples of whatever was being measured would be taken. Double sampling is the type of sampling used when there exists a linear relationship between previous data that have been collected and data that are now needed. An example of double sampling would be estimating the population of the United States in 1990 based on data collected from 1980.

Four classes or types of measurement are used to describe data. Nominal measurement classifies only the observations into groups or categories. For example, objects could be classified by color and counted; data could be reported as 34 red, 49 blue, and 115 white crystals in a growth chamber. Ordinal data are those grouped by size into groups, but no absolute value is placed on the differences between the groups. Snow crystals could be counted and placed into one of three groups: small, medium, and large. Crystals in the small group are clearly smaller, but are not by actual, measurable units from those placed in the medium or large categories. Interval measurement units have real differences between the units. Commonly, things such as length, 10 millimeters versus 20 millimeters, indicate a real difference of 10 millimeters between the two measurements. Ratio measurements are created by dividing one number by another to create a single number.

The assumptions or parameters required of each statistical method must be satisfied for the statistical procedure to provide an unbiased analysis of the collected data. Descriptive statistics organizes the individual observations and provides a concise summary of their character. Any set of data may provide descriptive statistics. The arithmetic mean and standard deviation are examples of descriptive statistics that may be compiled from a set of data. Inferential statistics test hypotheses about the population or universe from which the sample was drawn and often contrast that universe with other populations or universes. Statistical techniques allow scientists to measure the strength of their conclusions in terms of probability. The methods chosen to test hypotheses are often called either parametric or nonparametric statistical methods. Nonparametric statistical methods can be used on all measurement categories. Parametric statistical tests can be used only on those data having interval or ratio measurements, unless the nominal or ordinal measurements are mathematically transformed. If the data are collected and analyzed properly (that is, the assumptions of collection and analysis by a specific statistical method are met), then the statistics cannot lie. If the parameters are not met, than the statistician can make a mistake regarding what the data really mean.

Perhaps the most unique feature of twentieth century applied mathematics has been the invention and development of electronic computers. The growth of statistics, particularly in the applied area, has been phenomenal, and the development of statistical programs or packages has taken the drudgery out of solving problems. Statistical packages often available on mainframe and personal computers include BMPD, Minitab, SAS, and SPSS-X.

Applications

The notation of statistical works is often difficult to follow, but convention provides the reader with some guidelines. Usually, true value for the universe is represented by a Greek letter; the population parameter by a capital letter; and the sample or estimated value by a lowercase letter. For example, the Greek letter mu (μ) represents the actual mean (arithmetic average) for the universe, capital X with a bar over its top ($\bar{X}$) as the mean for a population, and a lowercase x with a bar over its top ($\bar{x}$) as an estimate of the mean based on data collected by random sampling from that population or universe.

Any set of data will have descriptive statistics computed for it. For interval data, these regularily include a mean, median, mode, range, standard deviation, standard error of the mean, a confidence interval for the estimate of the mean, and a coefficient of variation. The individual data points may be graphed to form a scatter diagram or a histogram to show the general shape of the distribution or frequency of the data points. Assumptions may be made about the shape of this frequency distribution, and these hypotheses can be tested using different statistical methods to see if, in fact, the graphed data do meet the hypothetical distribution proposed for that sample of data.

Besides the type of measurement that determines what statistical method is appropriate for use, the number of populations or universes being compared also determines the selection of a specific statistical "test" for use. If the assumptions for a parametric statistical method are thought not to be met, then an alternative nonparametric test may be used to compare these sets of data. Nonparametric tests generally require a greater difference among the sets of data being compared to reject the null hypothesis than are necessary for the difference needed by a parametric statistical test to reject it at any given probability level.

For the comparison of two samples, the t-test is often used. The t-test may compare a mean calculated from sample data with either the universe or a population mean, with another independent mean, or with related means. Related means are those that produce sample values from the same unit in a pre- and postmeasurement situation. If the calculated t-value exceeds the tabular value expected by random chance given in the student's t-table, then the null hypothesis of equality will be rejected. The Mann-Whitney U-test is a nonparametric statistical test that may be used if the assumptions required of the t-test are not met. Often, the t-test requirement of equal variability among samples is not met, so the Mann-Whitney test is often preferred over the t-test. Interval measurement must be used to use the t-test.

Nominal data—those observations placed in different categories—may be tested using the chi-square or G-test procedures. Theoretical distributions can be tested against actual observed frequencies of some category. Ratios of occurrence of some character may be tested or ratios between two groups may be also compared to determine if the ratios are the same. This latter is usually known as row by column ($R \times C$) contingency table analysis.

If more than two sets of data are to be compared, again assuming interval-type measurements or proper transformation of the measurement data, then a field of statistical testing known as analysis of variance is often used. The calculated test statistic is the F-value, and its calculation involves some complex calculations, unless a suitable computer program is available for use. The actual application of treatment levels to the experimental groups is called experimental design. The experimental design used determines the actual procedures used to calculate the F-value. Some types of experimental designs that may be used are: completely randomized design, block design, latin square, and split-plot applications. The nonparametric alternative for the randomized design is the Kruskall-Wallis test, while the Friedman test is a possible nonparametric alternative for the randomized block experimental design. Analysis-of-variance statistical methods are often used where the investigator can control what is happening, either in the laboratory or in field plots, such as agricultural test stations.

The relationship between two variables, usually designated x and y, is often used by statisticians in statistical procedures known as regression and correlation analysis.

The *x* variable is often called the independent variable, while the *y* variable is the dependent variable. In other words, the *x* variable governs what value *y* may have. If only one *x* variable is involved in the equation, the regression is termed "simple"; if more than one *x* value is used to estimate *y*, then the relationship is known as multiple regression work. Regression is prediction; how the variable(s) *x* relate to *y* determine the predicted value of *y*. The relationship of the *x* value to the *y* value is referred to as "correlation." If an *x*:*y* relationship is positively correlated, then as *x* increases or decreases, the value of *y* also increases or decreases. If the relationship between the *x* and *y* variables is a straight line or linear, then the correlation coefficient *r* that is calculated when the regression equation is solved is also valid. If the relationship between *x* and *y* is not linear, then the calculated correlation coefficient *r* has no valid meaning and cannot be used. Nonparametric correlation coefficients, such as Kendall's tau (τ) or Spearman's rho (ρ), may be used to show the strength and shape of the relationship between *x* and *y*.

Context

Statistics began with the study of probability as it applied to gambling and life insurance. Life insurance, first issued in 1583, utilized probability to determine the odds of the insured's death. Modern insurance rates today are based on the same approach to probability. Statistics have been used for more than 250 years, but only since the mid-1900's has its use become widespread in every area of study and investigation. Explosive development in statistical theory and methods since the 1950's has made this area of mathematics particularly exciting and fruitful.

The mathematical theory of probability began in 1654 when Chevalier de Mere, a wealthy French nobleman who liked gambling, asked Blaise Pascal to solve a problem involving the distribution of money in an unfinished game of chance. The Dutch scientist Christiaan Huygens published a book on games of chance in 1667, but the first book devoted entirely to probability was Jakob Bernoulli's *Ars conjectandi* (1713; art of conjecture). In 1662, John Graunt applied statistical probability in his study of death in London.

Thomas Bayes's 1763 essay is well known as the first attempt to use the theory of probability as an instrument of inductive reasoning. Pierre-Simon Laplace in his *Théorie analytique des probabilités* (1812; analytic theory of probability) stated that the most important questions of life were really only problems of probability. Laplace was unrivaled for his mastery of analytic technique and was perhaps the eighteenth century's best applied statistician. During the middle nineteenth century, only the Soviet mathematicians Pafnuty Lvovich Chebychev and Andrey Andreyevich Markov made significant advances in probability theory.

After 1900, interest in probability and statistics soared, and it became one of the most important and fruitful areas of mathematical investigation. In 1900, Karl Pearson determined the theoretical distribution of chi-square. In 1908, W. S. Gogget calculated the distribution of *t*, the probable error of a mean, and this began the study of exact sampling distributions in statistics. Sir Ronald A. Fisher developed much of the theory and applications of experimental design and statistical analysis in the early 1900's.

The true beginning of nonparametric statistics was in 1936, when H. Hotelling and M. R. Pabst published their paper on rank correlation. I. R. Savage's 1962 monumental *Bibliography of Nonparametric Statistics* had more than three thousand references dealing with nonparametric statistics in it, indicating the growth and importance of this area of applied statistics. It is now one of the most successful branches of modern statistics. Analysis of variance and regression theory and applications have also developed since the mid-1940's.

Bibliography

Cochran, William G. *Sampling Techniques.* 2d ed. New York: John Wiley & Sons, 1963. This book is the bible of sampling, with most of the variations of the five major sampling approaches discussed. The derivations of formulas are sometimes not very clear.

Draper, Norman, and Harry Smith. *Applied Regression Analysis.* New York: John Wiley & Sons, 1966. This book is devoted entirely to regression analysis. In spite of its title, it covers both the theoretical aspects and the applications of linear regression. Somewhat difficult to read unless one has a firm grounding in theoretical mathematics.

Siegel, Sidney. *Nonparametric Statistics for the Behavioral Sciences.* New York: McGraw-Hill, 1956. This commonly used textbook of nonparametric statistics is useful for all areas of study, not only the behavioral sciences. Contains most of the standard nonparametric methods and often has answers to questions that newer books do not.

Snedecor, George W., and William G. Cochran. *Statistical Methods.* 7th ed. Ames: Iowa State University Press, 1980. This book is used in many beginning statistics courses. Useful for basic assumptions for statistical methods. Contains a good chapter on sampling and explains random and stratified random sampling particularly well.

Sokal, Robert R., and F. James Rohlf. *Biometry: The Principles and Practice of Statistics in Biological Research.* New York: W. H. Freeman, 1981. Contains useful information about current statistical methods. Examples are plentiful but are often difficult to follow. The availability of computer programs is assumed, so advanced topics, such as multiple regression, are not shown with worked examples.

Steel, Robert G. D., and James H. Torrie. *Principles and Procedures of Statistics, with Special Reference to the Biological Sciences.* New York: McGraw-Hill, 1960. The best book explaining how to do the arithmetic. Since it is dated, the book lacks some of the newer nonparametric statistical tests and information on the G-test. Examples are especially easy to follow and use.

Weatherburn, C. E. *A First Course in Mathematical Statistics.* Cambridge, England: Cambridge University Press, 1968. This book provides a very readable introduction to the topic of mathematical or theoretical statistics. Anyone with an average background in mathematics may use this book and should gain an understanding of the formulas and validity of the methods that are commonly used by statisticians.

Zar, J. H. *Biostatistical Analysis.* 2d ed. Englewood Cliffs, N.J.: Prentice-Hall, 1984. This book is one of the best introductory books on applied statistics that are available. Step-by-step examples are shown, and it discusses considerable theory with the application of procedures. Contains an excellent coverage of nonparametric statistical procedures. Includes extensive statistical tables in the back of the book.

David L. Chesemore

Cross-References

Error Analysis, 831; Probability, 1936; Statistical Mechanics, 2367.

STELLAR OSCILLATIONS AND HELIOSEISMOLOGY

Type of physical science: Astronomy and astrophysics
Field of study: Stars

Helioseismology is the study of the oscillations that take place within the sun. These periodic vibrations are caused by sound waves, which originate within the convective zone of the sun. By analyzing the motion of these waves, scientists can image the interior of the sun and develop a more accurate model of the sun.

Principal terms

CONVECTION: the flow of heat in a liquid or a gas; the hot material is less dense and flows upward, and the cooler material flows toward the bottom of the convection cell

FUSION: a thermonuclear reaction that powers the sun and the stars; in the sun, hydrogen nuclei are fused into helium, generating energy

NEUTRINOS: chargeless, massless subatomic particles that are created during the fusion process; virtually unstoppable, they pass easily through the earth

OBLATENESS: a somewhat spherical object that is flattened at the poles

Overview

It was discovered in the early 1960's that areas of the sun's surface are periodically oscillating up and down. This discovery, like many other scientific discoveries, was made somewhat by accident. The evidence for solar quakes was obtained while astronomers were attempting to measure the oblateness of the sun to verify another theory.

It was known for quite some time that the orbit of the planet Mercury did not follow the precise path predicted by the solution to Sir Isaac Newton's laws of gravity and motion. In fact, as Mercury orbited the sun, its orbit would also revolve around the sun. Although some precession was predicted, a percession in Mercury's orbit of 43 arc seconds per century was not predicted. It was proposed by some scientists that there was another planet within the orbit of Mercury. Others suggested that, after so many accurate predictions, Newton's law of gravity needed some alteration. In 1916, Albert Einstein proposed that a massive object, such as the sun, warps the space-time around itself. Mercury's orbit would then follow the curvature in space-time caused by the sun. Einstein's general theory of relativity accurately accounted for the motion of Mercury. According to an opposing theory of gravity, the scalar-tensor theory of Carl Brans and Robert H. Dicke, the sun is not a perfect sphere but has a slight equatorial bulge caused by the rotation of its core. Dicke proposed that a distortion of 0.05 percent of the sun's surface would explain the observed behavior of Mercury's orbit.

At the University of Arizona, astronomer Henry Hill built a telescope that was designed specifically to detect a distortion in the shape of the sun. When the telescope became operational and the sun's surface was studied, no evidence of a distortion was observed. Upon further observation and an additional series of measurements, Hill and his colleagues discovered the periodic oscillations of the sun's surface.

Light from the sun is analyzed by the use of a spectrometer. This device, similar to a glass prism, breaks up sunlight into its component colors. Since the cooler gases of the sun's atmosphere absorb some wavelengths of light, a spectrum, which contains several dark lines, is produced. These dark lines in the sun's spectrum form the chemical signature of the various elements that compose the sun.

The discovery of surface oscillations was made by observing the Doppler shift of various spectral lines in the light from the sun. By analyzing light, astronomers can determine whether the source is moving toward or away from the observer. If the spectral lines have been shifted to the blue (or short-wavelength) end of the spectrum, the source of the light is moving toward the observer. If the shift is toward the red (or long-wavelength) end of the spectrum, the source is moving away from the observer. By observing various points on the surface of the sun, astronomers were able to show the periodic oscillation of those points, as spectral lines would be alternatively redshifted and then blueshifted.

It was first observed that these periods of oscillation were about five minutes in duration. Since then, it has been determined that the entire surface of the sun is in a state of constant oscillation, with periods varying between minutes and hours. It might be said that the sun is ringing like a bell. In this case, however, the bell is being struck continuously.

Astronomers believe that the origin of these waves is the convective zone beneath the photosphere or "surface" of the sun. These huge boiling columns of gas carry heat from the sun's interior to the surface. The tops of these convecting cells produce the granulation seen in solar photographs. These granules may each be hundreds of kilometers across. The gas rises toward the surface accompanied by a tremendous roar. These sound waves oscillate through the sun and cause its surface to rise and fall periodically.

As sound waves travel downward into the sun, they encounter higher temperatures and pressures. These changing physical conditions result in the wave's velocity being increased. Eventually, the waves begin to bend upward toward the surface of the sun. When they reach the bottom of the photosphere, they are reflected back into the interior of the sun. The depth that the wave travels depends upon its length. The wavelength also determines how far a wave will travel around the sun before it hits the surface.

The sun's interior is conducting waves with virtually millions of different wavelengths and frequencies. Some waves have the exact length necessary to make an even number of bounces before they return to where they began. Astronomers categorize these waves by the number of times that they strike the surface in one complete cycle of the sun. For example, a wave with the designation I-4 strikes the surface in three places before it bounces back to its starting position. Once it returns to its origin, it has struck the surface of the sun four times. Scientists have found that waves with low I numbers travel deep within the sun and may be a key to revealing physical characteristics there, while waves with higher I numbers may be used to probe the shallow zones of the sun's interior.

Applications

During an earthquake, various types of seismic waves are generated. These waves are received at seismic stations around the world and their arrival times are noted. By determining the path and velocity of these waves, scientists learn much about the interior of the earth. Similar methods are used by exploration geophysicists to study the subsurface when searching for mineral deposits or potential oil traps. In this case, the seismic waves are generated by explosive charges or other artificial means. The waves then travel into the earth and are reflected as they strike various rock layers. Scientists can then determine the depth of these layers.

Scientists hope to be able to use solar seismic waves to image the interior of the sun, just as geophysicists use seismic waves to study the interior of the earth. Prior to this new development in solar physics, the processes that occur within the sun and the locations of various boundaries within the sun were theoretical. It is hoped that future helioseismic studies will continue to increase human knowledge of the sun.

Theorists believe that the sun is a giant ball of gas that is sustained by a thermonuclear fire burning within its core. Because of the great pressure and high temperature within this region of the sun, the nuclei of hydrogen are fused together to form helium. During this process, which is known as the proton-proton cycle, 600 million tons of hydrogen are consumed each second and turned into some 590 million tons of helium. The tonnage that is not converted into helium is transformed into energy. Most of the energy leaves the core in the form of high-intensity radiation called gamma rays. The remainder is in the form of chargeless, massless, subatomic particles called neutrinos.

By studying the rate at which neutrinos are emitted from the core, scientists can gain some insight into the processes that are occurring within the core. In the late 1960's, an experiment was set up deep within an abandoned gold mine in Lead, South Dakota. Conducted by Raymond Davis of the Brookhaven National Laboratory, this ongoing experiment uses a 378,000-liter tank of a chlorine solution to detect the solar neutrinos. As the neutrinos pass through the solution and strike individual atoms of chlorine, argon nuclei are formed. Since these argon nuclei are radioactive, they can be detected easily and the neutrino flux can be calculated. The results of the Davis experiment have touched off one of the major debates in modern astrophysics. The experiment is detecting only about one-third as many neutrinos as the theory indicates should be detected. Astrophysicists are wondering what is happening to the missing neutrinos. Several theories have been proposed to solve the problem. It has been suggested that perhaps the Davis experiment has flaws. This possibility has been examined thoroughly; the high-technology equipment involved in this experiment has been checked several times and the data have been interpreted and reinterpreted. The problem of the missing neutrinos remains.

There is a possibility that the model of the solar interior is incorrect. The possibility exists that the core of the sun is not as hot as current theory indicates. If the core is cooler, then the number of neutrinos emitted from the sun would be less, and this would account for the missing neutrinos. A particle known as WIMP, or weakly interacting massive particle, has been proposed as a possible solution. According to modern theory, WIMPs were formed in the early universe, when matter was in a very dense state and temperatures were extremely hot. These particles migrated toward the center of newly forming stars, where they remain today. It is believed that WIMPs may circulate within the interior of the sun, carrying heat away from the core. As a result of this process of carrying away heat, the fuel cycle would be slowed somewhat, thus reducing the number of neutrinos emitted.

There are still other possibilities to be considered. One interesting possibility is that the sun's core is rotating much faster than its outer layers. If this is true, the core temperature would be lower and, thus, the rate of neutrino emission from the core would be lower. A potential problem with this solution is that a rapidly rotating core would cause an equatorial bulge, and no such bulge has been detected.

It is obvious that the only way to determine precisely the cause of the missing neutrinos and whether the core is rotating is to have images of the interior of the sun. Helioseismic waves may provide the answer. It has already been determined by studying solar seismic patterns that the sun's convection zone is deeper than previously believed. Apparently, it composes about 30 percent of the solar radius. New information has been found that links the rate of rotation of the interior of the sun to the formation of sunspots.

More extensive studies using helioseismology have been planned. The Global Oscillation Network Group (GONG) at the National Solar Observatory in Tucson, Arizona, was founded to begin such studies. Also, the European Space Agency (ESA) plans to install helioseismology instrumentation on its Solar and Heliospheric Observatory (SOHO), which will be able to take relevant data for years to come.

The GONG program has outlined several specific goals for their solar studies program. The first goal is to determine the internal temperature, pressure, and composition of the sun from the surface down to the core. A second goal is to determine the rotational rates of internal layers of the sun. In addition to determining rotation rates, scientists will again attempt to detect any solar oblateness. This test will allow the accuracy of the general theory of relativity to be checked again and quite possibly, between the oblateness and rotation tests, the problem of the missing neutrinos can be solved. Scientists also hope to use helioseismic studies to investigate how energy is transferred from the solar surface to the chromosphere and corona. It is currently believed that intense magnetic fields, along with acoustic shock waves from the tops of convecting cells, are responsible for temperatures of 400,000 Kelvins in the chromosphere and temperatures of 2 million Kelvins in the corona.

GONG's plans include building a chain of 50-millimeter refracting telescopes to be stationed at various locations on the earth. These locations will ensure that at least two telescopes will be gathering data from the sun at all times. The network will observe the sun continually for a three-year period. This large amount of data is necessary if astronomers are to determine the path of sound waves through the sun and convert that information into a model of the solar interior. Each of the telescopes that will be deployed by the GONG project contains an instrument called a Fourier tachometer. This device is capable of measuring extremely small Doppler shifts at more than sixty-five thousand different points on the surface of the sun. By observing these shifts, astronomers can determine oscillation periods of these various points and form a detailed model of the solar disk.

Context

The discovery that the surface of the sun is oscillating was made in the early 1960's. At the time, scientists were gathering data on the oblateness of the sun in an attempt to determine which theory—the scalar-tensor theory of Dicke and Brans or Einstein's general theory of relativity—could best explain the motion of the planet Mercury. As it turned out, there was no noticeable oblateness of the sun, but subsequent observations revealed a Doppler shift in the solar spectra taken from various points on the sun. This Doppler shift provided evidence for periodic oscillations. Further investigations have revealed that the sun is ringing as if it were a large bell that is continuously being struck.

The millions of different wavelengths and frequency combinations of waves are believed to originate within the sun's convective zone. In this area, the tremendous heat from the core is flowing outward toward the surface. This method of heat flow, convection, is found only in a gas or a liquid. Here, the material toward the base of the zone is extremely hot. Hot liquids and gases are also less dense than cooler liquids or gases. As a result, the less dense material rises toward the surface, carrying the heat. At the base of the zone, cooler material flows in to fill the void. This material will be heated subsequently and will begin to move toward the surface. The sound waves given off by this movement of huge amounts of hot gases are the waves that cause the sun to vibrate.

The discovery of helioseismic waves and, therefore, the science of helioseismology will enable solar scientists to map the interior of the sun and perhaps solve some of the perplexing problems in solar physics. For example, if it can be determined whether the core is rotating or if the sun has an equatorial bulge, it may be possible to solve the problem of the missing neutrinos. Helioseismic imaging will also make it possible for astronomers to determine the boundaries for such zones as the convective zone, radiative zone, and the core itself. In addition, the accumulation of data from the GONG program will, over a period of years, enable astronomers to form an accurate model of the solar surface and interior.

Bibliography

Gamow, George. *A Star Called the Sun*. New York: Viking Press, 1964. This very readable volume describes the sun, solar processes, and energy generation within the sun. Stellar evolution is also discussed. Some basic algebra is used in illustrations. Well suited for the layperson.

Lopresto, James Charles. "Looking Inside the Sun." *Astronomy* 17 (March, 1989): 20-28. A somewhat technical article discussing the origin of the study of helioseismology and its possibilities in solar research. Although very little mathematics is used, the article contains an abundance of technical terms. The reader should have a background in basic physics and astronomy.

Mitton, Simon. *Daytime Star: The Story of Our Sun*. New York: Charles Scribner's Sons, 1981. A nontechnical volume accessible to the general reader. Mitton discusses the sun, its structure, its processes, and its future.

Pasachoff, Jay M. *Astronomy: From the Earth to the Universe*. Philadelphia: Saunders College Publishing, 1991. A fairly technical volume covering topics in stellar and solar system astronomy. This volume is used as a text for college freshman-level astronomy courses but would be accessible to the informed reader. Contains an excellent unit on the sun.

Seeds, Michael A. *Foundations of Astronomy*. Belmont, Calif.: Wadsworth, 1990. This general astronomy textbook contains a section on the sun. Suitable for the general reader.

____________. *Horizons: Exploring the Universe*. Belmont, Calif.: Wadsworth, 1991. Although this volume is intended for use as a college-level general astronomy text, it is accessible to the general reader. Contains an excellent discussion on the sun.

David W. Maguire

Cross-References

Acoustics, 23; Electromagnetism, 750; General Relativity: An Overview, 950; Optics, 1673; Star Formation, 2360; Thermodynamics: An Overview, 2493; Thermonuclear Reactions in Stars, 2509.

STORAGE RINGS AND COLLIDERS

Type of physical science: Elementary particle (high-energy) physics
Field of study: Techniques

Storage rings and colliders are similar devices that high-energy physicists use to increase the range of energies that can be studied. Because experiments are conducted in the center-of-mass frame of reference, all the energy is available to produce new phenomena.

Principal terms

ANTIMATTER: a form of matter which has the same mass as normal matter (electrons and protons) but reversed charges

CENTER-OF-MASS FRAME: the (usually moving) coordinate system in which the interaction is at rest

COLLIDER: a device which brings energetic particles into close vicinity so that they may interact and their interactions may be observed; generally refers to devices in which both particles are in motion

DETECTORS: devices for observing the elementary particles produced in reactions

LUMINOSITY: a measure of the intensity of beam-beam interaction in a colliding-beam machine

QUARKS: the fundamental components of strongly interacting particles; a baryon is made up of three quarks, and a meson is made up of a quark and an antiquark

STORAGE RING: a device which employs magnetic fields to trap charged particles for extended periods, the particles generally moving in a closed path or a ring

Overview

High-energy physicists exploit Albert Einstein's relationship between matter and energy by using very energetic collisions to produce new forms of matter. The more energy available for a reaction, the more likely something new and unique can be produced. Since much of the matter studied by particle physicists is unstable and decays in a few billionths of a second or less, scientists are required to produce this matter in order to study its behavior. Physicists have employed particle accelerators, such as the synchrotron, to produce particle beams rivaling the most energetic particles found in nature. As scientists have learned more and more, there has been a push to higher energies. To achieve more energetic collisions, large accelerators have been built. Colliders provide a means to achieve very high-energy collisions at a modest cost. The fundamental idea behind a collider is to perform the experiment in the center-of-mass frame of reference. The maximum amount of energy available for a collision is the energy in the center of mass. In any other frame of reference, some of the energy of the colliding particles is needed to keep the reaction products moving. This reserved energy is not available to the reaction.

This concept of the relativity of the amount of energy is familiar. For example, if one is standing on the side of a road, then the cars that move past have a considerable amount of energy. A person riding in a car, however, is not aware of this energy. Consider a collision between a very massive object, such as a bowling ball, and a light object, such as a tenpin. In the collision, the direction of the bowling ball is not deflected very much; it retains most of its kinetic energy, and momentum and very little of this energy is transferred to the tenpin. A collision between a light object and a very heavy one has similar results. For example, when one bounces a ball off of the ground (off of the earth), the ball reverses its direction and rises almost to its initial height. Again, very little of its energy is transferred to the earth (though twice its momentum is transferred). Collisions between equal-mass objects maximize the amount of energy transferred between them. This concept of equal-mass collisions must be generalized because, for relativistic reactions, the mass is no longer constant. The maximum energy is available for the reaction if the sum of the momenta of all the particles in the reaction is zero. In a reaction of two particles, this criterion is met when the collision is head-on and the particles have equal but opposite momenta. If the two particles have the same mass, then they will also have the same energy if their momenta are matched.

Colliders are instruments to produce head-on collisions between elementary particles. If the momenta of the interacting particles are the same, then the maximum energy will be available to create new particles. The traditional way of carrying out high-energy physics experiments has been with "fixed targets." In fixed-target experiments, a very energetic beam of particles, from an accelerator, is directed at a target. The target is a piece of material at rest in the laboratory. Collisions then occur between the high-energy beam and the at-rest target particles. Relativity dictates that, for very high energies, the center-of-mass energy available for the reaction only increases as the square root of the beam energy. Therefore, if the accelerator beam energy were doubled, then the center-of-mass energy would only increase by a factor of 1.41. In a collider, if the accelerator beam energy were doubled, then the center-of-mass energy would increase by a factor of four, that is, a factor of two for each beam.

Colliders provide a very high center-of-mass energy per collision, but for careful studies, physicists need many collisions. In order to achieve many collisions between the two energetic particle beams, the beams are stored. A storage ring is a device very much like a particle accelerator, except that the beams are allowed to circulate for many hours or days. The circulating beams are then periodically, perhaps several times per cycle, brought into collision with each other. A measure of the collision rate is a quantity called luminosity. A machine's luminosity indicates how many collisions can be recorded for a given process. The luminosity is proportional to the number of particles in either beam. Clearly, if the number of particles circulating in one beam is doubled, then the chance that the beam will interact is doubled as well.

The luminosity is also inversely proportional to the cross-sectional area of the beams. If the particles can be squeezed down to a very small area, then there is a higher probability that they will interact with a particle in the other beam. Particles in storage rings are often stored in bunches. The luminosity also depends on the frequency with which the bunches cross. In a very large ring, the bunches may cross infrequently. Many bunches can also be stored, one behind the other in the ring, in order to increase the number of collisions.

In most colliders, two beams are brought to a head-on collision. The storage ring must be capable of storing beams of both particles for long periods of time, which means that the beams must be composed of such stable particles as protons and electrons, and their antiparticles (antiprotons and positrons). A substantial engineering benefit is realized if the collisions are between a particle and its antiparticle. The ring stores particles magnetically, so that if the momentum and charge of a stored particle are reversed, then the particle will follow the same path but in a reverse direction. If the two beams are particle and antiparticle, then both can be stored in the same ring. This is not always the optimal design choice, however, since one of the beams is composed of antimatter, which must be produced in high-energy collisions because it is not found in nature. This requirement limits the ultimate intensity of the antiparticle beam.

Applications

Storage rings and colliders have been successfully employed for the discovery and study of many new particles. Electron-positron colliders were the first to be widely exploited. The positron, which is the antiparticle to the electron, is relatively easy to produce. Since it is positively charged, it will circulate opposite to the electron in a storage ring and only one ring is needed. Electron-positron colliders produce reactions through the electromagnetic interaction. The fact that this interaction is well understood makes it much easier to design and carry out an experiment and to interpret the results.

Electron-positron colliders have been used to discover particles containing the quantum numbers of charm and beauty quarks. These are new unstable forms of matter that can be studied only as they are produced and allowed to decay in the laboratory. A very convenient feature of electron-positron colliders is that they are very effective at producing particles that have the same quantum numbers as the photon. Recall that the photon is produced in the collision of an electron and an antielectron (a positron), so that neutral quark-antiquark states should have comparable quantum numbers and be easily produced. In general, quark-antiquark bound states are meson resonances. Many new particles can be found in the decay of these easily produced resonances. For example, the excited states of a charm-anticharm quark bound state (known as the psi) decay copiously to charmed particles. Similarly, an excited bound state of beauty and antibeauty (known as the upsilon) is used as a "B factory," a convenient and copious source of beauty particles. Electron-positron storage rings have been most effective when the operating energy has been chosen to take advantage of these channels for the production of new particles.

Electron-positron colliders were also used in the discovery of the tau lepton. The tau lepton is a particle which is very similar to the muon and the electron. It differs from these particles only in that it has a much higher mass, about thirty-six hundred times that of the electron. It also has a new quantum number which restricts how it may interact or decay. It can be most easily produced in the reaction in which an electron and positron annihilate to make a tau and an antitau lepton. In that way, no net change in the tau quantum number is produced because the tau and antitau have opposite values.

Electron-positron storage rings have a number of limitations. As the energy needed to effect the collisions rises, the cost of building and running the collider rises very rapidly. Much of the added expense is the cost of storing such energetic beams. As the electrons and positrons circulate in the storage ring, they undergo acceleration to keep them moving in a circle. Accelerated charged particles radiate energy. In a storage ring, this energy is known as "synchrotron radiation." The amount of energy radiated depends on the accelerated particle's mass and energy. This synchrotron radiation is a unique, intense source of ultraviolet light and X rays. A number of electron storage rings have been built to be used as a source of this radiation, which is a major tool in the study of materials. Synchrotron radiation is the major source of energy loss in electron-positron storage rings.

To avoid the limitations of synchrotron radiation, the concept of the linear collider has been proposed. A linear collider (also called a single pass collider) would have the center-of-mass energy advantage of a storage ring but would not store the beams. The beams are accelerated toward each other. They collide and then are discarded. In order for such a method to produce the high luminosity needed to obtain enough collisions, the beams themselves must be very intense because they will only cross once and are not circulated or stored for reuse.

Proton-antiproton colliders are used to achieve energies five to five hundred times those that are accessible to electron-positron colliders. The proton is eighteen hundred times heavier than the electron, so synchrotron radiation, which depends on the particle mass to the inverse fourth power, is no longer a serious problem. On the other hand, since they are so heavy, antiprotons are very hard to produce and store. Antiprotons are generated by nuclear collisions, but they make up only a small part of what is produced. Because they are produced in energetic collisions, they come out with a very wide range of momenta. In order to be stored, all antiprotons must have a very similar momentum. A technology called "cooling" has been developed to collect and store a suitable number of antiprotons to make collisions reasonably intense. Because of the difficulty of collecting and storing antiprotons, a number of colliders have been built with two separate storage rings, so that two beams of protons may be collided. Since the two proton beams have the same charge but are moving in opposite directions, they cannot be contained in the same magnetic storage ring.

Proton-antiproton colliders have been used to discover the very massive particles responsible for the weak nuclear force. These particles weigh from eighty to ninety times the mass of the proton. Since the proton-antiproton interaction is dominantly the strong nuclear force, the collisions produce many particles, and sophisticated detectors are needed to sift through the products to find events of interest.

The weak nuclear force is the dominant interaction for electrons and positrons when they have a center-of-mass energy near the mass of the neutral weak boson, known simply as the Z. This feature has been employed to produce and study this neutral weak boson. In this case the benefits outweighed the problems associated with electron-positron colliders at high energy.

In spite of the gains in center-of-mass energy to be had from equal energy in each beam, it is sometimes useful to have beams of different energy. Such machines are called asymmetric colliders. Asymmetric colliders are used when it is important to push the particles resulting from the interaction in a specific direction in order to observe them better. For example, if one is studying the lifetime of unstable particles, then this push will time-dilate (slow) the lifetime so that it may be more easily observed. Colliders have also been built with dissimilar particles, such as an electron-proton collider. These provide the advantages of a collider but the ability to study the proton structure with an electromagnetic probe.

Context

The fundamental concepts involved in colliders are fairly basic in physics, but realizing the concept required much technology. The technology of particle accelerators themselves had to be developed to a high degree, as storage rings are basically a kind of accelerator. Any residual gas present in the ring would interact with the stored beam and eventually destroy it. Therefore, storage rings require a very high vacuum in order to operate reliably. The storage time, or beam lifetime, for conventional accelerators is on the order of seconds, while storage rings must maintain a stable beam for ten thousand times longer or more. The requirement of stability also makes severe demands on the focusing system of the accelerator. In order to keep a collection of particles together in a beam, all the particles must travel on nearly the same path for days. This places high stability requirements on the equipment and necessitates high sensitivity to engineering tolerances in the construction. The demand for high luminosity means that very high beam currents must be stored. Interactions between the colliding beams and within the bunches themselves lead to destabilizing effects for which the accelerator focusing system must compensate.

Storage rings became major tools in the 1970's and 1980's because physicists had a clear picture of the phenomena that they wanted to study and could identify the center-of-mass energy that would be most effective. These energies were beyond what could be conveniently reached with conventional, fixed-target devices, so there was a clear need for innovation.

Bibliography

Carrigan, Richard A., Jr., and W. Peter Trower, eds. *Particles and Forces at the Heart of Matter.* New York: W. H. Freeman, 1990. This is a collection of reprints from *Scientific American* on particle physics. Several articles extend the discussion of collider techniques and many present the motivation for and the results of collider experiments.

Cline, David B., Carlo Rubbia, and Simon van der Meer. "The Search for Intermediate Vector Bosons." *Scientific American* 246 (March, 1982): 48-59. This article tells the story of the first proton-antiproton collider and its use to discover the fundamental particles that mediate the weak interaction.

Jackson, J. David, Maury Tignier, and Stanley Wojcicki. "The Superconducting Supercollider." *Scientific American* 254 (March, 1986): 66-77. Describes the motivation and design of the Superconducting Supercollider, a very ambitious storage ring project. This device is designed to collide two proton beams at an energy of 20 trillion electronvolts each.

Mistry, Nariman B., Ronald A. Poling, and Edward H. Thorndike. "Particles with Naked Beauty." *Scientific American* 249 (July, 1983): 106-115. This article discusses the use of electron-positron collisions to produce particles with the beauty property. Since these particles are unstable, they can only be studied as they are produced. Electron-positron collision is the best method known for the production and study of beauty quarks.

Perl, Martin L., and William T. Kirk. "Heavy Leptons." *Scientific American* 238 (March, 1978): 50-57. Discusses the discovery of the tau lepton with electron-positron collisions. The tau lepton is a member of the third family of elementary particles.

Wilson, Robert R. "The Next Generation of Particle Accelerators." *Scientific American* 242 (January, 1980): 42-57. This article describes the evolution of the particle accelerators that are used for high-energy physics research. Explains the difference between fixed-target experiments and collider experiments.

John M. LoSecco

Cross-References

Antimatter, 123; Baryons, 195; Bosons, 252; Conservation Laws, 546; Detectors on High-Energy Accelerators, 643; Forces on Charges and Currents, 907; Mesons, 1374; Newton's Laws, 1509; Synchrotrons, 2464.

STRING THEORIES

Type of physical science: Elementary particle (high-energy) physics
Field of study: Unified theories

String theories are a set of theories which propose that subatomic particles are not hypothetically defined by points in space and time but by tiny, one-dimensional strings.

Principal terms

NINE SPACE: the concept that the universe is made up of nine spatial dimensions

PLANCK LENGTH: an exceptionally small subatomic measurement on the order of 1.6×10^{-33} centimeters

RENORMALIZATION: the process of canceling out absurd mathematical predictions in a physical theory by introducing mathematical factors in equations

RESOLUTION: in subatomic experimentation, the smallest resolvable distance that can be probed with a synchrotron

STANDARD MODEL: the theoretical model used to probe the subatomic structure since 1960

SUPERGRAVITY: a theoretical attempt to unify gravity with elementary particle theory

SUPERSTRING THEORY: a theoretical proposal that defines elementary particles as tiny strings in ten dimensions

SUPERSYMMETRY: a theory that extends the symmetry relationships of the standard model to the Planck scale

SYNCHROTRON: a particle accelerator used in theoretical physics to cause very high-speed collisions between subatomic constituents

TOPOLOGIES: descriptions of how strings are expressed in one dimension, typically as "lines" or as "circles"

Overview

The 1960's and 1970's were remarkable years in theoretical physics. Discoveries about the atomic structure were inundating physicists gathering data with the aid of the world's enormous high-energy particle accelerators (synchrotrons). Physicists used these machines to accelerate atomic parts to enormously high velocities and crash them, headon, into other atomic parts. What resulted was a shower of smaller particles as the atomic pieces shattered. Then, physicists would look at the tracings from their instruments and discover the nature of the new particles, later cataloging and naming them.

German physicist Albert Einstein (1879-1955), in his elegant and revolutionary theories of relativity, demonstrated unequivocally that matter and energy, under the right conditions, were interchangeable. Einstein also firmly believed that every law in physics could be joined into a single, all-encompassing theory he called the unified field theory (also wryly known in science as "the theory of everything"). Such a theory would unite the four forces of nature: gravitation, electromagnetism, and the two nuclear forces (strong and weak).

Armed with this knowledge, information coming in from particle accelerator experiments and observations of the universe, scientists developed an all-encompassing vision of creation that accounted not only for the formation of the stars and planets but also for all matter itself. This theory, called the big bang theory, stated that the entire universe began from a tiny pinpoint of energy, infinitely hot and infinitely dense. Ten billion years ago, this tiny point of energy exploded in an unimaginably violent eruption. As the universe expanded, it was at first so hot that it was absolute energy, and no matter could exist at all. As it expanded and cooled, however, tiny particles of matter began to precipitate out of the cloud of energy. These tiny particles came together and ultimately formed what are known as atoms. The first matter to precipitate was the simplest form of matter: hydrogen. Although the big bang theory dealt with the creation of the universe from a philosophical point of view, it did not embody a completed, unified field theory. As physicists began to look at the particles that they were creating with their accelerators, their ultimate aim was to construct a unified field theory that would also explain how matter itself formed out of the primordial explosion. They observed that it required more and more energy to divide atomic particles into smaller and smaller pieces. The act of completely disassembling matter seemed to require all the energy of the big bang itself—in reverse.

A synchrotron can be compared to a microscope, enabling physicists to peer ever deeper into an atom and even into the subparticles of atoms. With what is presently known about the atom, it is relatively straightforward to calculate how much energy it requires to look into its innermost parts. What physicists have found is that, using only the current from a common household outlet, they can look into the atom and distinguish the electron from the atomic nucleus. To look into the nucleus, however, requires far more energy. Furthermore, to look inside the proton and neutron requires the energy of a small city. Resolution is the smallest distance that can be probed with a synchrotron, and with the largest particle accelerator in the world, current resolution is approximately 1×10^{-16} centimeters (which enables scientists to look inside protons and neutrons). Scientists have calculated that to disassemble the atom completely would require a resolution smaller than 1.6×10^{-33} centimeters (called the Planck length) and a particle accelerator larger than tens of thousands of solar systems packing the energy of the stars.

In the 1980's, this calculation created something of a crisis in the physics community. Ideally, theoretical physics is seamlessly joined to experimental physics. Theorists develop theories that are ultimately proved by evidence from particle accelerators. Without supporting evidence, theories cannot develop beyond theories: They default to mere conjecture and philosophical musings. One American physicist, Howard Georgi, described this dilemma as having plunged physics into mere "recreational mathematical theology." Because the energy required to confirm theories seemed to surpass available resources and technology, the physics community embarked on untestable grounds of theory by the early 1980's.

The drive to develop a single grand unified field theory has been approached with partial theories that join two of the four forces of nature together. The joining of two of these theories has been a largely successful approach mathematically, but when physicists have attempted to go further or to join such physical concepts as relativity and quantum mechanics, glaring, seemingly absurd errors arise out of massively complex mathematical constructs. These errors are canceled out by a process called renormalization, which means that physicists introduce values into their calculations to cancel out the errors and fit the observations. Hence, these theories are incomplete because they cannot exactly predict observed processes without renormalization.

One such attempt was the effort to join quantum mechanics to general relativity (called a classical field theory as compared to a quantum field theory). Renormalization was not even possible. Finally, a concept was devised called supergravity, in which a theorized particle called the graviton (particles responsible for the expression of gravity) was united with other speculated particles into a single particle with widely different characteristics (dubbed a "superparticle") because it would have solved most of the renormalization problem. The term "super" resulted from a concept called supersymmetry, which defined particle interactions. Supergravity was a great theoretical construct, but it was seriously deficient in describing the particles that were actually observed in synchrotrons. Until 1984, it was widely believed that supergravity was still the prime contender for the grand unified theory itself.

In 1919, German physicist Theodore Kaluza, intrigued by Einstein's view of a four-dimensional universe (width, depth, height, and time) made one of the earliest attempts at a unified field theory by incorporating electromagnetism as a fifth dimension. Swedish physicist Oskar Klein later linked Kaluza's idea to quantum mechanics. The Kaluza-Klein theory stated that a fifth dimension, of which humans have no manifest awareness, was literally curled up in a tight circle and played no part in the observable universe. Yet, the aspect of this seemingly insignificant and deeply hidden fifth dimension allowed for a partial unification theory. It was such a bizarre concept that it was virtually ignored for more than fifty years and has since been proven not viable.

Just as the Kaluza-Klein five-dimensional unification concept languished on the shelf, quantum mechanics was born, which described the physics of the atomic interior. Quantum physics, like relativity, is a four-dimensional science. It describes a subatomic world so divorced from the cause-and-effect, easy-to-understand world of classical physics that even Einstein declared it a "stinking mess" and shunned it for all of his professional life.

Quantum mechanics clearly won the war over relativity's classical field theory relative to the subatomic world, and armed with its often bizarre philosophical notions, physicists began to peer deeper and deeper into the atomic core. Yet, quantum physicists occasionally needed Einstein's relativistic notions even at atomic distances, and they used them in an untidy mergence fraught with the constant necessity to renormalize and cancel out mathematical absurdities whenever the two disparate sciences were forced to converge.

The initial attempts to look into the atom's core uncovered discrete particles called the proton and neutron. Using what physicists called the standard model, particle physicists found that there were smaller particles still deeper. As they used synchrotrons with higher and higher energies, however, the mathematics of the standard model began to be less and less effective at explaining the smaller values. Furthermore, none of them could account for gravity at quantum distances and masses.

Theoretical physics, despite the strangeness of quantum reality, still had difficulty in imagining more than a four-dimensional universe. String theory can be traced back to 1960, when Gabriele Veneziano and Yoichiro Nambu attempted to use concepts involving more than four dimensions to support a discovery in the standard model that described discrete particles or events. These theories used a one-dimensional line or a one-dimensional line curled back in on itself to form a circle—hence the name "string theory."

In 1984, two of string theory's strongest proponents, John H. Schwarz of the California Institute of Technology and Michael Green of Queen Mary College, London, suggested that merging concepts in string theory with supersymmetry would free many renormalization problems and mathematical anomalies. This merger also neatly incorporated gravity into quantum physics for the first time, and they called it the superstring theory. These one-dimensional strings exist in two "topologies": open and closed strings. An open string has the topology of a line, and a closed string has the topology of a circle. String theory defines the distinct particles discovered by the standard theory as states or modes (conceivably vibrations) of a single string. Presumably, string states that are smaller than the Planck mass should be observable as distinct, experimentally derived particles.

There are three leading string theories: type I, type II, and heterotic. All these theories share some major attributes. They involve a theory of the universe encompassing ten dimensions, nine spatial and one in time. All incorporate quantum gravity, free from arbitrary mathematical renormalization. The three theories are distinct because of how they deal with internal symmetry between the grouping characteristics of standard theory particles and how they approach symmetry between the particles. The string theories offer superior answers to questions posed by the limitations of the standard model at resolutions at or smaller than the Planck length, but they also pose other questions. Only one of the string theories will ultimately survive, but whether it will be a combination of the three theories, two of them, or a yet unknown competitor is still uncertain.

The primary complaint against string theory has been that it is untestable because of the extraordinary energies required to look into the deepest part of the atom. Yet, by the late twentieth century, those complaints were being addressed. Researchers announced that they had developed an extension of string theory that allows for experimental verification at currently available synchrotron energies.

Understanding the concept of extra dimensions has been a challenge to physicists from the outset of modern physics. The three dimensions of width, depth, and height are obvious realities of the physical world, and as such require little deliberation to understand just how they fit into the physical world. Einstein tied time into the physical world through his theories of relativity and formally introduced it as the fourth dimension. Although such a linkage was all but conspicuous with minimal thought, Einstein successfully demonstrated the linkage of time and space mathematically. When the Kaluza-Klein theory of a universe of more than four dimensions was first contemplated, the ability of physicists or anyone else to look at the universe as it is and clearly understand it through visual observation alone began to break down. Although a fifth dimension could apparently unify electromagnetism with relativity in mathematical theory successfully, the concept of an extra hidden dimension simply could not be visualized in a rational sense outside the enigmatic mathematical paradigms used to described it.

When supersymmetry and superstring theory overtook the Kaluza-Klein theory in the 1970's, it brought with it not one extra dimension, but six, and the difficulty in understanding the idea of extra dimensions increased dramatically. Physicists began to refer to the new multidimensional universe as nine space (referring to the nine spatial dimensions besides time). The simplest question that can be asked about nine space is why it cannot be observed directly. The answer is relatively straightforward: Although humans live in those nine spatial dimensions which make up all of matter, the extra six dimensions are curled up and embedded in the subatomic world, in a region of space so tiny that they simply cannot be observed directly. Theory predicts that, if these extra dimensions do indeed exist, then they are confined to regions on the order of the Planck length.

To observe anything with particle accelerators directly at such a small scale would require energies that are 1 billion times greater than have ever been produced. Consequently, humans observe the universe on a large scale and see the three-dimensional, smoothed surface of matter that is, in fact, made up of a subsurface with an extraordinary, fine veneer rich in dimensional texture.

The closest look at the texture of all matter reveals that the tiniest details of substance are made up of points in space that open up into highly curved regions with nine spatial dimensions. While these tiniest regions of space can be observed, however, it can only be seen that they open and close back in on themselves (they are curled up into these tiny regions) in a distance of 10^{-33} centimeters. These tiny points of space and time are so very highly curved that, if it were possible to inject a person into these tiny universes, then that person would strike out in a given direction and end up back where he or she started in only 10^{-33} centimeters.

The question that physicists are asking is why these six tiny dimensions are so radically curved and, hence, are so tiny. They also ask why the other three spatial dimensions that one can directly observe so comparatively are flat and spread out for light-years to the edge of the observable universe. Such questions are at the very frontier of theoretical physics, and outside of sheer philosophy, they do not have clear answers.

Applications

Unification theory has come to dominate the thinking of physics. String theory has not only offered the hope of achieving physics' greatest objective but also opened up wide new avenues for experimental physics. With this new theory, scientists have been finding ways to use available synchrotrons to peek inside regions of the atom heretofore thought impossible. By altering traditional techniques, they may be able to verify the existence of the strings indirectly. This has energized the community of experimental physicists, who are busy designing such experiments.

As physicists delve more deeply into the string models, new approaches continually suggest themselves. Hence, the string theories continue to evolve into more accurate representations of the actual atomic core. Current models predict not only new strings further in the atom (which, again, may also be related as new particles) but also gravitons and even an unsuspected form of invisible matter being dubbed "cryptons."

Vitally important information is also being reexamined by the new string theories. The standard model theory suggested that protons (hence all matter itself) was unstable and would eventually decay. Experiments have not supported this theory, but the theoretical evidence is substantial. String theory also supports proton decay but declares it to occur at a period significantly longer than standard theory predicts. With the new string theory in hand, researchers will be able to reestablish the watch for decayed protons.

String theory is a staggeringly complex mathematical construct. Such theories usually begin with simple conceptual notions, then the appropriate mathematics are applied to them. Einstein's theories of relativity are examples, and string theory is at just such a stage. To incorporate string theory into the appropriate geometries, however, will probably require the development of a new mathematical methodology.

Context

String theory may represent the final thrust to develop an experimentally supported grand unified field theory, or it may merely represent another step to the ultimate theory in physics as physicists peer ever deeper into the atom. In either case, string theory has become one of the most intensely developed theories in physics since quantum mechanics opened the doors to the center of the atom in the early twentieth century.

String theory will ultimately represent a radical new way of viewing the universe. Just as relativity opened doors to new concepts never before dreamed of by classical physicists, so then does string theory. String theory has hinted at a universe of ten dimensions. How six extra dimensions in the universe could lie hidden in the depths of all matter is an extraordinary, meaningful philosophical question that may ultimately touch every aspect of science as a whole. Verification of such invisible matter as cryptons (often called shadow matter) would have important implications for cosmology. If its mass is as large as string theory predicts, and its stable lifetime is 10^{18} years, as string theory suggests, then it would probably account for enough "missing or dark matter" to allow the universe ultimately to contract back into the big bang state, which is a fundamental question of contemporary science.

The idea that proton decay is longer than was expected also has important implications for science. If string theory can accurately predict this decay constant, it may also provide important alternative clues as to why this event has not yet been observed. A new round of experiments may support this decay, which would also verify aspects of string theory through experimentation.

Developing methods of testing string theory has vital ramifications for both experimental and theoretical physics. Without the ability to test its theories, even with such elegant representations as string and superstring theory, the very essence of science is rendered ineffective. The methods that are underway to test string theories will ultimately mean much for physics and its future. Throughout the history of science, humanity has sought the fundamental questions about the tiniest parts of all that there is. Some of these questions may lie in the multidimensional, supersmall world of superstrings.

Bibliography

Crease, Robert P., and Charles C. Mann. *The Second Creation*. New York: Macmillan, 1986. In this magnificent book, Crease and Mann follow the making of twentieth century physics from its nineteenth century roots to today's most enigmatic mysteries. It is a book that microscopically examines characters and personalities as well as the issues of physics. Strings are discussed alongside the other pioneering work of the mid- to late 1980's, and they are linked clearly to the unified field theories. This book is highly readable and is probably the most complete book ever written on the personalities and work of twentieth century particle physicists.

Hawking, Stephen W. *A Brief History of Time*. New York: Bantam Books, 1988. In this eminently readable work, one of the most prominent physicists of the twentieth century examines the universe from his view of creation to the present. Hawking examines the far-flung reaches of space and time, from black holes to the interior of the atom, and discusses the elementary particles of the atomic nucleus. Hawking also discusses the concept of string theory, its relationship to grand unified theories, and what it means to physics. This book is written for the lay public, illustrated, and easy to understand.

Pagels, Heinz R. *The Cosmic Code*. New York: Bantam Books, 1982. This book describes quantum physics as "the language of nature." Pagels embarks on a literary quest to explain some of the most profoundly difficult topics in quantum physics to the layperson in a clear manner. He succeeds and opens up the interior of the atom for a clear view of what is inside. The book reads clearly and is illustrated, and one need not be a physicist to enjoy the material.

____________. *Perfect Symmetry*. New York: Bantam Books, 1985. In this follow-up book to *The Cosmic Code*, Pagels delves even deeper, providing a layperson's perspective. The aspect of the atom's core is covered and Pagels supports a lively discussion of the grand unified field theories and other frontier topics in physics. A very readable book that has become a classic of the genre.

Schwarz, John H. "Superstrings." *Physics Today* 40 (November, 1987): 33-40. This article, written by one of the leading theoretical physicists and proponents of string theory, is an excellent and widely referenced article on the theory of strings. Provides a superior insight into the history of strings and its promise. It is of moderate difficulty, but well worth the effort. Illustrated.

Sutton, Christine. *The Particle Connection*. New York: Simon & Schuster, 1984. Sutton, a former physicist turned reporter, explains the details behind the tools of the physicist's trade: the particle accelerator. She follows up the discussion with stories of how the machine is used and the nature of the particle chase at CERN, the European laboratory for particle physics. Illustrated, and reads so that the student with a reasonable grounding in science can grasp its message.

Dennis Chamberland

Cross-References

The Structure of the Atomic Nucleus, 182; The Big Bang, 210; General Relativity: An Overview, 950; Grand Unification Theories and Supersymmetry, 985; Nonrelativistic Quantum Mechanics, 1564; Nuclear Forces, 1580; Special Relativity, 2333; The Standard Model, 2353; Synchrotrons, 2464; The Unification of the Weak and Electromagnetic Interactions, 2635.

SULFUR COMPOUNDS

Type of physical science: Chemistry
Field of study: Chemical compounds

Sulfur, the thirteenth-most-abundant element in the earth's crust (before carbon and chlorine), forms a wide variety of compounds of importance in industrial chemistry, in inorganic and organic laboratory chemistry, and in the chemistry of living organisms.

Principal terms

AMIDE: an organic compound in which the acidic hydrogen atom of a carboxylic or sulfonic acid is replaced by an $-NH_2$ group ($RCOOH$ to $RCONH_2$, or RSO_3H to RSO_2NH_2); when more than one acid group is attached to the nitrogen, the compound is an imide

CROSS-LINKS: connections between two polymer chains, or between different parts of a single polymer chain; the connections may be direct bonds, or bonds through one or more other atoms

HETEROCYCLE: an organic ring compound (a compound in which a group of atoms forms a closed, cyclic structure) in which the ring is composed of atoms of more than one element

OXIDATION: in inorganic chemistry, removal of electrons from an atom or ion; in organic chemistry, removal of hydrogen atoms from a compound, or addition of oxygen atom(s)

POLYMER: a long chain compound whose molecules are hundreds or thousands of atoms in length and are composed of small groups of atoms that repeat like links in a chain

PROTEIN: a biological polymer in which the links are amino acids

REDUCTION: in inorganic chemistry, addition of electrons to an atom or ion; in organic chemistry, addition of hydrogen atoms to a compound, or removal of oxygen atom(s)

THIO-: a prefix in chemical nomenclature that indicates that one or more atoms in a compound, usually oxygen atoms, have been replaced by sulfur atoms (for example, a thioester, RCOSR, from the regular ester, RCOOR)

Overview

Sulfur is a group VI element, which means that it has six electrons in its valence shell. Thus, in its inorganic compounds, it can either pick up two electrons to form a -2 ion (sulfide, S^{2-}) or give up electrons, usually in twos, to form various oxides and oxyanions, among them sulfur dioxide (SO_2), sulfur trioxide (SO_3), sulfite (SO_3^{2-}), thiosulfate ($S_2O_3^{2-}$), and sulfate (SO_4^{2-}) ions, the last frequently in the form of sulfuric acid (H_2SO_4). Except for the sulfides, these inorganic compounds are covalently bonded about the sulfur atom, with the entire oxyanion acting as a building block in an ionic crystal. The organic compounds of sulfur are entirely covalent, with sulfur bonded to carbon or hydrogen atoms with two shared electron pairs, or to oxygen by double bonds. Examples include mercaptans (R–SH), sulfides (R–S–R), disulfides (R–S–S–R), sulfonates ($R-SO_3H$), and sulfoxides (R–S(=O)–R) (in all cases, R– stands for an organic molecular structure of almost any kind). Many of these compounds are found in plant and animal chemistry. Sulfur can also enter into polymeric structures, as part of the primary polymer chain or as a cross-linking agent. Finally, sulfur is an integral part of many molecules in physiological chemistry, where it forms a cross-linking bond between protein chains, as well as part of the ring structure of some pharmaceuticals.

Among inorganic sulfur compounds, the sulfides include a number of minerals that are ores for both sulfur and a metal, such as galena (PbS, lead sulfide), cinnabar (HgS, mercuric sulfide), or iron pyrites (FeS_2). The metal is recovered by roasting the ore in air, which oxidizes the sulfide to sulfur dioxide. This can be recovered and used to make sulfuric acid. Mercuric sulfide is also used as a pigment, as are zinc and cadmium sulfides (ZnS and CdS). The last two compounds give light when struck by electrons, and are used as phosphors in cathode-ray tubes and some laboratory instruments. Hydrogen sulfide (H_2S) is a highly toxic gas with the odor of rotten eggs.

Sulfur dioxide and sulfur trioxide are made by burning elemental sulfur in air. Sulfur dioxide has the sharp, acrid odor observed in fumes from burning matches and fireworks. It can be dissolved in water and neutralized with soda to make sulfites (Na_2SO_3) and bisulfites ($NaHSO_3$). Sulfites, bisulfites, and the parent sulfur dioxide are used as disinfectants, preservatives, and bleaches, particularly for foodstuffs and natural products. Textile fibers, wicker, gelatin, and beet sugar are bleached in this way. Sulfur dioxide has been used as a disinfectant in breweries and as a preservative in table wines.

When sulfur trioxide reacts with water, the product is sulfuric acid:

$$SO_3 + H_2O \rightarrow H_2SO_4$$

Sulfuric acid is by a considerable margin the largest-volume material produced by chemical industry, and has been for at least two centuries. Production in the United States in the 1970's and 1980's averaged nearly 40 million tons per year. Some 70 percent of sulfuric acid production goes into fertilizers, as ammonium sulfate or calcium acid phosphates. The rest is used in other manufacturing processes.

Some metal sulfates are of historical interest or industrial importance. Glauber's salt, sodium sulfate decahydrate ($Na_2SO_4 \cdot 10H_2O$), the "sal mirabile" of alchemy, is still used in dyeing and printing, and as an ionic bulking agent in some detergents. Epsom salt, magnesium sulfate heptahydrate ($MgSO_4 \cdot 7H_2O$), has a place in the pharmacopoeia. Gypsum, calcium sulfate dihydrate ($CaSO_4 \cdot 2H_2O$), can be dehydrated to the hemihydrate, $CaSO_4 \cdot \frac{1}{2}H_2O$, which is plaster of Paris, or to anhydrite, $CaSO_4$, which is a useful drying agent. Plaster of Paris hardens with water by returning to the gypsum structure. Barium sulfate ($BaSO_4$) is the opacifying agent used in X rays of the gastrointestinal system and is used industrially as a pigment base.

Some other inorganic sulfur compounds are important. The ions thiosulfate ($S_2O_3^{2-}$) and dithionite ($S_2O_4^{2-}$) are valuable reducing agents in the laboratory, usually as the sodium salts. Sodium thiosulfate pentahydrate ($NA_2SO_4{\cdot}5H_2O$) is the "hypo" used in photographic development. Its purpose there is as a complexing agent, to remove unreacted silver ions, which, if left in the negative, would cause overall blackening.

Organic sulfur compounds can be grouped as reduced or oxidized sulfur compounds, the structural difference being that the reduced compounds contain only single bonds to the sulfur atom(s) while the oxidized have double bonds, usually to oxygen. Many of the reduced compounds have pungent odors, some quite unpleasant, or are components of flavor in vegetables such as onions and garlic. This is particularly true of the thiols, or mercaptans (R−SH), or of compounds that contain the −SH group. We may take the mercaptans in ascending order of complexity of the carbon-containing group: Methyl mercaptan (CH_3SH) has been identified in radish root. Ethyl mercaptan (CH_3CH_2SH) is not found in nature but is used as an odorant in natural (piped) and cylinder gases; it can be detected at levels as low as 0.02 part per billion. The compound n-propyl mercaptan ($CH_3CH_2CH_2SH$) is found in onions. Isoamyl mercaptan, $(CH_3)_2CHCH_2CH_2SH$, and trans-2-butenethiol, $CH_3CH{=}CHCH_2SH$, are components of the spray of skunks. Dimercaptans are found in asparagus and cabbage:

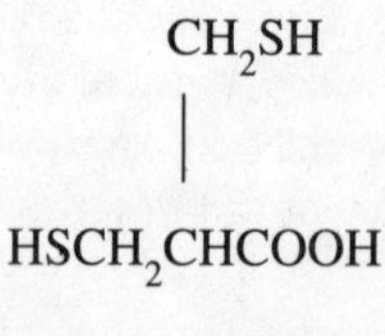

from asparagus

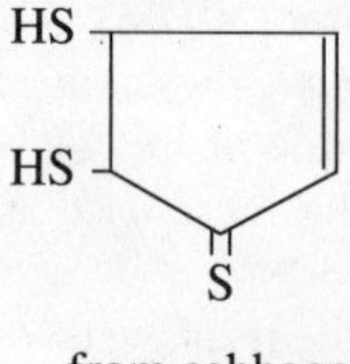

from cabbage

The compound found in asparagus gives rise to two pungent metabolites that appear in the urine after one eats asparagus: $CH_3{-}S{-}CH_2CH_2COSH$ and $CH_2{=}CHCOSCH_3$.

Sulfides (R−S−R) and disulfides (R−S−S−R) include dimethyl sulfide, $(CH_3)_2S$, isolated from paper-pulp-processing liquors. Various sulfides are found in plants of the *Compositae* family, such as marigolds and chrysanthemums. The principal component of garlic oil is diallyl disulfide, $(CH_2{=}CHCH_2)_2S_2$. Pineapple juice contains the mercapto ester $(CH_3)_2CHCOSCH_3$. The seeds and oil of mustard plants contain complex sulfur compounds whose molecules also incorporate a molecule of the sugar glucose. The vesicant (blister-producing) mustard gas, used in war, is the simple sulfide $ClCH_2CH_2{-}S{-}CH_2CH_2Cl$.

Various biochemical compounds are sulfides, many of them heterocycles (ring structures containing carbon with other elements) that also contain nitrogen. Coenzyme A, abbreviated CoA or CoA-SH, is a large molecule with a mercapto group that can be converted into a thio ester. Acetyl CoA has the structure $CoA-S-COCH_3$, and can transfer the acetyl group to other biological molecules. The natural penicillins are sulfur-nitrogen heterocycles. The compound luciferin, which on biological oxidation provides the light of glowworms and fireflies, is also a sulfur-nitrogen heterocycle.

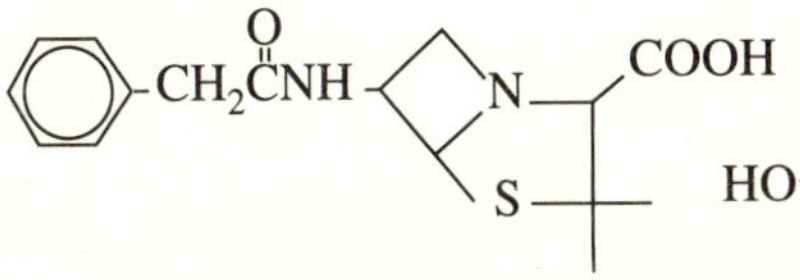

penicillin G

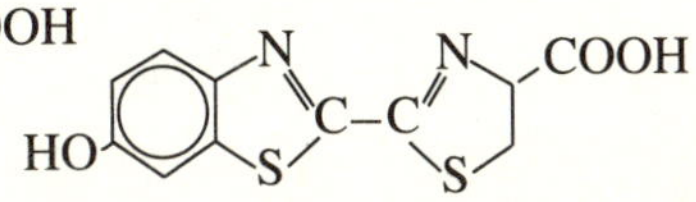

luciferin

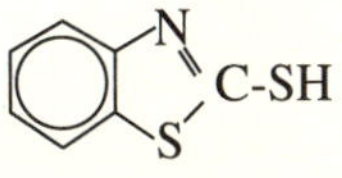

mercaptobenzothiazole

Sulfide and disulfide cross-linkages play a part in both biochemistry and industrial chemistry. Disulfide links are found in protein molecules; mercapto groups in the amino acid cysteine form a disulfide bond that links one part of the long protein chain to another, to hold it in a particular shape required for its job as an enzyme or other biologically active compound. Presence of many disulfide cross-links produces the rigid structural proteins found in hair, horn, hoof, and nail materials, which cannot be broken down by most animal digestive systems. This is why owls cough up pellets containing feathers, claws, and so forth of creatures they have eaten whole. It is, conversely, why moth larvae can eat wool fibers; they have the enzymes to handle the disulfides.

Sulfide linkages are of immense importance in rubber chemistry. The isoprene polymer that is rubber (as found, for example, in rubber cement) has poor structural properties. It can be firmed up by mixing with elemental sulfur and heating, a process called vulcanization. The sulfur reacts with residual double bonds in the polymer chains, attaching them to one another by sulfide cross-links. Depending on the number of cross-links, the rubber can be made elastic, or firm, or even totally rigid. The chemicals used in the rubber industry to accelerate vulcanization are sulfur compounds; mercaptobenzothiazole is only one of many such compounds.

Oxidized sulfur compounds include sulfoxides, RS(=O)R; sulfonic acids, RSO_3H; and sulfate esters, $R-OSO_3H$. The most important of the sulfoxides is dimethyl sulfoxide, or DMSO, $(CH_3)_2S=O$. It is a powerful solvent in both laboratory and industrial chemistry. In the latter, it is used to dissolve the synthetic polymer Orlon; the solution is extruded through fine jets into a water bath, where the DMSO disperses into the water and leaves a polymer strand. DMSO has the interesting property of crossing the skin barrier of the body, carrying certain dissolved drugs with it.

Although sulfonic acids and sulfate esters have important laboratory uses, their most common application is as detergents. Detergent molecules have the form $R-SO_3^-Na^+$ or $R-OSO_3^-Na^+$, where R is a long hydrocarbon chain, typically $C_{12}H_{25}$. They act by combining the hydrocarbon end of the molecule with greasy dirt, while the ionic sulfonate end combines with water, forming a bridge that allows the dirt to be rinsed away in the water. The virtue of sulfonate detergents over soaps, which act similarly but have a carboxylate ion as the water-soluble end, is that the sulfonate ion does not precipitate with the cations of hard water, remaining effective even in seawater.

Two other sulfonates should be mentioned; both appear as amides. The first is an example of the group of antibacterial agents called sulfa drugs, all of which have the *p*-aminobenzenesulfonamide structure of sulfanilamide, as shown. The differences lie in the nitrogen-containing structures (some quite complex) that replace the $-NH_2$ group at the right of the structure. The second is saccharin, a cyclic imide of benzene-1-carboxylic-2-sulfonic acid.

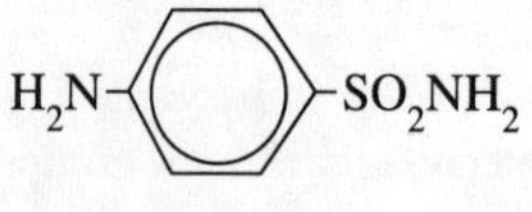

sulfanilamide

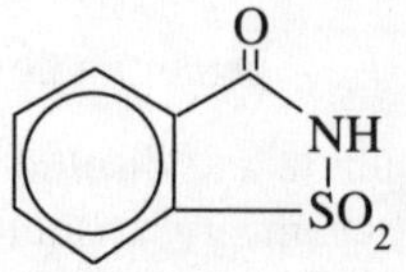

saccharin

Applications

Sulfur compounds are ubiquitous in chemistry. Both inorganic and organic sulfur compounds are found everywhere in laboratory chemistry, industrial chemistry, and physiological chemistry. For all that, however, they are little in evidence in ordinary applications. Most people know, for example, about chlorine compounds as hazards to the environment; about heavy metals as toxins; about acids in general, even in household chemistry. Yet, except for sulfuric acid (which most people never have seen as such) and possibly sulfites, people are not likely to know which sulfur compounds affect them and their surroundings, and how.

This is in part because the compounds do not advertise themselves as containing sulfur. One must look closely at the label of lawn and garden fertilizer to find that it contains ammonium sulfate; one must know some chemistry to realize that the hydrogen ion in the acid phosphates came from sulfuric acid. The alum used to crisp up homemade pickles is ammonium aluminum sulfate, but the can says so only in small letters. Similarly, small print identifies California wine as containing sulfites. As for pharmaceuticals, the system adopted for condensing their long chemical names more often than not conceals all information about the chemical elements they contain. Certainly, onions and garlic (or the friendly skunk at the back door) carry no labels at all to identify their flavor- and odor-producing materials. It is easy to overlook the sulfur compounds that affect everyday life.

It is even easier to overlook, or know nothing of, the larger volumes of sulfur compounds that are consumed internally by the chemical industry or in other manufacturing processes. As one example, the important industrial base soda ash (sodium carbonate, Na_2CO_3) was at one time manufactured by the Leblanc process, using sulfuric acid as a starting material. Salt (the mineral halite) was treated with sulfuric acid to produce salt cake (sodium sulfate), which was then reacted with limestone (calcium carbonate) to precipitate calcium sulfate and free the sodium carbonate. Thus, the presence of sulfuric acid in soda ash manufacture was concealed, and as the soda ash was generally used in further manufacture, its presence was not found in the final products, either.

The fertilizer and rubber industries have already been mentioned as internal consumers of sulfur products. The paper industry consumes sulfite and sulfate pulp. In cloth dyeing and finishing, enormous quantities of alums—sulfates of iron (III), chromium (III), or aluminum (III)—are used as mordants in the dyeing process. Aluminum alum is used in the manufacture of dyes, to make "lakes." It is also used to clear water of clay and other colloidal material in public water systems; in sugar manufacture, to clarify the product; in tanning; in marble and porcelain cement; and in a dozen other applications that are seldom seen because the products used by consumers show no traces of the chemicals used in manufacture. Sulfur compounds are truly ubiquitous.

Context

Sulfur is one of the handful of elements known as elements since ancient times. This fact is reflected in the name itself, which comes to us from Sanskrit via Latin, and refers in meaning only to the element itself, not to its origin or chemical properties (as do "francium" and "germanium," for example, and "silicon"—from "flint," *silex*). Another clue to its age is that each language has its own name for sulfur, just as for iron or gold; this is in contrast to the recently discovered elements like ruthenium or argon, whose names are identical in all tongues.

Compounds of sulfur have long been known, as well. Some of the ancient names are still familiar. "Oil of vitriol," meaning (more or less) concentrated sulfuric acid, comes from alchemical times, as do "blue vitriol" (copper sulfate) and "green vitriol" (ferrous sulfate), and the various "alums" mentioned above. Other examples of sulfates as vitriols can be given, as well as of sulfates named for their discoverer or place of origin (Glauber's salt, Epsom salt). "Orpiment" (antimony [III] sulfide), "liver of sulfur" (mixed polysulfide and thiosulfate of potassium), and "flowers of sulfur" (elemental sulfur as a finely divided powder) are other old names, still current, that show our long familiarity with sulfur compounds.

Sulfur is found in elemental form at the rims of volcanoes (the "brimstone" of the Bible) and, more important, in underground deposits. In the nineteenth century, the most important source was Sicily, where the sulfur-bearing earth from the mines was mounded up and burned. The heat from burning part of the sulfur melted the rest, which flowed out and solidified as it cooled. In 1894, the extensive deposits in Texas and Louisiana were first tapped by the Frasch process, which melts the sulfur in situ by pumping superheated steam down to the underground bed, then forces the molten sulfur to the surface with compressed air. Today, this is the principal source of elemental sulfur.

As already noted, sulfuric acid is the leading heavy-industrial chemical. (Other inorganic contenders are nitrogen and oxygen, from liquid air, lime, ammonia, and caustic soda.) It can be made from elemental sulfur or from the sulfur dioxide obtained when sulfur-bearing ores are roasted. Without sulfuric acid, much of chemical industry would come to a halt. This is the most obvious sense in which sulfur is vital in our contemporary world. The countless other appearances of the element in everyday chemistry and physiology, however, of which the examples given above are only a sampling, show that sulfuric acid is merely center stage in a cast of thousands, every member of which is important to humans.

Bibliography

Baum, Stuart J. *Introduction to Organic and Biological Chemistry*. 4th ed. New York: Macmillan, 1987. One of the few texts, introductory or advanced, that groups the reduced organic sulfur compounds together and discusses their descriptive chemistry.

Embree, Harland D. *Organic Chemistry*. Glenview, Ill.: Scott, Foresman, 1983. Deals with reduced and oxidized organic sulfur compounds under relatively few headings, but very lightly.

Fessenden, Ralph J., and Joan S. Fessenden. *Organic Chemistry*. 4th ed. Pacific Grove, Calif.: Brooks/Cole, 1990. A thorough text (one of many that might be cited), but one that epitomizes the problem in trying to find material about sulfur compounds in the textbook literature: Most of the (organic chemical) material is here, but scattered through the text by compound type (thiol, disulfide, sulfonate, and so on), and interlarded with information about reaction mechanisms, nuclear magnetic resonance spectra, and so forth, that contributes little to the descriptive chemistry.

Greenwood, Norman N., and Alan Earnshaw. *Chemistry of the Elements*. Elmsford, N.Y.: Pergamon Press, 1984. Technical and advanced, but with absolutely exhaustive coverage of the inorganic compounds of sulfur. Slender on applications and descriptive chemistry. Illustrates the other shortcoming of the textbook literature: Everything is here, but only on the chemistry of the inorganic compounds. For organic compounds and applications, consult another source.

Haynes, Williams. *The Stone That Burns: The Story of the American Sulphur Industry*. New York: Van Nostrand, 1942. Ancient, but covers every aspect of the industry that produces sulfur by the Frasch process along the Gulf Coast, including the sociology of the company villages around the industrial complexes. Some information on sulfuric acid production, and on sulfur production in Italy, Sicily, Japan, and Java. Haynes was an authority on American chemical industry.

Kutney, Gerald W., and Kenneth Turnbull. "The Sulfur Chemist: The Bearer of Ill Wind?" *Journal of Chemical Education* 61 (1984): 372-375. One of only two references that discuss sulfur mining and organic and inorganic sulfur compounds in one place. Necessarily a selection of information, like the present article, but contains other material. Gives name for sulfur in forty-three languages, for those who need it.

Linstromberg, Walter W., and Henry E. Baumgarten. *Organic Chemistry: A Brief Course*. 6th ed. Lexington, Mass.: D. C. Heath, 1987. Another popular short-form organic text in which the material about sulfur compounds is scattered by compound type: mercaptan, sulfonate, and so on.

Pratt, Christopher J. "Sulfur." *Scientific American* 222 (May, 1970): 63-72. The other source that discusses sulfur in all aspects. Stronger on production than on compounds and applications, but a fair account of each.

Radel, Stanley R., and Marjorie H. Navidi. *Chemistry*. St. Paul, Minn.: West, 1990. A general chemistry text with a good treatment of the inorganic chemistry of sulfur, but with the usual incompleteness of textbook accounts.

West, James R., ed. *New Uses of Sulfur*. Washington, D.C.: American Chemical Society, 1975. A collection of articles, somewhat advanced, discussing current uses of sulfur in coatings, structural materials, asphalt paving mixes, and lithium-sulfur and sodium-sulfur electrical cells.

Robert M. Hawthorne, Jr.

Cross-References

The Chemistry of Air Pollution, 61; Biological Compounds, 224; Inorganic Compounds, 1125; Organic Compounds, 1689; Oxygen Compounds, 1738; Soaps and Detergents, 2214.

SUNSPOTS AND STELLAR STRUCTURE

Type of physical science: Astronomy and astrophysics
Field of study: Stars

The sunspots, the accompanying magnetic fields within and around them, and their periodicity are among some of the important observables associated with the internal structure, differential rotation, and convection zone of the sun.

Principal terms

CORIOLIS FORCE: deflective force caused by solar rotation on moving matter

FACULA: bright spots or streaks on the photosphere associated with the magnetic field

KINETIC PRESSURE: the average force per unit area produced by atoms and molecules

PENUMBRA: the lighter outer region surrounding the darker umbral region on sunspots

PHOTOSPHERE: a relatively thin surface layer of the sun from where the majority of photons (light considered as particles) are emitted

PROMINENCES: arches of glowing gases often seen in the vicinity of sunspots

SOLENOIDAL FIELD: the magnetic field similar to that of a bar magnet found when an electrical current is passed through a cylindrical coil of wire

UMBRA: the darkest inner region in the sunspot

ZEEMAN EFFECT: the broadening or splitting of spectral lines of light emitted by atoms caused by the presence of magnetic fields

Overview

The sun, a massive gaseous sphere with a radius of 696 million meters, some 150 billion meters away, is the only life-sustaining source of energy available to Earth. With 333,400 times the mass of Earth, 99.9 percent of the mass of the solar system resides in the sun. Almost 94 percent of the solar mass consists of hydrogen atoms; 5.9 percent is helium, and the remainder is composed of traces of the remaining ninety elements.

The most important properties of the sun are its great mass and the fact that one-half of this mass is inside a central spherical core of one-fourth the solar radius, which produces the energy emitted by the sun. The core temperature of the sun is 15 million Kelvins, and its pressure is estimated to be 250 atmospheres. At this temperature, despite the inordinately high density, the nuclei remain gaseous. Solar energy results from the fusion of four hydrogen nuclei to form helium, and in each case, 0.70 percent of the mass entering the reaction is converted into energy. The sun generates 400 trillions of trillion watts of power, which amounts to burning 5 million tons of mass per second. The sun is 5 billion years old and is likely to shine at the present rate for several billion years to come. In due course, with depleted hydrogen supply, the core will consist mostly of helium.

The energy generated in the solar core is transported toward the surface through a series of radiative transfers. Thus, nuclei absorb the gamma rays, re-emitting them as X rays, which are then converted to ultraviolet rays and finally to visible light. The zone of radiative diffusion, however, extends to only 85 percent of the solar radius, at which point the pressure of the gas drops so that a turbulent convective process takes over and transports the energy to the surface. The convective cells that form are responsible for the observed boiling and bubbling on the solar surface. The temperature at the outward surface of the convective zone—the photosphere—is closer to 6,000 Kelvins, and it is in the photosphere that the sunspots are found.

The sun was assumed to be a perfectly radiant sphere until 1610, when Galileo (1564-1642) reported sighting dark spots—sunspots—on the photosphere. In 1843, Heinrich Samuel Schwabe (1789-1875), a German amateur astronomer, announced after twenty years of observation that the sunspots appeared and decayed in an eleven-year cycle. It was realized later that, in reality, the cycle consists of twenty-two years, if the magnetic field reversal is taken into account. Furthermore, it was also discovered that there are considerable variations throughout the observed period. Since the beginning of the twentieth century, the last four periods have averaged only 10.4 years, with the rise time, which lasts only four years, being quicker than the duration of decay. Also, superimposed on the solar cycle may be an eighty-year cycle with episodes of waning and waxing sunspot incidences.

The sunspots appear at the beginning of the cycle at latitudes ±40 degrees (+, north; −, south of the solar equator), progressing down to ±5 degrees when the next cycle starts. An individual spot survives about twenty-seven days, roughly a solar rotation period, if it is small in size, while a larger one lasts more than a rotation. Sunspots are regions of powerful magnetic fields, and it will be seen that most of the surface activities of the sun, such as flares, solar wind, prominences, and the like, are related to intense, time-varying, and transient fields of photosphere and corona. Within the central region of the sunspot, a magnetic field on the order of 0.30 tesla is strong enough to exert pressure that exceeds the kinetic pressure, thus inhibiting the energy transporting gas flow from the lower level of the convective zone and making it darker by contrast to the photosphere. Also, the lack of flow of hotter gas facilitates expansion, aided by fanning out magnetic fields, leading to some 2,000 Kelvins cooler than the surrounding photospheric temperature.

The diameters of the umbrae, the darkest regions of the sunspots, seldom exceed 20,000 kilometers, limiting their depth to about the same distance and large-scale mass movement to about half their size. Sunspots and their associated fields appear to be anchored at this depth. Vertical flow of matter in sunspots is found to be limited to 25 meters per second, while the horizontal flow does not exceed 50 meters per second within the umbrae. Sunspots also show irregular patterns of bright points called "umbral dots" or "umbral granulation," in addition to solar filigree caused by delicate small-scale movement of magnetic field elements. Such movements are studied by spectral analysis, just as the strength of the magnetic field is determined by Zeeman spectral shift and the velocity fields are determined by the Doppler effect.

Umbrae in sunspots are surrounded by lighter, and therefore hotter, regions with complex structures known as penumbrae. The penumbral magnetic field has horizontal fine structures, giving the region a filamentary overlapping white and gray appearance. The average field strength is assumed to be nearly one-half the umbral region. Penumbral matter flow is nearly horizontal, unlike that in umbrae, with progressively decreasing velocities.

Regions known as "pores" with dark umbrae also occur, having magnetic fields on the order of 0.20 to 0.25 tesla. A variety of structures occur near bright chromospheric regions, which are called faculae or, if more intricate, filigree, with field intensities on the order of 0.10 to 0.20 tesla. Also occurring are compact magnetic structures known as magnetic knots, having fields of up to 0.1 tesla.

The sun rotates eastward like Earth. It was discovered that the adjacent sunspots exhibit opposite magnetic polarities; that is, the field coming out of one returns into the other. Further, the polarity of the field of the leading sunspot in one solar hemisphere is always opposite to the polarity of the field of the leading sunspot in the other hemisphere. George Ellery Hale discovered in 1924 that the solar magnetic field reversed once every eleven years so that the sunspots in reality had a twenty-two-year cycle. The intense fields at adjacent sunspots in either solar hemisphere were also found to be solenoidal, so that they fanned out of or into the sunspots vertically to the plane of the photosphere.

Thus, solar activity is marked by the manifestation of periodic occurrences of sunspots, accompanied by complex variations in associated magnetic field intensities. The solar magnetic field is copious, dissipative, and forever rising to the surface, only to be diffused and dispersed through numerous active regions. Characteristically, the surface magnetic fields, weak or strong, are not continuous and occur in conjunction with the active regions of the sun. For example, among the granular boundaries, a basic feature of the photosphere, concentrated tubular fields emerge and disperse into the solar atmosphere. A global magnetism that is somewhat akin to Earth's or poloidal fields appears to coexist and evolve during the activity cycle, along with toroidal fields, which are found at the deeper levels of the convective zone, that surface through the sunspots at higher latitudes at the beginning of the periodicity.

The dynamo theory of solar magnetic fields developed by Eugene N. Parker, Horace W. Babcock, and others in the 1950's provides a framework of explanations to the observed sun's magnetism, the sunspot activity cycle, and other major features such as flares, prominences, and solar winds. There are compelling reasons to believe that the convective zone consists of highly ionized gas—a mixture of ions and electrons—which is called the solar plasma. The rotating plasma induces both a current and an intense magnetic field. The sun does not rotate like a solid body. The rotational velocity of the outer layers increases with latitude. The solar equatorial rotational period of twenty-five days progressively increases to thirty days at about latitude 60 degrees. This differential rotation of the convective zone is assumed to be responsible for driving the dynamo and amplifying the solar magnetic fields. Another aspect of the solar dynamo is the radial differential motion to a depth of 30 percent of the sun's radius. The implication is that a greater solar mass motion is involved in the generation process of the field, extending to deeper levels than previously believed. Another important velocity component affecting the field-generating mass movement is the Coriolis force, which imparts a westward drift to longitudinal motion. The Coriolis force thus modifies north-south movements of sunspots and other field-oriented large- and small-scale surface features.

At the beginning of an activity cycle, solar magnetic field lines at a depth of one-tenth of the sun's radius are considered dipolar, or poloidal, with field lines emerging from the south pole and entering at the north. The differential rotation of the sun, after a few rotations, stretches these flux tubes constrained to move along the plasma surface so that they wrap around parallel to the equator on either side. As the density of the field lines increases, the magnetic buoyancy of the field is elevated. The convective motion of the plasma further twists the flux lines like strands of rope. The enhanced tension in the magnetic field lines at higher latitudes finally breaks through the solar surface as sunspots roughly along the east-west direction. The mechanism satisfactorily explains the incidence of sunspots and their geometry in either solar hemisphere and the bipolar nature of the fields of adjacent spots. As the flux tubes weaken, as a result of diffusion of energy through the sunspots at higher latitudes with the progressing cycle, the spots migrate toward the equator until it is time for a new cycle to begin. The dispersal, eruption, and random walk processes of the field are brought on by the large-scale movement of the plasma in the convective zone below, weakening and finally leading to the reversal of the polarity of the poloidal field at the start of a new cycle. The complexity of the envisioned process, which leads to the twenty-two-year magnetic cycle, is a matter for further investigation. In view of the numerous physical processes involved in the complete breakup of poloidal fields and the weakening of the toroidal fields, the possibility of nonreversal of the polarity at the end of a definite time frame exists.

Applications

The sun is Earth's nearest star. Everything scientists know about the sun is applicable to most other stars. In addition to the obvious properties—namely, the mass and luminosity—the sunspots and the attendant magnetic cycle are aspects that can be observed readily. The sunspot cycle is related to long-term temperature fluctuations on Earth; variations in the charge-related properties of the interplanetary medium caused by the changing magnetic field and the consequent irregularity of the ejected mass in the form of flare, solar wind, and the like; and drastic variations in the electromagnetic induction properties of terrestrial installations brought on by so-called magnetic storms (which occur at peak sunspot activity cycle and are caused by a rapidly dissipating solar magnetic field). There appear to have been long episodes of solar inactivity, with the consequent drop in Earth's temperature. An accumulated body of information indicates that the sun may be a variable star, possibly with complex periodicity. Such a prospect for the solar luminosity changes would have far-reaching ramifications for the history of Earth's temperature.

The occurrence of the magnetic field is common in the universe, in the Milky Way galaxy, and in the solar system. The dynamo mechanism or its variant appears to be the most plausible mode of large-scale amplification of magnetic fields. A thorough knowledge of the solar dynamo will add to an understanding of the sunspot cycle and a host of other surface mass ejection processes. It is believed that the magnetic fields of the major planets, including those of Earth, are also products of the dynamo mechanism.

The study of sunspot cycles has necessitated sophisticated and precise instruments that help unravel the mysteries of smaller and more varied magnetic elements of the sun. The field of helioseismology has helped astronomers to probe in greater depth the convective and the radiative zones of the sun, enabling them to reexamine and refine the dynamo theory, which in turn has led to progress in the knowledge of planetary magnetic fields.

Spectroscopic studies of magnetic fields in other stars have revealed that the majority of stars possess both toroidal and poloidal fields similar to those found on the sun, with intensities ranging from 0.20 to 2.00 tesla; the observed poloidal fields appear to have periodicities ranging from one-half of a day to decades; and among many of the lower main sequence stars, "starspots" are suspected to exist.

Magnetic field activity, presumably similar to that of the sun, appears to be stronger in many younger stars, which rotate more rapidly in comparison with older and lower-angular-velocity stars. The younger stars in Orion have field intensities three times that of similar-sized older stars. It is evident from observations that starspots in relatively younger stars, which radiate intense X rays and ultraviolet rays, actually make them appear dimmer. It is plausible that the sunspot and the associated magnetic field intensity may have been stronger at an earlier epoch of the solar system, when presumably the sun was spinning more rapidly. Closer to home, the sunspot activity has left a permanent record, dating back thousands of years, in concentrations of the carbon 14 radioactive isotope, produced by the modulating effect of the solar magnetic field on cosmic rays reaching Earth. The reduced number of sunspots implies a weaker magnetic field. This, in turn, leads to a greater cosmic-ray flux that reaches Earth and produces higher concentrations of carbon 14. A measurement of concentrations of carbon 14 at various levels of the soil is easily correlated with sunspot activity.

Context

Humans have worshiped the sun from the dawn of history. With Galileo's discovery of the sunspots in 1610 began a new era of inquiry into the causes of the blemishes seen in the otherwise perfect photosphere. It was during the middle of the nineteenth century that Schwabe announced the discovery of the eleven-year sunspot cycle; a coherent picture of the phenomenon emerged much later. The current solar research is concerned with many broad areas, such as sunspots, their emergence, and the solar magnetic activity cycle, as well as the generation of thermonuclear energy, its transport mechanism to the surface, and the sun's atmospheric physics.

It has become possible to explore the eleven-year cycle in the ultraviolet region of the spectrum with the aid of the Nimbus 7 and Solar Mesosphere Explorer satellites. In the 1980's, the Solar Maximum Mission satellite carried out delicate experiments and simultaneously collected a vast amount of data on the solar atmosphere, enhancing the understanding of the structure and variable radiations that occur there. Research on Earth's radiation belt, combined with the deep-space planetary probe data, leads to the conclusion that the pervasive solar wind that creates a dynamic interplanetary medium is yet to be understood with respect to its interactive properties.

Solar physics, solar observation, and the complex aspect of the subject started with a chance sighting of sunspots by an amateur astronomer. The current high-resolution solar telescopes that use clusters of mirrors are able to provide considerably more detailed information on ever-smaller areas of the sun. Scientists have been able to scan the surface and the atmosphere of the sun from high-energy gamma rays to low-frequency radio waves of the electromagnetic spectrum. With the advent of supercomputers and the availability of space shuttles to deploy specialized instruments, such as high-resolution solar telescopes, high-energy X-ray and gamma-ray telescopes, and pinhole/Occulter facilities, the coming decades will witness spectacular progress in solar physics, which will be aided by advances in related fields. Historically, the source of solar energy could be explained only in terms of thermonuclear fusion; the lessons of plasma physics and magnetohydrodynamics should help scientists to explain the solar dynamo mechanism and, hence, the related activity cycle, including sunspots.

Bibliography

Foukal, Peter V. "The Variable Sun." *Scientific American* 262 (February, 1990): 34-41. Written for the general reader, the article describes the history of sunspots and the variation in the solar luminosity. Presents the current findings, pointing to the possibility that the steady sun is more than likely a variable star. An interesting article for all levels of readers.

Gibson, Edward G. *The Quiet Sun*. NASA SP-303. Washington, D.C.: Government Printing Office, 1973. Contains many photographs, illustrative graphs, and charts. Although technical, this volume is written so that a nonscience general reader will not experience any difficulty with the text.

Giovanelli, Ronald G. *Secrets of the Sun*. New York: Cambridge University Press, 1984. This book, which contains many photographs of sunspots and their progression in the eleven-year cycle, is specially written for nonscientists, scientists, and lay readers alike. A rare example of an author's serious attempt to introduce the reader to a rather complex set of ideas.

Jordan, Stuart, ed. *The Sun as a Star.* NASA SP-450. Washington, D.C.: Government Printing Office, 1981. Although technical, this volume is a complete source of solar physics. Includes a large number of references at the end of each chapter. A nontechnical reader can skip over the occasional mathematical equations without experiencing a sense of loss in logic or continuity.

Parker, E. N. "Magnetic Fields in the Cosmos." *Scientific American* 249 (August, 1983): 44-54. Parker describes the process of amplification of the existing fields. The solar field and the attendant sunspot cycles are described adequately by the dynamo mechanism insofar as the features are concerned. Describes the longitudinal or poloidal magnetic fields of major planets in the solar system. The article is directed toward an interested general reader.

Wentzel, G. Donat. *The Restless Sun*. Washington, D.C.: Smithsonian Institution Press, 1989. Written primarily for the general reader, this excellent volume contains photographs, illustrations, and information on all types of solar activities. Even a casual reader can get engrossed in reading this work because of the easy style and depth of material covered. Geared for a beginner.

V. L. Madhyastha

Cross-References

The Measurement of Magnetic Fields, 1289; Solar Flares, 2220; Solar Wind, 2228.

SUPERCONDUCTORS

Type of physical science: Condensed matter physics
Field of study: Solids

Superconductors have zero resistance to the flow of electrical current at temperatures below a critical point. Once a current is established in these large-scale quantum systems, it persists indefinitely.

Principal terms

CONDUCTIVITY: the measure of a material's ability to carry electrical current

ELECTRICAL CONDUCTOR: a substance that permits easy flow of electrical current

ELECTRICAL CURRENT: the flow of electrical charges through space or through a material

KELVIN TEMPERATURE SCALE: no temperature can be below 0 Kelvin, where classical motion ceases; water freezes at 273 Kelvins and boils at 373 Kelvins

QUANTUM: one of the indivisible units into which energy and other physical quantities can be subdivided

RESISTANCE: a measure of the obstruction of flow of electrical current in a material

Overview

In 1908, the Dutch physicist Heike Kamerlingh Onnes (1853-1926) liquefied the rare gas helium at 4 Kelvins above absolute zero, or equivalent to −269 degrees Celsius. In 1913, he would receive the Nobel Prize in Physics for this pioneering work in low temperatures; meanwhile, Kamerlingh Onnes discovered a remarkable occurrence that opened a new field of science. Shortly after his discovery of liquid helium, Kamerlingh Onnes used this freezing liquid to carry out low-temperature studies of the electrical resistance of metals. He was working with mercury, and, as is common to metals, the resistance of mercury decreased slowly as the temperature dropped. Below the temperature of 4.2 Kelvins, the resistance plummeted and Kamerlingh Onnes was unable to detect any resistance. Mercury turned into a superconductor of electricity below the critical temperature of 4.2 Kelvins.

Many experiments by Kamerlingh Onnes and others convinced the scientific community that superconductors have not merely a low resistance but zero resistance to the flow of a steady electrical current. Once a current begins circulation within a superconductor, the current persists. This notable ability of superconductors does not fit the classical scientific model of electrons moving through a dense forest of atoms in a metal; it was not fully explained until 1957, when John Bardeen, Leon N Cooper, and John Robert Schrieffer used quantum theory to describe the process.

Bardeen, Cooper, and Schrieffer shared the 1972 Nobel Prize in Physics for their work on the so-called BCS theory.

In the years immediately following Kamerlingh Onnes' discovery of superconductivity, the unbearably low critical temperatures rose in fits and starts. Within two years, the poor metallic conductor lead demonstrated a critical temperature of 7 Kelvins; it was not until the mid-1930's that superconductivity appeared in niobium, another poorly conducting metal, at 9 Kelvins. Within another few years, a version of niobium oxide passed the 10-Kelvin barrier. In the early 1950's, the commercially important intermetallic superconductor Nb_3Sn displayed a critical temperature of 18 Kelvins. In 1973, scientists passed the 20-Kelvin barrier. The superconductor was a combination of niobium and the semiconductor germanium, Nb_3Ge, which has a critical temperature of 23 Kelvins for transition into the superconducting state.

In 1986, Karl Alexander Müller and J. Georg Bednorz discovered the class of high-temperature superconductors, based on the copper oxides. Their superconductor, called LSCO, is a ceramic formed by the combination of lanthanum, strontium, and copper oxides. LSCO has a critical transition temperature around 37 Kelvins, more than 50 percent higher than the previous superconductor. As a result of their work, Müller and Bednorz shared the 1987 Nobel Prize in Physics. Within two years, the critical temperature tripled, to more than 120 Kelvins. In 1987, C. W. Chu, M. K. Wu, and their coworkers discovered superconductivity at 92 Kelvins in yttrium-barium-copper oxide, $YBa_2Cu_3O_7$, often called YBCO or 1-2-3. This astounding increase in the critical temperature triggered the rapid discovery of other variants of the copper oxide superconductors. These copper oxides show conduction mainly in parallel layers formed by the copper and oxygen atoms. The new discoveries raised the critical temperature to 110 Kelvins in a copper oxide of bismuth with strontium and calcium, and 122 Kelvins in TBCCO, a copper oxide of thallium with barium and calcium. For the first time, high-temperature superconductors, or HTSCs, existed that possessed critical temperatures above that of liquid nitrogen, a common, inexpensive commercial refrigerant obtained by cooling the nitrogen of the air.

The experimental progress in the discovery of higher-temperature superconductors emphasizes two facts about superconductors: The superconductors must be conductors; paradoxically, they should be poor conductors, and the poorer the better. The search for high critical superconducting temperatures moved scientists inexorably away from good conductors and toward materials that are closer and closer to the edge of electrical conduction.

The three best conductors among the common metals are silver, copper, and gold, but superconductivity exists only below 0.3 Kelvin in these metals, and even may be absent. The base material for Müller and Bednorz's copper oxide superconductor is La_2CuO_4, which is an insulator. In other words, it does not conduct electricity. Strontium must be added to the composition in place of lanthanum to provide conduction, in a manner similar to dopants that provide conduction in semiconductor devices. YBCO is more akin to a normal conductor; like LSCO and the other copper oxide superconductors, however, it is only a two-dimensional, layered conductor. To understand why such unusual and poor conductors make good superconductors, it is necessary to study the mechanism that produces superconductivity.

Electrons are the basic carriers of electrical charge and provide electrical current in most common materials. Quantum mechanics, which was pioneered by Max Planck (1858-1947), explained much about the nature of electrons. First, electrons spin on their axes eternally. The spin may be up or down, but the spin is constant, at a fixed angular momentum of half the quantized value, or Planck's constant divided by 6.283. Second, electrons, and all particles with half-spin, are exclusive. Called fermions for the American physicist Enrico Fermi (1901-1954), these particles occupy only one energy state at a time. The state describes the motion of a particular spinning electron in space. A maximum of two electrons, one with spin up and the other with spin down, can occupy a given state and share the energy and motion that characterize that state. Two more electrons at the same velocity must move in a different direction. If more electrons are added, they must enter different energy states and directions, filling the energy states like water filling a jug.

Normal metals have twice as many states as electrons and act as half-filled jugs. If the jug is tipped, the water runs up the sides. If voltage is applied to a metal, the electrons run through the metal producing the current. In contrast, insulators and semiconductors act like completely filled jugs. If the filled jug is tipped, the water has no place to move. Similarly, a voltage produces no appreciable current in an insulator or semiconductor. The strontium in the LSCO superconductor acts to absorb some of the electrons from the filled La_2CuO_4, leaving a hole in the electron sea, like a bubble in the filled jug. If the jug is turned over, the bubble moves up, opposite gravity. In fact, water is flowing down around the bubble. A voltage applied to LSCO causes the holes to move opposite to the electrons, producing a current as if the carriers were positive. This hole motion dominates the behavior of all the copper-oxide, high-temperature superconductors.

Quantum mechanics helps explain one more aspect of superconductors. Integral spin particles, such as light photons with spin 1, are not exclusive. Called bosons for the Indian scientist Satyendra Nath Bose (1894-1974), these particles may share one energy state. For example, a typical helium-neon laser has more than a hundred billion photons in one energy state, which shapes its narrow red beam. If electrons in a conductor could pair together, so that every two half-spins acted as a single 1-spin, then all the paired electrons would move together in the same energy state and could be beamed in one direction. When a voltage is applied, the result is superconduction; when no voltage is applied, the result is a persistent superconducting current. Such paired electrons, or paired holes, do form below the critical temperature of a superconductor and are called Cooper pairs. A related pairing turns very cold liquid helium into a superfluid. The forces responsible for the electron pairing are also present at higher temperatures and act to produce both resistance and low conductivity in the normal conducting state. Thus, poor conductors favor strong Cooper pairs and good superconductivity. Cooper pairing seems to defy the rule that electrical charges repel. In a conductor, any electron feels the repulsions of all the other electrons and the attractions of the bound positive ions that compose the fixed lattice of the conductor. The other electrons and ions are on all sides of the electron and the net forces cancel. As a result, the electrons or holes in a conductor are, on average, free. There are two ways in which carriers of the same charge may attract each other when one looks beyond the average.

The first method considers the situation in which the carriers are electrons; this method equally applies to the analogous situation in which the carriers are holes. As an electron moves through the latticework, it attracts the opposite fixed charges, deforming the lattice slightly in the carrier's wake. Another electron, which has opposite spin and, normally, is moving in a direction opposite to the first electron, will be attracted to the lattice deformation. Quantum physics allows electrons of opposite spin to pass easily through each other. Thus, both carriers will be attracted to the same region of the latticework and, effectively, to each other. The deformation of the lattice sets up sound waves, or phonons, in the conductor; thus, the pair attraction is the result of phonon interaction. Below the critical temperature, the phonon interaction binds Cooper pairs and yields superconductivity. The BCS theory fills in the details and describes traditional low-temperature superconductors in which phonon pairing predominates.

A second method that may yield pair binding takes place when an electron repels the bound negative electron cloud surrounding the positive ion latticework. The effect is to leave a positive wake that attracts the second electron in the Cooper pair. This binding has several variations under the heading of excitonic interaction. Conventional phonon binding is not strong enough to explain the high critical temperatures of the copper oxide superconductors. Whether excitonic binding can explain the superconductivity in these high-temperature superconductors is not clear. Superconductors are one of the few examples in nature where quantum effects dominate large-scale wonders.

Quantum mechanics produces further unusual effects in superconductors. The Cooper pairs appear experimentally when a voltage is present between two superconductors separated by a thin insulating barrier. Quantum theory states that particles can tunnel through thin barriers. When a voltage V appears between two superconductors, Cooper pairs tunnel between the superconductors at the frequency f given by $hf = (2e)V$, where h is Planck's constant and e is the electron's charge. The separation region between the two superconductors forms a Josephson junction; tunneling is known as the Josephson effect, which has been verified experimentally to an extraordinary degree. The presence of $2e$ rather than e in the detected frequency proves the existence of the Cooper pairs.

A superconductor expels any magnetic field present when it makes the transition from the normal conducting state to the superconducting state. Known as the Meissner effect, this process renders superconductors diamagnetic. Since any current present in the superconducting state generates a magnetic field, this field is also expeled from the superconductor's interior and penetrates only through a thin layer of the superconductor. The thickness of this layer is known as the London penetration depth.

All current within the superconductor moves in this thin layer at the surface of the superconductor, rather than throughout the volume as in a normal conductor. Many applications require only a thin thread or layer of superconductor to carry current. The layer may be deposited on another material for strength, while the thread, or threads, may be buried within a support material.

Superconductors vary in the manner in which they expel magnetic fields. Type I superconductors expel magnetic fields completely from their interior, whereas type II superconductors expel magnetic fields by segregating the fields into thin filaments that form a vortex pattern within the body of the superconductor. Type II superconductors are the more useful superconductors, since they can withstand higher magnetic fields by increasing the number of vortex filaments, thereby multiplying their internal surface area as the magnetic field rises. High-temperature superconductors are of type II. Quantum physics dictates the magnetic flux and magnetic field time area trapped in the vortices. The magnetic vortices are quantized in multiples of $h/2e$.

Applications

The most important large-scale application for conventional low-temperature superconductors is the production of large magnetic fields. Superconducting magnets are used in laboratories that study high-energy physics or seek to imprison the sun's energy in plasma fusion, in hospitals that use magnetic-resonance imaging for diagnosing patients, and in the removal of magnetic materials from garbage and wastes.

Large currents that course through low-temperature, superconducting windings generate the large magnetic fields in these applications. Liquid helium is the conventional coolant for the low-temperature, type II superconductors. An important characteristic of any superconductor used to generate large magnetic fields is the maximum magnetic field that the superconductor can withstand at the coolant temperature, before losing its superconductivity. Common low-temperature superconductors are Ni-Ti, which can withstand modest magnetic fields of 9 teslas; Nb_3Sn, which can withstand fields up to 15 teslas; and V_3Ga, which will withstand 20 teslas. (One tesla is about twenty thousand times the earth's magnetic field.)

For strength, handling, and safety, the superconductors used for magnetic field generation are embedded within a metal matrix. For example, Nb_3Sn filaments are often encased in a bronze conductor and Nb-Ti filaments in copper. If cooling fails during operation, the overlying conductor must assume the current carried by the superconductor. This action prevents explosion of the generator coils when the superconductor switches to its normal conducting but resistive state and increases in temperature.

The maximum magnetic field that a superconductor can withstand increases with the critical temperature of the superconductor. Thus, the maximum magnetic fields of the high-temperature superconductors are immense, up to 100 teslas at low temperatures. These maximum fields depend on the orientation of the copper oxide planes and on the actual current that the superconductor carries. High currents, as well as magnetic fields, act to suppress superconductivity.

In magnetic applications, bulk high-temperature superconductors must carry at least 400,000 amperes per square centimeter at a magnetic field of 1 tesla, in order to compete with copper wire in liquid nitrogen. Good-quality, crystalline thin films of the new superconductors easily exceed this specification. Yet, bulk samples of these ceramic superconductors have minute crystal grains, separated by disorganized boundaries. These grain boundaries lower the ability of the bulk superconductors to carry current. In addition, movement of the magnetic vortices trapped in these type II superconductors produces heating. The success of the low-temperature superconductor Nb-Ti depends on normally conducting titanium precipitates that pin the vortices in the superconducting composite. The new high-temperature materials require similar inventive techniques, which will come, to eliminate grain boundary effects, to pin magnetic vortices, and to produce bulk samples that are easy to fabricate into useful and inexpensive superconducting coils.

A second application of low-temperature superconductors is to measure extremely small magnetic fields using SQUIDs, or superconducting quantum interference devices. The device consists of two Josephson junctions placed within a superconducting wire loop. As with magnetic vortices, the magnetic flux within the superconducting loop is quantized in multiples of $h/2e$. The current within the SQUID, along with the loop area, gives a precise measurement of the magnetic field. SQUIDs can measure very small voltage differences, currents as minute as several electrons per second, and magnetic fields that are one ten-billionth of the earth's magnetic field. This precision enables scientists to map the magnetic field generated by brain activity. Development of high-temperature superconductor SQUIDs would replace cooling with bulky liquid helium with more compact and portable liquid nitrogen or electrical cooling systems, and would provide more widespread use of these precise quantum measurement devices.

There are many possible uses for high-temperature superconductors when they are cooled to liquid nitrogen temperatures. In addition to magnetic field coils and SQUIDs, these uses include magnetic levitation of trains, magnetic energy storage, electrical transmission lines and their surge protectors, hybrid semiconductor-superconductor computers and computer interconnects, and low-loss, microwave circuits and lines. These possible uses depend strongly on the success of material scientists in fabricating, shaping, and coercing the conductors that have resulted in the creation of a new area of science.

Context

In 1913, Kamerlingh Onnes described his 1911 experiments on mercury. "Mercury has passed into a new state which . . . may be called the superconductive state. . . . The behavior of metals in this state gives rise to new fundamental questions." The answers to these questions required quantum theory, initiated by Max Planck in 1900 and culminating with the BCS theory four decades later. The classical mechanics of Sir Isaac Newton and electromagnetism of James Clerk Maxwell cannot describe the large-scale events involved with superconductivity. While a steady electric field exists inside the superconductor, the carriers continually accelerate, as if in a resistance-free vacuum instead of buried in a maze of atoms. Only quantum theory explains this seemingly unbelievable behavior.

The carriers of superconductors are Cooper pairs, bound together at temperatures below the critical temperature. In the superconducting state, all the bound pairs condense into the lowest energy state available, mimicking bosons, and cannot give up energy. Resistance heats a conductor when current flows and the heat comes from energy lost by the carriers. Since the superconducting Cooper pairs have no energy to give, no heating is possible and there is no resistance. Above the critical temperature, thermal energy breaks apart the Cooper pairs, spreads the electrons in energy, and generates the resistance of the normal conductor. Quantum mechanics explains how persistent currents in the superconducting state stay together while streaming through the atom maze.

Particles such as the Cooper pairs also behave like waves. Cooper pairs behave in the same manner as the boson waves of light. A light beam moves through flat glass without spreading. The beam does experience the dense atoms in its path, but the wave phenomenon of constructive interference continually reforms the beam in the direction of motion, and only in that direction. The atoms slow the light beam and bend its path at the surface of the glass, but the light travels straight once it is inside the glass. When the beam exits the glass, it bends parallel to its original direction and resumes its original speed. If trapped inside a glass fiber in a phone network, the light travels at undeterred speeds along the fiber guide. Like the light beam, persistent supercurrents are quantum beams of supercarriers that are guided within the superconductor, forever moving, and never able to lose energy and rest.

Bibliography

Cava, Robert J. "Superconductors Beyond 1-2-3." *Scientific American* 263 (August, 1990): 42-49. Discusses HTSCs. The readings are lucid and the layouts are appealing. Illustrations are attractive and informative.

Clarke, John, and Roger H. Koch. "The Impact of High-Temperature Superconductivity on SQUID Magnetometers." *Science* 242 (October 14, 1988): 217-223. Gives a detailed account of the operation of superconducting SQUIDs and the possibilities of operating high-temperature superconductors in liquid nitrogen.

Forsyth, E. B. "Energy Loss Mechanisms of Superconductors Used in Alternating-Current Power Transmission Systems." *Science* 242 (October 21, 1988): 391-399. An understandable account of results from engineering tests on the design, use, costs, and promise of electrical transmission systems that used the common, low-temperature superconductor Nb_3Sn. These tests provide important background data for assessing the use of high-temperature superconductors in power transmission and similar applications.

Ginzburg, Vitaly L. "High-Temperature Superconductivity: Past, Present and Future." *Physics Today* 42 (March, 1989): 9-11. Written by a well-known Russian theoretical physicist who was working on high-temperature superconductors as far back as 1971. Fascinating reading.

Tinkham, Michael, et al. *Physics Today* 39 (March, 1986): 22-80. This special issue on superconductivity contains six articles detailing the history, science, and applications of low-temperature superconductors, written on the seventy-fifth anniversary of Kamerlingh Onnes' discovery and before HTSC.

Peter J. Walsh

Cross-References

The Absolute Zero Temperature, 1; Conductors and Resistors, 540; Electrons and Atoms, 778; The Exclusion Principle, 860; Insulators, 1132; Liquefaction of Gases, 1256; Nonrelativistic Quantum Mechanics, 1564; Superfluids, 2425.

SUPERFLUIDS

Type of physical science: Condensed matter physics
Field of study: Liquids

Superfluidity is that state of a fluid, such as principal isotopes of helium and hydrogen at extreme low temperatures, in which it has zero viscosity and exhibits other properties of macroscopic quantum order not predicted from classical atomic and hydrodynamic theory. Although superfluidity is primarily an area of theoretical and experimental study in cryogenic condensed matter physics, the electronic gas in a superconducting metal is also a superfluid, and theories of superconductivity and superfluidity are closely linked.

Principal terms

BOSE-EINSTEIN STATISTICS: a form of quantum mechanical statistics used to describe an assembly of "bosons," which are particles whose spin is an even multiple of Planck's constant, *h*, in a noninteracting configuration at low temperatures

DISPERSION CURVE: a plot of the energies and frequencies of permitted excitations of an acousto-thermo-mechanical system, usually shown as energy versus inverse frequency (wavenumber)

FERMI-DIRAC STATISTICS: a form of quantum mechanical statistics used to describe an assembly of "fermions," which are particles whose spin is a half-integral multiple of Planck's constant

LIQUID CRYSTALS: liquids which exhibit many physical properties characteristic of solid crystals, such as nonisotropic (directional) molecular orientation and optical birefringence

NUCLEAR MAGNETIC RESONANCE: a phenomenon of atomic nuclei whereby nuclei in a static magnetic field absorb radio frequency energy at characteristic (resonant) frequencies, the precessional behavior of which can be used as an accurate measure of temperature and system order

PHONON: the quantum of acoustic-vibrational energy in a crystal or in low-temperature fluids; a "quasi-particle" which travels at the local velocity of sound

ROTON: a quantum of acoustic-vibrational energy, similar to but distinct from phonons, which occur near the minimum of the dispersion curve for low-temperature helium 2

SPECIFIC HEAT: the quantity of heat required to raise a unit mass of uniform material one degree in temperature

SUPERCONDUCTIVITY: the state of zero electrical resistance in a material, frequently but not always at low temperatures

SUPERLEAK: a phenomenon in superfluids such as helium whereby a superfluid can easily pass or leak through a medium with fine pores, or flow in a fountain-like fashion through capillaries which are resistive or impermeable to any normal fluid with viscosity

Overview

Helium 4 was first liquefied by the Dutch physicist Heike Kamerlingh Onnes in 1908. In 1924, Kamerlingh Onnes and Josef Dana continued the cryogenic studies of helium that first revealed behaviors not consistent with then-current theories of solid and liquid matter. Just below 2.2 Kelvins, they found values of specific heat so abnormally high as to suspect major experimental error. In 1930, Willem Keesom and Klaus Clusius of the Leiden Low Temperature Laboratories confirmed Kamerlingh Onnes and Dana's results. In addition, when they plotted their measurements of density and specific heat versus temperature for helium 4, they discovered a tripartite transition whose specific heat peak occurred exactly at the same temperature as the density maximum. Because of its visual similarity to the Greek letter λ, this transition is known as the lambda point. Above the lambda point, helium 4 behaves as a dense classical gas; at lower temperatures, however, its behavior is radically different and unexpected. Keesom distinguished two forms of helium 4: Above the lambda point, he called it helium I, and below the lambda point, he called it helium II.

Between 1930 and 1936, extensive research discovered additional unusual low-temperature thermal fluid behaviors. As Kurt Mendelssohn's book, *The Quest for Absolute Zero* (1977), recounts, almost all key experimental and theoretical results were published in volume 141 of the British journal *Nature* in 1939. University Of Cambridge physicist Rudolf Peierls and others not only reconfirmed the Leiden results but also discovered that net heat flow through helium II is not proportional to temperature difference; in other words, the lower the temperature difference, the greater the heat conductivity, which suggested that when supplying heat to helium II, liquid flows in (not away from) the direction of the heat source. It was also noted that below the lambda point, helium II's fluid viscosity decreased significantly. In the same *Nature* issue, Soviet physicist Pyotr Kapitsa and John Allen and Austrin Misener of the University of Cambridge described viscosity experiments that measured liquid helium II's friction by flow through fine capillary tubes and narrow plates and showed less resistance the narrower the orifice through which it had to pass. Allen subsequently found that net flow in helium II is nonlinear in relation to temperature, with pressure gradients existing proportional to heat input into the fluid. The latter effects are called "fountain pressure" or "superleaks." Kapitsa suggested the name "superfluidity" for those collective phenomena, as an analogy to previously discovered superconductivity.

The theoretical physicist Fritz London proposed the first theory of superfluidity. In the 1920's, Albert Einstein had shown that an ideal gas, obeying the alternative quantum mechanical statistics of the Indian physicist Satyendra Nath Bose, undergoes unusual change when cooled to sufficiently low temperatures. Because fluid particles are subject to quantum as well as classical mechanical forces, kinetic energy does not vanish at absolute zero even in theory. This so-called residual zero-point energy is a result of the fact that both positions and momenta of fluid particles are subject to Werner Heisenberg's uncertainty relation: $\Delta x \Delta p = h/2\pi$. In quantum mechanics, there are two statistical species of particles. Bosons have symmetric wave functions, such that any number can occupy the same quantum state simultaneously. By contrast, fermions have antisymmetric wave functions with a rigid exclusion principle (developed and named for Wolfgang Pauli) that precludes a given state from being occupied by more than one fermion. Examples of bosons include photons, helium 4 atoms, and phonons. Examples of fermions include electrons, neutrons, and helium 3 atoms. London revived the Bose-Einstein theory to assert that a temperature eventually is reached where some gas particles condense, not in the usual position space to form a solid, but in the quantum-mechanical space of velocities, thereby forming a superfluid.

Shortly thereafter, Hungarian physicist Laszlo Tisza, after having examined experimental results of disks oscillating in superfluid helium II (which, unlike capillary flow tests, showed definite finite viscosity), suggested that helium 4 be considered a mixture of two fluids—normal and superfluid—analogous to the two-conductor superconductivity model. Tisza also predicted, prior to later experimental confirmation, that heat impulses injected into superfluid helium II would travel through the liquid as phonons, rather than as ordinary acoustic pulses. London then demonstrated that superfluids do not carry entropy. In several modified forms, notably that of John Bardeen, Leon N Cooper, and John Robert Schrieffer (known as the BCS theory), this two-fluid model of superfluidity became a basic framework for many experiments in low-temperature physics for decades afterward.

In the early 1940's, Soviet physicist Lev Landau developed a different approach to theoretical superfluidity. Landau described normal fluids via quantized phonons and rotons, representing local thermal equilibrium states. In developing specific hydrodynamic equations to describe superfluid motions, Landau concluded that, strictly speaking, superfluids are incapable of rotation. Landau further predicted what became known as the Landau critical velocity, which is a characteristic speed limit for objects moving through a superfluid. For speeds below the Landau critical velocity, the liquid retains its superfluid properties; above this velocity, it behaves in many respects like a normal fluid. Landau's theory predicted other subsequently observed phenomena, such as the slower wave known as "second sound" (a temperature-entropy wave of constant pressure and density) in contrast to ordinary sound waves. Normal and superfluid components of a liquid oscillate in opposite antiphased relations, such that the fluid density remains basically constant.

In 1941, Anton Bijl, Jan de Boer, and Antonious Michels demonstrated that the maximum velocity possible for superfluid flow in a thin film of helium apparently obeys the quantum critical velocity condition of $\mathbf{V}_c = (h/2\pi)\, md$, where m is helium's atomic mass and d is film thickness. London later proposed that the angular velocity of helium II is subject to similar quantum conditions, such that $\Omega < (h/2\pi)\, mr^2$, where Ω is the angular velocity and r is the helium container's radius. Further confirmation of Landau's theory was developed by Kapitsa in 1941, when he observed large temperature discontinuities between a copper cryogenic heater and its superfluid helium bath. This so-called Kapitsa resistance, a thermal boundary resistance arising from the acoustic impedance mismatch between the contacting materials, was associated with the anomalously high phonon transmission from solid metal into liquid helium.

In contrast to the descriptive phenomenological two-fluid model, the excitation model developed by Landau for superfluid helium considers the liquid below the lambda point to be analogous to a crystal at low temperatures. Superfluids are conceived as an inert background medium in which moves a gas of phonons and rotons. In terms of the dispersion relation, which plots helium excitation temperature versus its momentum, phonons and rotons correspond to absolute and local dispersion curve minima associated with the Landau critical velocity. Theories of superfluids usually seek to reproduce this dispersion curve as a form of verification.

The fact that critical velocities measured for helium II are usually much smaller than sound velocities in normal liquid has been used as evidence for the existence of quantized microscopic vortices. Quantized vortices also play a major role in explaining the apparent rotation of theoretically irrotational (nonrotating) helium II in a rotating container. For sufficiently large container angular velocities, both normal and superfluid components of helium 4 appear to rotate. What actually occurs, however, is that many quantized vortices, parallel to the container's rotation axis, cause the superfluid component to simulate solid-body rotation. Henry Hall and William Vinen first detected these vortex lines in 1956. Rayfield and Reif in 1964 examined vibrations of a fine submerged wire parallel to the rotation axis, thereby demonstrating circulatory vortex quantization. In 1979, Yarmchuk and others developed an optical imaging technique for photographing individual vortices.

A major prediction of Landau's superfluid theory was the proposed existence of another superfluid transition in helium at ever lower temperatures. Between the late 1940's and the end of the 1960's, many experiments sought to confirm or disprove Landau's prediction of a superfluid transition temperature of helium 3. Variously thought to occur at 1.5 Kelvins, 0.1 Kelvin, and 0.005 Kelvin, the transition temperature of 0.0026 Kelvin for superfluidity in helium 3 was accidentally discovered by Douglas Osheroff, Robert Richardson, and David Lee of Cornell University in 1972.

Liquid helium 3's specific heat undergoes an unusual finite discontinuity at its superfluid transition that differs from the lambda point transition of liquid helium 4, yet is similar to the normal/superconducting transition for a metal outside magnetic fields. Theoretically, helium 3 superfluidity arises through the pairing of electrons into groups known as Cooper pairs. The nonzero internal angular momentum predicted by Landau's quantum fluid theory of Cooper pairs has been confirmed as the cause of many of helium 3's unusual properties. Another related theoretical concept for helium 3 superfluidity is that of the "order parameter." The macroscopic order parameter is the large-scale analogue of a single-particle quantum mechanical wave function. In helium 3, the superfluid order parameter comprises nine individual quantum states (in contrast to only one in superfluid helium 4). This order parameter results in two helium 3 superfluid phases, designated helium 3A and helium 3B.

In contrast to superfluid helium 4, superflow circulation in the helium 3A phase is not quantized. As a result, rotating samples of helium 3A exhibit rotary motions intermediate between those of ordinary classical viscous liquids and superfluid helium 4. The high-pressure superfluid helium 3A phase has been shown to be similar to ordinary liquid crystals. The lower-pressure helium 3B phase lacks the coupling and liquid-crystal properties of the A phase. For example, experimental studies using nuclear magnetic resonance (NMR) have shown that helium 3B vortices are hundreds of times larger than those in superfluid helium 4. NMR studies have also shown that at temperatures equal to 0.6 of the superfluid transition temperature for helium 3, both temperature and gyromagnetic discontinuities occur as a result of quantum properties of helium 3B's half spin. Both helium 3A and helium 3B have characteristic frequencies that are associated with oscillations of the macroscale order parameter around its equilibrium position. These resonant ringings, measured by John Wheatley of the University of California and Richardson's Cornell University group, are designated the Leggett effect for the British theoretical physicist Anthony Leggett, who first predicted them. Several other unexpected vortex properties observed in both the A and B phases of superfluid helium 3 differ from those of helium 4. A-phase vortices are multiply quantized, possessing continuously distributed values of vorticity. Vortices in the B phase are inherently magnetized, spontaneously possessing a dipole moment and unusual vortex properties.

The superfluid phase transition and behavior of thin films differ notably from those in bulk superfluid helium 3. Because the long-range order parameter cannot exist in two-dimensional samples, a new kind of quantum-mechanical order arises from binding quantized vortices of opposite charge. Thin films of superfluid helium 3 have offered physicists the chance to examine "gapless" anisotropic superfluids. In investigations of superfluid helium 3 thin films, local disorders often result from crystalline defects and roughness in the helium container walls. These problems can destroy Cooper pairing, as well as effectively lower the superfluid transition temperature.

In superfluid helium 3 and superconducting metals, Cooper pair formation produces a gap in this excitation spectrum, so that a specific energy is required to break up a Cooper pair. This gap, which is isotropic in superconductors, is directionally dependent (anisotropic) in superfluid helium 3A. This gap is evident in the A phase's specific heat behavior, which does not decrease exponentially with temperatures as superconductors. A rough surface, however, notably alters the number and energy distribution of helium 3 Cooper pairs. For roughness sufficiently large with respect to Cooper pair dimensions, Cooper pairs can be broken using almost no energy. This gapless superfluid phase is characterized by helium 3's linear specific heat decrease, as in ordinary helium 3.

Applications

Unlike cryogenic and, more recently, higher-temperature superconductors, superfluids have not been found to possess any direct or immediate industrial application. In order to achieve, maintain, and measure milli- and micro-Kelvin temperatures, however, numerous advances have occurred in cryogenic refrigeration, thermometry, and materials.

Soviet physicists Isaak Pomeranchuk (in 1950) and Dmitri Anufriyev (in 1965) developed a very-low-temperature cooling method for helium 3, using the adiabatic compression of a liquid-solid mix along its melting curve. Using suitable precooling, such as that provided by dilution refrigerators, and applying pressure, liquid helium 3 is brought to the solid phase, with the resulting energy release known as Pomeranchuk cooling. The cooling power of this method is the total amount of heat absorbed when a given quantity of liquid helium 3 is converted into solid at constant temperature. Pomeranchuk cooling successfully achieves temperatures of about 0.001 Kelvin (1 milli-Kelvin).

Further cooling via adiabatic demagnetization employs spin states of salts and metals. Applying a magnetic field to weakly interacting magnetic dipoles yields reduced entropy, since magnetized states have greater order. During magnetic field application under constant temperature, released heat is removed by placing the spin system in thermal contact with a helium 4 bath. When steady-state conditions are achieved, the magnetic state is isolated and the field is reduced under adiabatic conditions. Temperatures of less than 0.1 milli-Kelvin have been achieved with this method, and even lower temperatures (approximately 10^{-7} Kelvin) are possible. Successful nuclear adiabatic demagnetization requires extremely large magnetic fields (exceeding 10 teslas), however, which are not always readily achieved even when employing superconducting SQUID (superconducting quantum interference device) magnetometry.

The vapor pressure of liquid helium 3 itself can be used to measure its temperature down to 0.3 Kelvin. For measurements at lower regimes, the small heat capacities of most materials preclude techniques which themselves introduce significant heat to disturb the original system state. Semiconductor resistance thermometers can be used accurately to about 0.01 Kelvin. Paramagnetic and nuclear orientation methods can be employed at even lower temperatures. In NMR thermometry, which is accurate between 2 Kelvins and 300 nano-Kelvins (or 300×10^{-9} Kelvins), continuous wave or pulse magnetic fields are applied to platinum or copper nuclei. Although spin-lattice thermal relaxation times can be very long in the milli- and micro-Kelvin regime, NMR relaxation measurements accurately record temperatures existing immediately prior to measurement. Another milli- or micro-Kelvin-applicable thermometer is gamma-ray anisotropy, which measures directional differences of cobalt and manganese radioactivity counts.

There has been considerable development in improving cryogenic properties of diverse materials such as stainless steel, copper, epoxies, graphite, dielectric grease, and varnishes. Perhaps the most direct spin-off application of superfluidity is in high-resolution ultrasonic microscopy (cryo-ultra-microtomy). Because sound wave attenuation increases notably for short wavelengths of higher frequencies in conventional fluids, superfluid helium 4 at temperatures below 0.2 Kelvin permits wavelength resolutions as small as 30 nanometers, which is twenty times shorter than red light (5,000 angstroms). Superfluid helium 4 has negligible sound attenuation as a result of decreasing residual scattering of phonons by other thermal vibrations. For finer resolution, superfluid helium 3 is used for temperatures to 0.3 Kelvin. Still finer resolution is possible using dilution refrigeration for cooler, less attenuating superfluids. In the milli-Kelvin temperature and gigahertz frequency regimes, sound scattering from helium zero-point motion is the last remaining source of attenuation, reducible by maintaining temperature and increasing superfluid pressure.

Context

The origins of the discovery, theory, and subsequent refinements of scientists' understanding of superfluidity have often involved complex and not always congruent interrelations of statistical and quantum mechanics, fluid dynamics, and solid-state physics, as well as recent mathematical theories such as the renormalization group. The fact that superfluidity underscores the reality of macroscopic quantum ordering is an important verification of quantum mechanics. Superfluidity has provided yet another example how semiformal or phenomenological theories often not only precede, but also supplement more formal rigorous mathematical explanations.

Further progress in explaining extant phenomena and extending observations to lower temperatures and other fluids has both conceptual and technological prerequisites. Unless new ultra-low temperature cooling methods are practically devised, further improvements in cryostat construction materials and magnetometry are needed. In addition, almost all numerical models for low-temperature fluid systems such as vortices require extensive computer time and will remain of limited predictive utility until made less computer-bound through new algorithms, computer hardware, and symbiotically conceived algorithms optimized for given hardware architectures.

Another theoretical development in superfluidity, which is likely to continue as a focus of further study, is spin-polarized superfluid hydrogen. Scientists believe it is possible to achieve a mesoscale-ordered liquid, with electron spins antiparallel to magnetic fields, by milli-Kelvin cooling superfluid hydrogen under very high magnetic fields. In contrast to normal hydrogen 2 so-called hydrogen ¶ is expected to have Bose-Einstein statistics, a gas at all temperatures, and no liquid phase. Both bulk and thin films of hydrogen ¶ are predicted to exhibit superfluid transitions at 0.3 Kelvin.

Bibliography

Bennemann, K. H., and J. B. Ketterson, eds. *The Physics of Liquid and Solid Helium*. 2 vols. New York: John Wiley, 1976-1978. The most general introduction to pre- and post-superfluid helium 3 scenarios in low-temperature physics. The review paper by D. Lee and R. Richardson is particularly recommended for its conceptual and experimental background, helping readers who are confused by the variety of phenomenological and mathematical models for superfluidity.

Betts, David S. *An Introduction to Millikelvin Technology.* New York: Cambridge University Press, 1989. Provides what is probably the best introduction to refrigeration and thermometry in experimental superfluidity and superconductivity. Many sections consider the historical development of low-temperature scientific and engineering instrumentation. Extremely well referenced to applications as well as theoretical and experimental cryogenic literatures.

Galasiewicz, Zygmunt M., ed. *Helium 4*. New York: Pergamon Press, 1971. These collected papers are notable largely for their historical value. Covers the many rapid successions of discoveries about superfluid helium 4, London and Landau's theories, and the protracted search for the superfluid helium 3 transition. Includes a valuable editor's introduction that provides a good overview of the papers' content and their historical import.

Keesom, Willem H. *Helium*. Amsterdam: Elsevier Scientific Publishers, 1942. Although originally conceived as an intermediate-level (undergraduate science) monograph documenting the then-present state of the art, it can be read profitably as a very detailed account of the history of cryogenics from the late nineteenth century through 1940. A good balance between recounting experimental and theoretical results. Supplemented by Lifshits and Andronikashvili's work, *A Supplement to "Helium"* (1959).

Khalatnikov, Isaak M. *An Introduction to the Theory of Superfluidity.* Translated by Pierre C. Hohenberg. New York: W. A. Benjamin, 1985. A very comprehensive if rather advanced (undergraduate) account of all phenomenological and formal theories of superfluidity. Contains many useful illustrations, figures from original Soviet papers, and an extensive bibliography. Requires caution as a result of nonstandard nomenclatures and a biased technical emphasis. Should be read after McClintock, Meredith, and Wigmore's work and Richardson and Smith's work.

Kidnay, A. J., and M. J. Hiza, eds. *Cryogenic Properties: Processes and Applications.* New York: American Institute of Chemical Engineers, 1986. A good general audience treatment of most normal and some selected superfluid applications to physical chemistry and allied research. Offers a sketch of a number of industrial and applied research areas that may benefit from spin-offs of present and future superfluidity research results.

Lifshits, E. M., and Elevter L. Andronikashvili. *A Supplement to "Helium."* New York: Consultants Bureau, 1959. Largely of historical interest, this book updates Willem H. Keesom's account of the state of the art of physical knowledge about superfluidity and superconductivity for the post-World War II period, concentrating on Soviet theoretical and experimental progress.

McClintock, P. V. E., D J. Meredith, and J. K. Wigmore. *Matter at Low Temperatures.* New York: John Wiley & Sons, 1984. Although more advanced than David Betts's work, this text gives the most complete and current English-language account of solid and fluid normal and superstate properties. Provides many illustrations of experimental results, and initial efforts at finding industrial applications for these results.

Mendelssohn, Kurt. *The Quest for Absolute Zero*. 2d ed. New York: Halsted Press, 1977. The classic historical account, written for the general reader, for most theoretical and experimental cryogenic work up to the late 1960's (prior to superfluid helium's discovery). Many illustrations and special sections help make technical aspects readily grasped without prior background.

Richardson, Robert C., and Eric N. Smith. *Experimental Techniques in Condensed Matter Physics at Low Temperatures*. Redwood City, Calif.: Addison-Wesley, 1988. Although an intermediate-to-advanced level (undergraduate) reference, this text recapitulates the conceptual and technical development of low-temperature refrigeration, thermometry, and adjunct materials science beyond that given in previous standard accounts by Edward White and Oli Lounasmaa. Includes accounts of the accidental discovery of the superfluid helium 3 states by Richardson, Osheroff, and Lee and their subsequent work.

Gerardo G. Tango

Cross-References

Acoustics, 23; Crystal Symmetries, 612; Entropy, 800; The Exclusion Principle, 860; Nuclear Magnetic Resonance Imaging, 1587; The Interpretation of Quantum Mechanics, 1986; Resonance, 2107; Superconductors, 2417; Thermodynamics: An Overview, 2493; Ultrasonics, 2615; The Uncertainty Principle, 2628; Viscosity, 2685.

SUPERNOVAS

Type of physical science: Astronomy and astrophysics
Field of study: Stars

Supernovas, or exploding stars, are spectacular events emitting light at rates up to hundreds of millions times greater than the sun. They have been recorded throughout history and are an important mechanism in the synthesis of new elements as well as triggering new star formation.

Principal terms

ANGULAR MOMENTUM: momentum caused by the rapid spinning of a star about its axis of rotation

CHANDRASEKHAR LIMIT: the limiting size of a white dwarf star that could support its own weight, which is about 1.4 times as massive as the sun

GRAVITATIONAL CONTRACTION: the inward pull on a star because of its mass

GUEST STAR: another term for a nova or new star

ISOTOPE: a variety of an element that has a different mass number because of variations in the number of neutrons

NEUTRINO: a tiny particle that has no apparent mass or electric charge, no effect on ordinary matter, and travels at a speed nearly that of light

NEUTRON STAR: a very dense star composed of neutrons, typically with a radius of 10 kilometers and a mass equal to or slightly greater than the sun

NUCLEOSYNTHESIS: the process leading to the formation of new elements

SHOCK WAVE: a zone of compression and heating of matter traveling faster than the speed of sound

SUPERNOVA REMNANT: gaseous material expelled by a supernova explosion

Overview

A supernova, an explosion of a star, is one of nature's most spectacular events. In the first ten seconds, as much energy radiates from a tiny core 30 kilometers across as from all of the other stars and galaxies in the entire visible universe. The energy of this short burst is one hundred times more than the sun will radiate during its 10-billion-year lifetime. The dramatic supernova of 1987 (Supernova 1987A) discovered in the large Magellanic Cloud, although one of the fainter supernovas, emitted light at 100 million times the rate of the sun. Supernovas of the more massive stars may emit light at up to 600 million times that of the sun.

Supernovas are very rare. The last one observed in Earth's galaxy was Johannes

Kepler's in A.D. 1604. According to estimates, up to five supernovas may occur in the Milky Way galaxy each century but most escape detection because of obscuration by interstellar dust in the galactic plane. In other galaxies, the rate varies from a few each century in the largest spirals to one every few centuries in the smallest spirals.

Supernovas in the night skies were noticed in ancient times. Although these civilizations had little scientific knowledge of their observations, the records are still an important resource for modern astronomy. Supernovas are dying stars, and observations on them provide estimates of the death rates among stars. Data on the time at which the starlight reached Earth may be provided in the historical records, along with perhaps information on the brightness, color, and length of time visible.

If a bright stationary star was observed for several months, it was more likely a supernova rather than a less spectacular and more common nova. Chinese records have provided reliable reports of various celestial events. There were particularly noteworthy events called "guest stars" sighted in the years 185, 386, 393, 1006, 1054, 1181, 1572, and 1604. The supernova of 1006 was perhaps the brightest to have appeared in the past thousand years. Because of its position in the sky far south and below the horizon from northern Europe, the best observations came from Arabic, Chinese, and Japanese sources. This supernova was described as round in shape, two and one-half to three times the size of the planet Venus, twinkling considerably, and illuminating the horizon with a brightness greater than one quarter of the Moon.

The guest star that was sighted in June, 1054, was described as bright as the planet Jupiter, visible in daylight for three weeks and in the night sky for two years. Interestingly, this bright object, although surely visible, went unrecorded in European annals. The split in Christianity between the Roman Catholic church and the Greek Orthodox church in the east may have caused Church officials to have deleted such a "portent event" in the heavens from European historical records. The event was noticed in the Middle East and in North America by native Americans. The sighting is recorded as a petroglyph incised on sandstone at Navajo Canyon, Arizona. The date has been estimated from association with Pueblo Indian dwellings and pottery by archaeological methods.

Tycho Brahe, a Danish astronomer, won fame for his observations of the supernova of 1572. The supernova is said to have inspired him to devote a lifetime to collecting measurements of the positions of the stars and planets. Based on careful observations and measurements over an eighteen-month period, Brahe concluded that the phenomenon was not a kind of comet or meteor but a star that had never been observed before.

Kepler, who was formerly Brahe's assistant, sighted a supernova on October 17, 1604. He described the star as similar to Jupiter in brightness, with the color of a diamond. Visibility continued through October, 1605, although the star briefly disappeared behind the sun from November, 1604, until January, 1605. Sightings were also recorded in Chinese and Korean records; comparing their observations to those of the Europeans allowed scientists to determine the accuracy of the records.

As an aftermath of the star's explosion, hot gases expand outward with speeds approaching that of light, forming a cloud of luminous material called a supernova remnant. Within the past forty years, astronomers have identified the remnants of almost all of the supernovas known from history and more than a hundred other remnants. The Crab nebula in the constellation Taurus was studied extensively since the 1920's, when photographs taken eleven years apart showed outward expansion of gaseous material. Knowing the expansion rate of the gases from Doppler shift studies, Edwin Powell Hubble, in 1928, concluded that to reach its present size would take nine hundred years. The prediction agreed with sightings of 1054 and established the Crab nebula as the first confirmed supernova remnant.

The remnant of the brilliant 1006 supernova was first detected as a radio source in the early 1970's, but it was not photographed until 1976. Kepler's supernova of 1604 was first photographed in 1947 by Walter Baade after determining is location from Kepler's position. The position of the remnant of Brahe's supernova was first confirmed as a strong radio source in 1952 and photographed shortly after with the George Ellery Hale 5-meter telescope.

Stars obtain their energy from hydrogen fusion for the greater portion of their life spans. The sun will continue to use its supply of hydrogen for about 5 billion more years, but stars more massive will burn their supplies more rapidly. A star equal to 20 solar masses will shine with a brightness ten thousand times that of the sun but will exhaust its fuel in only a few million years. Whatever its mass, a star will eventually run out of hydrogen.

The end of the first stage in the star's life cycle is reached when most of the hydrogen in the core has been fused into helium. The star was stable, generating energy with the pressure of hot gases outward and compensated for by the pull of gravity inward. When the rate of energy production slows down, outward radiation diminishes, and gravity compresses the core more compactly. Temperatures rise through the star, igniting unburned hydrogen in regions surrounding the core. The outer layers of the star expand to immense proportions, while the surface cools; the star now becomes a red giant. The core continues to contract until the temperatures reach 100 million Kelvins; then helium atoms fuse to form carbon atoms, releasing greater energy. The star becomes differentiated, with the heavier elements located in the centermost zones and the lighter elements fusing in shells farther out. Carbon fusion releases still more energy and fusion produces oxygen nuclei. The fusion may progress, creating the heavier elements, neon, magnesium, silicon, and iron in a stepwise process known as nucleosynthesis.

Most stars are not massive enough to reach the stage of carbon fusion and end up with a core of carbon, becoming white dwarf stars. The more massive stars can fuse heavier elements, but further fusion in any case cannot proceed past the isotope of iron, which has an atomic weight of fifty-six. If additional particles were added onto this iron nucleus, energy would be removed from the surrounding shells rather than released from the core as in the previous fusion process. A massive star that ends up with an iron core is surrounded by shells of progressively lighter elements in onion-like layers.

At this stage in its life cycle, the massive star can no longer provide sufficient outward radiation pressure to balance the crushing pull of gravity inward, because the iron core is no longer releasing energy. The next surrounding layer of silicon, however, continues to burn, adding more iron onto the core. If the mass of the core reaches a critical value called the Chandrasekhar limit, collapse of the star begins in a matter of minutes. The compression by the collapse increases the core temperature; however, the iron cannot fuse at the higher temperatures but instead starts to disintegrate. The process removes energy from the core and in turn lowers the pressure of the surrounding gas, allowing additional compression. Electrons are forced tightly into the nucleus, combining with protons, and release tiny particles called neutrinos. The neutrinos remove energy from the center of the star very quickly, accelerating the collapse process and permitting gravitational contraction to incredible densities, reaching that of an atomic nucleus. A shock wave is formed, but it takes several hours to reach the surface. Great numbers of neutrinos remove enormous energy from the exploding star and generally continue their journey outward uninterrupted.

There are two possible scenarios for the next stage of the star, depending upon its mass. For stars of 8 to 10 solar masses, the explosion leads to the formation of a neutron star, which is merely the remnant of the collapsed core compressed to the density of an atomic nucleus, or 100 million times that of water. These stars are very tiny and hot, with core temperatures exceeding 1 billion Kelvins. Neutron stars rotate rapidly because of conservation of their original angular momentum, and their magnetic fields are strongly concentrated into two beams at the poles. Neutron stars that blink on and off rapidly when this beam of radiation is observed are called pulsars. Very massive stars equal to 30 solar masses or more are so massive that the gravitational forces overcome the core strength of even a neutron star (known as neutron degeneracy), and the star collapses to a singularity point known as a black hole.

Applications

Telescopic surveys of galaxies have expanded the number of supernovas discovered from 20 in 1934 to more than 650 by 1978. The number of discoveries increased dramatically after 1950, with the development of the Schmidt telescopic camera used for surveying large areas of the sky. Detailed analysis on smaller regions was performed with telescope apertures as large as 5 meters.

High-altitude observatories were able to study radiation emitted by Supernova 1987A that would otherwise be absorbed by the earth's atmosphere. Information on the size of the original star was obtained from the ultraviolet flash recorded by the International Ultraviolet Explorer satellite. Observations followed for several months, revealing emissions from a shell of gas surrounding the supernova. Radiation characteristics led to the conclusion that the initial light emitted from the supernova was from material at a temperature of a half million Kelvins. The high temperature, combined with the strong ultraviolet radiation, is expected whenever a powerful shock wave breaks through the surface of the exploding star, ejecting gaseous material.

Measurements of the expansion velocity led to the determination of the total energy of the explosion.

Several months after the initial outburst of Supernova 1987A, an infrared telescope flown at 12,000 meters on a jet transport revealed many elements from the supernova core, including iron, nickel, cobalt, oxygen, neon, sodium, potassium, silicon, and magnesium. The intensity of the infrared lines proved that the quantities of these elements was larger than what could have been present initially in the star at birth.

Neutrino observation has been used in the analysis of the explosion of Supernova 1987A, as it has given astrophysicists the precise time of the explosion. Neutrinos, once emitted from the shocked core, travel virtually untouched outward through the remainder of the star. Electromagnetic radiation or light, however, is absorbed and reradiated by matter while on its way to the star's surface, emerging several hours later with the shock wave. As a result, neutrinos released from the star arrived at Earth before the initial light burst after traveling for 160,000 years. By measuring the time difference from the detection of the neutrinos to the first observed light from the exploding star, it was concluded that the original star was relatively small, placing it in the blue rather than the red supergiant class. The neutrino detectors are typically large tanks of water placed in deep underground mines as a shield from cosmic background radiation and other undesired events. The tanks are surrounded by electronic amplifiers called photomultipliers that detect rare collisions of the neutrinos with electrons and positrons (which result from direct neutrino-proton collisions).

The light emitted by Supernova 1987A followed an unusual light variation called a light curve. Supernovas are categorized into either Type I or Type II based on their light curves. Type I has a rapid brightening, followed by a quick decline of more than 3 magnitudes from maximum brightness in one hundred days, followed by a more gradual decline of 7 magnitudes after three hundred days. Type I supernovas evolve from less massive stars and are observed in both spiral and elliptical galaxies. Type II supernovas have a rapid decline of more than 4 magnitudes in the first one hundred days, followed by a more gradual decline of 7 magnitudes. Type II supernovas evolve from the more massive stars and are observed only in spirals and irregular galaxies.

The light curve of Supernova 1987A was expected to follow that of the Type II Supernova but was observed approximately 2 magnitudes dimmer at maximum brightness. Since blue supergiant stars are smaller than Type II precursors, the explosion had to transport the gases farther to remove them from the surface of the star. Because of the greater amount of energy needed to break up the star, there was less energy to appear as visible light, making the supernova appear dimmer.

The triggering shock wave is thought to originate not at the center of the star but at the Chandrasekhar mass limit, or about halfway out of the iron core; inside this distance, only the pressure wave propagates. Computer-generated models on the propagation of the shock wave indicate a more complicated mechanism than previously believed. All the energy of the shock wave may be used up initially in dissociating nuclei, such as iron, into component nucleons. Perhaps the shock wave may be revived because of the absorption of large amounts of neutrinos. An additional process of convective transport may be involved in transferring the energy liberated by neutrino absorption to the shock wave.

Context

Supernovas, or massive explosions of stars, are very rare events; only one has appeared visible to the eye in the Milky Way since 1604. By studying supernovas and the synthesis of elements that they produce in their unimaginably high temperatures and pressures, astrophysicsts can model the final stages in the life cycle of a star. Theoretical models, for example, indicate that stars such as the sun do not end in a violent explosion but pass through a more quiescent phase in late stages. These stars can expect to end up as tiny white dwarfs, which are merely the burned out remnants of a carbon core.

Stars of greater size may have sufficient mass to exert greater gravitational force on the core and compress it to higher temperatures and progressive nucleosynthesis. The cores of these stars become so dense as to end up with pure neutron cores of incredible densities that rebound with shock waves that blow them apart. Stars of even greater solar mass literally collapse upon themselves to form singularity points called black holes.

Supernovas are significant in the universe for a number of reasons. Interstellar clouds, the birthgrounds of new stars, may get their thrust for gravitational contraction and star formation by the triggering effect of a nearby supernova explosion, sending a shock wave into the region.

Extraterrestrial events such as supernovas may be responsible for major extinctions of life-forms on earth. Such an event would also set the stage for the development and radiation of new life-forms. Gene changes or mutations may result from excess radiation striking Earth from a nearby supernova. It is the process of mutation that results from the imperfection of the replication process that moves evolution forward with the development of advanced species. If it were not for mutations, then all organisms today would be merely replicas of the initial simple forms of life.

Since supernovas are so rare, it is not known when or where the next one is likely to occur or what that star will look like. The star would be in its last stages before collapse and would be a red giant star. Concentrating on nearby red giants, there are several possible candidates, including Ras Algethi in the constellation Hercules, Antares in constellation Scorpius, and Betelgeuse in constellation Orion. Betelgeuse is unstable and pulsates, which results in a stellar wind. Another possibility is the star in the Eta Carina nebula, which has a stronger stellar wind than Betelgeuse and is losing significant mass.

Overall, supernovas are important in providing an enrichment of heavy element supply to instellar space; the formation of the sun and Earth would not have been possible without the explosion and death of other stars.

Bibliography

Asimov Isaac. *The Exploding Suns.* New York: Dutton, 1985. This famous author of major science books covers many aspects of supernovas, from the observations of earlier civilizations to modern theories. An intriguing chapter outlines the formation of elements in a supernova and describes how life may have been affected by such periodic catastrophes.

Cooke, Donald A. *The Life and Death of Stars.* New York: Crown, 1985. Spectacular photographs and diagrams surperbly illustrate all chapters. The Crab nebula is among several case studies highlighted. The bright star in the Eta Carinae nebula is mentioned as a possible future supernova, along with several other stars.

Kippenhahn, Rudolf. *One Hundred Billion Stars.* New York: Basic Books, 1983. Stars, according to Kippenhahn, have lifetimes similar to human beings—they evolve from birth to death. The reader is taken through the final stages of a massive star leading up to the preliminary stages of a supernova through imaginary experiments. The style, written with humor, will captivate the general reader.

Marschall, Laurence A. *The Supernova Story.* New York: Plenum Press, 1988. The search for supernovas in the night sky is presented from the early records to those discovered in modern times. A documentary chapter on Supernova 1987A discovery is included, with emphasis on the observational techniques used.

Meadows, A. J. *Stellar Evolution.* New York: Pergamon Press, 1978. Evolution of main sequence stars is explained for various sized stars, including double stars. Hertzsprung-Russell (H-R) diagrams show the evolutionary course for each star's lifetime. The chapters are condensed, but adequately cover the basic theory.

Murdin, Paul, and Lesley Murdin. *Supernova.* New York: Cambridge University Press, 1985. An excellent collection of photographs and diagrams is included, with discussions of the ancient sightings. Tables are compiled for the magnitudes of Brahe's and Kepler's supernovas. A unique energy flow diagram for a supernova illustrates the nuclear processes of the evolving star.

Time-Life Books. *Voyage Through the Universe. Stars.* Alexandria, Va.: Time-Life Books, 1988. A chronology of the discovery of Supernova 1987A by Ian Shelton and its significance highlights the third chapter. Excellent color diagrams of the interiors of a supernova star illustrate time sequences from a few days until milliseconds before the final explosion.

Michael L. Broyles

Cross-References

Black Holes, 238; Main Sequence Stars, 1316; Neutrino Astronomy, 1489; Neutron Stars, 1503; Nuclear Synthesis in Stars, 1621; Optical Astronomy, 1650; Star Formation, 2360; White Dwarf Stars, 2742; X-Ray and Gamma-Ray Astronomy, 2755.

SURFACE CHEMISTRY

Type of physical science: Chemistry
Field of study: Chemical reactions

Surface chemistry is the study of chemical processes that occur in the vicinity of surfaces. The majority of both man-made and naturally occurring objects that exist in either the liquid or solid phase have surfaces and, thus, such studies have an unlimited number of applications.

Principal terms

ADSORPTION: a process in which a molecule or an atom becomes trapped on a surface

CATALYSIS: a process whereby a substance called a catalyst increases the rate of a chemical reaction; the catalyst does not become part of the products of the reaction

CHEMICAL BOND: the sharing of electrons to link atoms together to form molecules

ELECTRON CLOUD: the orbiting electrons that surround the positively charged nucleus of every atom; each chemical element is defined by the number of electrons (and their energy state); when an atom is part of a molecule, the local electron cloud surrounding that atom is different from the electron cloud surrounding that atom when it is isolated

HETEROGENEOUS CATALYST: a catalyst that is not of the same phase as the reactants

SPECTROSCOPY: a field in which the absorption and emission of different frequencies of radiation by molecular species are studied

Overview

Surface chemistry is the study of those chemical processes that take place on or in the neighborhood of a surface, which is usually also an interface between two phases (for example, a solid and a liquid). In order to realize the importance of surface processes, one has only to consider the human brain. The brain of a human has a volume approximately seven times that of an ape, whereas the surface area is approximately ten times greater. A good argument can be presented that supports the idea that it is the processes that occur at the surfaces of the brain that allow humans to consider themselves superior to the ape.

Numerous objects exist in nature in which the surface-to-volume ratio is large. (A value of one is the largest possible value and corresponds to the case in which all molecules are at the surface.) Biological systems provide many examples. Membranes have large surface-to-volume ratios, and processes occurring at the surfaces of membranes are fundamental in regard to the function of, for example, cells and organs. The discussion presented here is focused more on the chemical reaction processes that occur at solid surfaces. Chemical researchers and industry have learned to use such surfaces to speed up chemical reactions so that they occur at relatively faster rates.

It is required that either a solid or a liquid be present in a system under study in order to have a surface. The forces acting between the molecules (or atoms) of the adjacent phase and the molecules in the surface (for example, between molecules in a gas and molecules in a solid surface) determine the way in which the surface and the adjacent phase interact. As in most molecular interactions, it is the shape and character of the electron clouds around the different molecules that determine the forces between them. In general, surface chemistry can be divided into five areas: solid-gas, solid-liquid, liquid-liquid, liquid-gas, and solid-solid interfaces. The discussion presented here, however, is related more to the surface chemistry of solid-liquid interfaces and solid-gas interfaces. This does not mean that the chemistry associated with the other interfaces is unimportant. For example, the chemistry of solid-solid interfaces determines the mechanical properties of solids, and such interfaces are important in the manufacture of solid-state devices.

An important chemical surface process occurs in the catalytic converter of an automobile. The exhaust that moves through the pipe that extends from the exhaust manifold, which is located near the automobile's engine, contains carbon monoxide, nitric oxide, and oxygen. This exhaust enters the catalytic converter, where chemical reactions occur on the surface of beads that convert much of the exhaust into carbon dioxide and nitrogen gas. Eventually, the treated exhaust exits through the tail pipe. It is important to note that the molecules and atoms that comprise the surface of the beads remain there and do not become part of the products of the reaction.

In order to expand on the fundamental concept that led to the design of catalytic converters, some startling observations are reviewed from experiments. To do this, consider a closed vessel that contains a two-to-one mixture of hydrogen gas (H_2) and oxygen gas (O_2); that is, there are twice as many hydrogen gas molecules as there are oxygen gas molecules. The potential for the gases to react and form water is high. In fact, it can be shown that the state obtained after the reaction in which only water is present in the vessel is much more stable than the state in which the vessel contains only oxygen and hydrogen gases. Nevertheless, the vessel containing the two-to-one mixture of gases will remain as an apparent stable state for an indefinitely long time. On the other hand, if a platinum surface is located in the vessel, the reaction producing water occurs rapidly. At the end of the reaction, the platinum surface essentially remains unchanged. The results of such experiments represent a fascinating component of chemistry. Yet, the results are not at all magical. The platinum surface acts as a catalyst for the reaction; a catalyst is a substance that increases the speed at which a reaction occurs without being consumed by the reaction. More specifically, the platinum surface is classified as a heterogeneous catalyst because it is in a phase (a solid) different from the reactants (gases). The reaction between oxygen and hydrogen provides an example of why a catalyst is necessary and gives an insight into how a catalyst works. First of all, it is not very surprising that the mixture of oxygen gas and hydrogen gas apparently remains stable. The chemical bonds in hydrogen gas and oxygen gas are strong. Although the transformation to water through a chemical reaction is favored, these chemical bonds must be broken in order for the reaction to proceed. The breaking of these bonds in the absence of the platinum surface takes so long that, from a practical point of view, the reaction does not proceed at all. When a platinum surface is present, hydrogen and oxygen gas molecules stick to the surface. The interaction with the platinum atoms breaks the bonds in the gas molecules, producing hydrogen and oxygen atoms. These atoms are much more reactive; they can react with the gas molecules and with other atoms. The role of the platinum surface is to provide additional reaction pathways.

How does the platinum surface provide additional reaction pathways? In order to obtain some insight into this question, properties of surfaces should be analyzed. One might be tempted to think that, when a gas or liquid is in the presence of a solid surface, the only interactions between the two phases involve molecules (or atoms) approaching the surface and the deflection of the molecules by the surface through collisions. Nevertheless, there is always a force of attraction between an approaching molecule and the solid surface. Consequently, there is a finite probability that the molecules will become trapped on the surface. The process by which an atom or molecule becomes trapped on a surface is called adsorption. Heat is released through this process, and thus it is known that the system, after the molecules are adsorbed, is in a lower energy state. The amount of heat released, which can be measured and is called the heat of adsorption, reveals the relative strength of the attractive force. The fact that molecules are adsorbed is witness to the fact that the driving force that pushes the system to a lower energy state is usually very strong because of the forces of attraction between the atoms or molecules of the solid surface and the molecules in the gas or liquid phase.

An atom or molecule adsorbed on the surface is in a microscopic state that is different from the state it was in when it was a surrounding gas or liquid phase; that is, the electron cloud of the atom or molecule can be drastically different. Consequently, the interaction between other species in the phase surrounding the surface is also quite different, and this new interaction can have a considerably higher probability of leading to a chemical reaction.

Additional properties of surfaces that are helpful in the understanding of adsorption are roughness and porosity. If one uses the eye to judge roughness, a surface may look very smooth. Yet, after magnification to the molecular level, a surface will appear very rough even though it appears very smooth at the macroscopic level. An atomic model of a surface includes the presence of several different types of sites. The surface contains terraces, flat areas, and steps that lead from one terrace to either a lower or a higher terrace. The steps very seldom form straight lines, but rather have indentations called kinks. Other sites exist on a surface called detects, which include adatoms (atoms lying on top of a terrace) and vacancies (atoms missing from the flat area of a terrace). The environment of each type of site is unique. For example, a molecule (or atom) contained in a terrace will have the most molecules surrounding it, a molecule located in a step will have fewer neighbors, but more than a molecule in a kink. An adatom will ordinarily have the least number of neighbors. The different sites will interact differently with molecules of a different phase in the vicinity of the surface. The interaction with molecules of the surrounding phase might be strong enough that chemical reactions occur at some sites, but not others. Chemical products from reaction at one site may react at a different site.

Even if a surface appears to be smooth and dense, it is, to some degree, porous. For example, hydrogen atoms are known to penetrate the surfaces of palladium and platinum. When this occurs, the process is called adsorption. This phenomenon is known to affect catalytic properties of the surface and, hence, the outcome of catalytic reactions, especially in the case of platinum and palladium surfaces, reactions in which hydrogen is either added to or subtracted from a molecule.

Two kinds of adsorption can occur. One kind is called physical adsorption and the other is called chemisorption. In physical adsorption, the interaction between the molecules that are adsorbed and the molecules of the surface is relatively weak; that is, no strong chemical bonds are formed. Physical adsorption is closely related to condensation and, in fact, a special type of physical adsorption occurs when gas molecules fill tiny capillaries on the "surface" of a solid and condense to form liquid. Only one layer of atoms or molecules may result as a consequence of physical adsorption. Such a layer is called a monolayer. Physical adsorption may also involve several layers. The number of layers and character of the adsorption can change with respect to changes in conditions such as temperature and pressure. Adsorption can be characterized by what are called isotherms, plots of the amount of substance adsorbed on a surface versus the density (or pressure) of the substance. Physical adsorption involving only a monolayer allows the implementation of a method for accurately measuring the surface area of a solid. By assigning a value to the area covered by one adsorbed molecule (this can usually be calculated), the area of a surface can be calculated (for example, when a monolayer completely covers the surface), by counting the total number of molecules adsorbed; nitrogen gas is often used for this purpose.

Chemisorption is relatively strong and specific. It involves chemical reactions and as a result of the reactions, chemical bonds are formed between the molecules of the surface and the adsorbed molecules. Usually, chemisorption can be distinguished from physical adsorption. The heat of adsorption in physical adsorption has a value that is on the same order of magnitude as the heat released (per mole) when a gas is liquefied, whereas in chemisorption, the value is much larger and is in the same range as typical heats of reactions (heat released or adsorbed during chemical reactions). The two types of adsorption can be differentiated by examining the effect of temperature and pressure. Physical adsorption, in general, decreases rapidly with increases in temperature, whereas chemisorption first increases and then decreases with increases in temperature. (Unfortunately, there are many exceptions to the "rule.")

Physical adsorption of gases increases gradually with pressure and reaches a limiting value only at very high pressures. Generally speaking, chemisorption increases rapidly with increasing pressure at low pressures and soon reaches a maximum. Under certain conditions, however, the two types of adsorption processes will occur simultaneously, and it is often difficult to differentiate between the two.

Chemisorption occurs quite readily at certain temperatures when gases, such as hydrogen, oxygen, and carbon monoxide, are in the presence of a solid surface. Examples include the formation of a film of oxygen on a hot filament of carbon and the bonding of carbon monoxide to a surface of tungsten. Chemisorption also occurs readily when a liquid is in the presence of a solid surface; however, the conditions for optimal chemisorption are more varied. For example, the concentration of solute and the type of solvent play determining roles. As in physical adsorption, chemisorption can often, but not always, be characterized by plotting isotherms. Chemisorption is also divided into two types, molecular chemisorption and dissociative chemisorption. In molecular chemisorption, the whole molecule that approaches the surface bonds to the surface, whereas in dissociative chemisorption, the chemical reaction results in only part of the molecule bonding to the surface; the other part or parts dissociate during the reaction.

Chemisorption is important primarily because of the use of surfaces as catalysts. Much of the past research on catalytic reactions at surfaces has been empirical. It is, however, generally agreed that the forces involved are of the same type as those that are active in ordinary chemical reactions. Solid catalysts almost always combine chemically at the surface with one or more reactants. Yet, there are large differences between the activity of different catalysts and the type of activity. In some catalytic reactions on surfaces, it seems there is a uniform interaction over the whole surface between the reactants and surface molecules. Other reactions occur only at certain sites on the surface; these are called active sites.

The details of the reaction can be very complicated. The oxidation of small organic reactions at a platinum surface usually involves breaking the chemical bonds in the organic molecules that connect hydrogen atoms with carbon atoms. Chemical bonds form between hydrogen atoms and surface platinum atoms, and also between the surface atoms and the carbon atoms of the dehydrogenated organic molecules. Subsequent reactions depend on the constraints of the system and the phase (liquid or gas) in which the organic molecules originate. In both liquids and gases, pairs of neighboring hydrogen atoms combine to form hydrogen gas. In liquid solutions, if the platinum surface is charged, the hydrogen atoms can be transformed into protons. (A hydrogen atom minus an electron is a proton.) If the organic compound contains one or more oxygen atoms, or if the fluid (liquid or gas) contains oxygen or water, carbon dioxide can be a final product. Under different conditions, the reverse reaction can take place; using carbon dioxide as a reactant, organic compounds can be synthesized on platinum. Much research has been conducted on the latter reaction because this is how petroleum can be made. Both the forward and reverse reaction have been found to be affected by the crystal structure of the platinum, metal adatoms placed on the platinum surface, and the roughness of the surface. Many desired organic reactions do not proceed on a bare platinum surface. An oxide layer can be added to the surface by exposing the surface to oxygen gas or exposing water to a charged surface (reactions occur in which oxygen atoms and oxygen gas molecules form chemical bonds with the surface platinum atoms), and it has been found that additional organic reactions are catalyzed by the platinum-oxide surface.

In order for reactions that are catalyzed by solid surfaces to proceed to completion, there must be desorption of the products. Nevertheless, some reactions produce intermediates that remain on the surface and cannot be removed. These intermediates poison the catalysts, and either some method is required to remove the poison or the surface has to be replaced. On the other hand, chemically bonded species on a metal surface can protect that surface from unwanted chemical reactions. For example, oxide layers on aluminum protect it against corrosion.

Applications

In addition to looking at chemical processes at surfaces from a microscopic view—that is, the interaction of molecules—one can look at certain aspects of these processes from a macroscopic point of view—that is, the collective interaction of a tremendously large number of molecules. Surface processes, as a macroscopic process, were originally looked upon in this way, and many applications are still considered from this point of view. One important macroscopic quantity is surface energy; the corresponding force is called surface tension. A molecule in the bulk phase (for example, in the middle of a solid) interacts with molecules from all sides. At a surface, however, a molecule has fewer neighbors. Because of less molecular interaction, surface molecules have a higher energy. There is a tendency for the molecules at surfaces to behave collectively in such a way as to minimize the surface energy. This is especially noticeable at liquid surfaces because this process leads to surfaces with curvature.

The macroscopic view of some adsorption processes of liquids by solid surfaces is called wetting. Here, surface energy plays an important role. If a liquid is brought into contact with a surface that adsorbs it and if the liquid forms a film over the surface, the liquid is said to wet the solid. Surface forces determine whether a liquid substance spreads over a solid surface as well as into crevices and pores. Wetting usually means that a liquid spreads over a solid suface easily, whereas nonwetting usually means that the liquid tends to form a ball that runs off the surface easily. In order for a liquid to spread, it must increase its surface area, and hence, the adsorption must lower the energy more than the lowering of energy obtained when the liquid minimizes its surface by forming, say, a ball. Additives, called surfactants, are often necessary so that the liquid will possess a satisfactory degree of wetting. These additives consist of molecules with a polar and nonpolar part. The polar part (which can be thought of as an electrically charged part) sticks to the surface, whereas the nonpolar part protrudes away from the surface into the liquid. The nonpolar part must be sufficiently inert with respect to interactions with other molecules so that it does not interfere with the flow of the liquid.

The processes of adsorption and wetting of a solid by a liquid are important in the lubrication of surfaces and sprays for plants. Lubrication requires maintaining an adsorbed layer of a deformable material on each of two solid surfaces so that they do not come into contact. Liquids used as horticultural sprays for insects must have the property of wetting. Thus, they must be easily adsorbed and have low surface energy.

Many applications are described best by the microscopic point of view, especially adsorption of gases. Because of the fact that different compounds are adsorbed to a different degree, both physical and chemical adsorption can be used to separate species from gas mixtures. At 1,000 degrees Celsius, charcoal adsorbs oxygen, nitrogen, and argon gases from air, but does not adsorb helium and neon. Close to the temperature that air becomes a liquid, however, charcoal adsorbs almost all the neon gas, but not the helium. Charcoal was also employed in the gas masks of World War I to adsorb toxic gases. Charcoal is also used by industry to adsorb, and thus save, valuable volatile solvents during processes in which rapid evaporation of the solvent is necessary.

Almost all chemical technologies use surface catalysis. An important catalytic industrial process is the hardening of vegetable oils. The process causes the substance to change from a liquid oil to a solid. For example, margarine is produced using this process. The product is more stable against malodorous chemical derivatives that can form through chemical reactions that occur when vegetable oil is exposed to air. The process is carried out on a nickle surface. The catalytic properties of the surface enable chemical reactions, in which additional hydrogen atoms are added to the molecules that make up the vegetable oils, to occur at a sufficiently rapid rate to make the process economically feasible.

Surfaces with special molecular properties can be prepared for use as catalysts in specific processes. One such catalyst is employed in the production of linear polyethylene, which is used for making plastics. The catalyst is prepared by sticking a salt containing chromium into a glasslike silica gel. The mixture is heated and then cooled. The resulting surface, which is rather rough, consists of chromate (a chromium atom surrounded by four oxygen atoms) dispersed on a layer of silica. Isolated chromium centers provide sites at which, beginning with a solution containing ethylene, polymers (very long molecules) grow by way of chemical reactions. This is also an example of where the catalyst changes structure during the reaction; the particles of the catalyst become fragmented during the polymerization process. The polymer chains grow outward from the sites on the catalyst, and these sites may be within pores. Hence, particles on the surface can be shattered. In typical industrial reactors, the polymers precipitate out of solution and become separated from the catalytic particles.

Context

Many aspects of surface chemistry have been known since the earliest development of the physical sciences. One can follow some parts of the historical development of surface chemistry by noting the era in which applications were initiated. For example, it apparently was known in the fifteenth century that the carbon obtained by heating wood in a closed container removes coloring from solution. In 1791, this knowledge led to the process in which charcoal was used commercially to remove colored matter from sugar solutions.

In the first part of the nineteenth century, a patent was obtained for employing spongy platinum as a catalyst for the reaction in which sulfur dioxide is transformed to sulfur trioxide. (At that time, the process failed because the catalyst was easily poisoned.) Techniques for liquefaction of gases through physical adsorption on solids were developed around 1884. At approximately the same time, the procedure of using adsorbent carbon to remove gases in order to obtain a vacuum was developed.

One of the more famous chemical reactions, the synthesis of ammonia from nitrogen gas and hydrogen gas, was realized shortly before World War I by the chemist Fritz Haber. An iron catalyst is used in the process and today this process is so developed that ammonia is quite inexpensive. The development of the Haber process also calmed the fear that a world disaster would occur in the middle of the twentieth century because of a predicted shortage in the amount of fixed nitrogen (nitrogen is an atom in ammonia).

Measurements of macroscopic quantities, such as surface tension and the amounts of gases adsorbed by a solid surface, were possible around the middle of the nineteenth century. Experimental measurements could be related to macroscopic theoretical parameters; most of the theory was developed by Josiah Willard Gibbs. Microscopic measurements of chemical surface processes could not keep pace with the development of such measurements in other branches of chemistry until well into the 1950's. At that time, space exploration promoted the development of methods for the preparation of clean surfaces and reproducible surface studies. Important advances for microscopic investigations of surface chemistry are the developments of surface spectroscopies. Surface spectroscopies have enabled chemists to determine the elemental composition and structure of the surface layer, the state of surface adsorbates, and the determination of intermediate species of surface catalyzed reactions. Much research remains to be done on determining the mechanism of surface reactions, and it is expected that spectroscopy will play a key role in this endeavor. Research will also continue on the development of modified surfaces that will increase the efficiency of, for example, nuclear energy conversion, and both fossil-fuel conversion and generation.

Bibliography

Adamson, Arthur W. *Physical Chemistry of Surfaces*. New York: John Wiley & Sons, 1976. A somewhat technical account of surface chemistry written at the level of an undergraduate junior or senior. Yet, many of the descriptive parts are accessible to those with less background but who have a good intuitive approach to the physical sciences. A good starting point for those who want to examine a particular aspect of surface chemistry in detail.

Campbell, Ian M. *Catalysis at Surfaces.* New York: Chapman and Hall, 1988. This book almost exclusively focuses on catalytic surface reactions. Details are presented on the mechanism of such reactions. Campbell presents an excellent account of applications, which include catalytic craking of crude oils (the breakdown of crude oil to small hydrocarbons, which are then used as fuel), synthetic gasoline production, and the hardening of vegetable oils.

Somorjai, Gabor A. *Chemistry in Two Dimensions: Surfaces.* Ithaca, N.Y.: Cornell University Press, 1981. This is a detailed descriptive account that covers almost all topics of surface chemistry. Although there are mathematical explanations of various aspects of surface chemistry, these are mainly useful formulas that can often be skipped without losing the basic points. Chapters on adsorption and the surface chemical bond are particularly well written and are enjoyable to read.

Soriaga, Manuel P. *Electrochemical Surface Science: Molecular Phenomena at Electrode Surfaces.* Washington, D.C.: American Chemical Society, 1988. Contains many articles on the surface chemistry of electrode-electrolyte interfaces. The majority of articles describe applications of surface spectroscopies. For example, chapter 24 presents the results of an investigation using surface spectroscopy of the process in which a current is produced when carbon dioxide is obtained after methanol reacts at a platinum surface.

Weiser, Harry B. *A Textbook of Colloid Chemistry.* New York: John Wiley & Sons, 1949.

Hauser, Ernst A. *Colloidal Phenomena: An Introduction to the Science of Colloids.* New York: McGraw-Hill, 1939. Although dated, these books are good sources on colloid chemistry, which is considered by many colloid chemists as equivalent to surface chemistry. Highly descriptive. One might think that later developments would prove some of the theory discussed in these books as false. It is incredible, however, how much of a somewhat intuitive approach to the subject in these books was later proved correct.

Mark Schell

Cross-References

Adsorption, 46; Catalysis, 317; Colloids, 435; Electrochemistry, 721; Electrokinetics, 728.

SYMBOL MANIPULATION PROGRAMS

Type of physical science: Computation
Field of study: Artificial intelligence

Symbol manipulation techniques are used in high-level computer programming, in contrast with programs that focus on efficient manipulation of numbers. Logical and systematic design allows the construction of large programs that achieve complex reasoning tasks and are easy to understand.

Principal terms

ABSTRACT DATA TYPE: a class of structured objects described abstractly in terms of the predicates, operations, and functions that can operate on it

FRAME: a structured object consisting of <attribute, value> pairs

KNOWLEDGE: the result of processing, refining, and structuring data, extracting important aspects

PREDICATE LOGIC: a generalization of propositional logic, with symbols representing predicates and variables that may be quantified

PROPOSITIONAL LOGIC: a logic in which symbols denote propositions and are combined into formulas using connectives representing "and," "or," "implies," "not"

Overview

Symbol manipulation programs perform high-level computations on well-structured and organized data. Unlike number-crunching programs that perform arithmetic operations on numerical data, symbol manipulation programs manipulate and reason about knowledge represented as structures and organizations of symbols. Representations may be organized in various ways—for example, abstract data types, frames, semantic networks, and logical formulas.

In abstract data types, each object has a standardized structure and is accessed only through a restricted set of operations whose meaning is well understood. For example, a "queue" may be specified as a structured object whose components can be accessed using only operations, such as "first," "rest," and "add-to-Q"; the state of the queue may also be checked via the predicate "is empty;" a special operator to generate new (empty) queues may also be available. In such a specification, there is no mention of how queues are implemented on a specific computer or in a specific programming language. Cells and pointers may be used in one implementation, and arrays of some kind may be used in another implementation. One may choose to modify the underlying implementation at any time, perhaps to improve efficiency, without having to modify the high-level program itself. This is possible because there is no reference to implementation details in the high-level program that manipulates queues. Separation of concerns between the high-level program and the underlying implementation implies that programs are more easily understandable, maintainable, modifiable, and portable. Some other common examples of abstractly specified data types are lists, stacks, dequeues (double-ended queues), trees, and relations. All these have in common the property that their instances are accessed only through well-known and clearly defined operations. Variables in a program can be declared to belong to any of these classes of objects, just as variables can be declared to belong to one of the types (integers, characters, reals, and the like).

Frames are structured objects in which information is stored as a set of pairs. The first element in each pair is an "attribute" or "slot"; the second element is a "value" or "filler" associated with that attribute. The following is an example frame: {<name, David's>, <class, Restaurant>, <me, eastern>, <average—entree—price, $7>, <location, M.Street>, . . . }. Different frames may belong to the same schema, with the same attributes but different values. For example, many frames including the above "David's" frame may be instances of a "Restaurant" schema.

Schema may be organized into an inheritance hierarchy, with lower-level schema being subclasses of the schema above them in the hierarchy. One advantage of this methodology is that knowledge representation is more compact. Instead of separately specifying the attributes and values of each schema and frame, one can assume that every frame and schema inherits attribute-value pairs from the nodes above it in the hierarchy. For example, if the schema "birds" has an attribute "method of movement" with the value "flying," and "eagles" is a subschema of "birds," then one need not separately specify that eagles fly, since this property is inherited from the "birds" schema.

In specific cases, it is possible to override the inherited property by specifying an alternate value. For example, "penguins" may be a subclass of "birds" for which the value "walking" is associated with the attribute "method of movement." Now, if someone is told only that Tweety is a bird, one would infer that Tweety can fly; but if someone is told that Tweety is a penguin, the more specific result applies and one can conclude that Tweety walks rather than flies, although Tweety is a bird. This kind of inference mechanism is an important aspect of commonsense reasoning.

Symbol manipulation programs contain various reasoning and inference mechanisms. Symbolic logic is used to represent knowledge, and the first step in problem-solving is to translate expressions and sentences from English (or some other natural language) into the language of logic.

The simplest logic is called "propositional logic": Each symbol stands for an entire proposition in this logic, and the legal expressions or well-formed formulas are obtained by using logical connectives ($\&, \vee, \rightarrow, \neg$) to combine these symbols. For example, if the symbol P stands for the proposition "molecules contain atoms" and Q stands for "crystals contain molecules," then $P \& Q$ is a well-formed formula whose meaning is intended to be "molecules contain atoms, *and* crystals contain molecules." Each of P, Q, $P \vee Q$, $\neg P$, $\neg Q$, $P \rightarrow Q$ is also a well-formed formula.

The meaning of each formula can be extracted uniquely from the meanings of the symbols and the connectives in the formula. For example, "$P \vee Q$" stands for "*P or Q*," "$P \rightarrow Q$" stands for "*P implies Q*," and "$\neg P$" stands for "*not P*."

In classical propositional logic, there are only two truth values: true (T) and false (F), and the way in which connectives modify the meanings of formulas can be captured by "truth tables." Each truth table is a set of rows of truth values, where each row corresponds to one possible assignment of truth values to the individual "atomic" proposition symbols. In the following truth tables, each column represents the truth values of a formula for different combinations of truth values of the component atoms *P*, *Q*. The table needs to have only four rows of truth values, exhausting all possible combinations of truth vales of *P*, *Q*.

P	Q	$P\&Q$	$P \vee Q$	$P \rightarrow Q$	$\neg P$	$(P\&Q) \rightarrow (P \vee Q)$
T	T	T	T	T	F	T
T	F	F	T	F	F	T
F	T	F	T	T	T	T
F	F	F	F	T	T	T

The last column illustrates the computation of truth values of complex formulas containing several connectives. The connectives behave in the same way even if other formulas are substituted for *P* and *Q*: For example, the truth value of "$P \rightarrow Q$" when "*P*" is false and "*Q*" is true is exactly the same as the truth value of "$(P \& Q) \rightarrow (P \vee Q)$" when "$P \& Q$" is false and "$P \vee Q$" is true. The truth values of "$(P \& Q) \rightarrow (P \vee Q)$ " depend uniquely on the truth values of "$P \& Q$," "$P \vee Q$" in the same way as the truth value of "$P \rightarrow Q$" depends on the truth values of "*P*," "*Q*." The formula "$(P \& Q) \rightarrow (P \vee Q)$" is called a "tautology" because it is always true, as shown in the last column of the above table, irrespective of the truth values of its component propositions. Some other formulas are called "contradictions," always false irrespective of the truth values of propositions in the formula. For example, $P \& \neg P$ is a contradiction, because both *P* and $\neg P$ cannot be simultaneously true, irrespective of the intended meaning of the symbol *P*. Not every formula is a tautology or a contradiction.

Propositional logic cannot be used to express and reason about formulas with some inherent structure. For example, there is no way to conclude "crystals contain atoms" from representations of the statements "crystals contain molecules" and "molecules contain atoms" in propositional logic. Such reasoning is expressed naturally in a more complex formalism called "predicate logic," which also contains predicates, variables, and quantifiers. A predicate is something that can be applied to some number of arguments to yield a truth value. For example, one may use a predicate called *Contain* with two arguments, and consider *Contain*(*crystals*, *molecules*) to be true, and similarly consider *Contain*(*molecules*, *atoms*) to be true, whereas *Contain*(*crystals*, *monkeys*) is expected to be false. In order to conclude *Contain*(*crystals*, *atoms*) from those formulas, the idea must be expressed that if *x* contains *y*, and *y* contains *z*, then *x* contains *z*, for every set of values denoted by *x*,*y* and z. Formally, this is expressed as the formula "$\forall x, \forall y, \forall z.([Contain(x,y)\&Contain(y,z)] \rightarrow Contain(x,z))$," read as "for-all x, for-all y, for-all z, $Contain(x,y)$ and $Contain(y,z)$ imply $Contain(x,z)$." The symbols x,y,z are used here as variables, which can take on values from some domain of objects, and "$\forall$" is a special symbol called the universal quantifier; a formula "$\forall x.F$" holds if the formula F holds for every value of the variable x.

Variables may take value from an infinite number of objects, such as integers. For example, the formula "$\forall x.[odd(x) \vee even(x)]$" may be interpreted on the domain of integers to be equivalent to the following infinite set of formulas: $\{odd(0) \vee even(0), odd(1) \vee even(1), odd(2) \vee even(2), \ldots\}$. This means that it is not possible to verify predicate logic formulas by constructing finite truth tables for the formulas. The preferred way of reasoning with formulas in predicate logic is by using axioms and inference rules to derive theorems. Axioms are a set of formulas assumed to be truths, from which reasoning begins; all axioms are theorems, by definition. Inference rules are syntactic rules that tell which new theorems can be generated from other theorems. For example, if $P \rightarrow Q$ is a theorem, and if P is also a theorem, then an inference rule called *modus ponens* indicates that Q is also a theorem; the definition of this inference rule is not concerned with the details of what P and Q are, nor with the intended meaning of these formulas. Well-known formulations of axioms and inference rules for predicate logic are such that all and only tautologous formulas are provable theorems. This relates the proof techniques with the semantics of formulas, allowing one in practice to work entirely by manipulating symbols, rather than by trying to reason about their meanings.

Applications

The development of large computer programs is a difficult and complex task. Errors that appear to be minor can cause huge programs to crash, resulting in disasters and expenses that can be avoided by using more understandable symbol-manipulation programs. A large proportion of the effort in computer programming is spent in maintaining existing programs, rather than writing new programs. To tame such tasks, formal specifications using methodologies such as abstract data types have been found to be useful.

The more abstract and the higher the level of a program, the easier it is to understand its logic clearly, making it easier to modify it, port it to other machines, or add to its capabilities. Entire programming languages and computers have been developed on the basis of the symbolic or logical processing paradigms. These have become increasingly popular, because the statement of a problem in a logic programming language can itself be directly executed to give the solution, instead of first going through a complicated translation into a computer program in a conventional programming language. So the task of the programmer essentially consists of specifying the problem in clear, logical, unambiguous terms, rather than encoding it in a language that is difficult to comprehend.

Another motivation for symbolic processing techniques is that they are capable of being used to describe cognition and reasoning tasks performed easily by humans. It is natural to describe qualitative reasoning in qualitative terms, using symbols to denote various objects and concepts, rather than in quantitative or numeric terms. Automated reasoning is an active research topic in artificial intelligence. Many computer programs that mimic human performance in various reasoning tasks have been built and successfully used to prove complex mathematical theorems. For purposes of simulating human commonsense reasoning, however, classical predicate logic appears to be inadequate.

Unlike classical logic, human reasoning commonly appears to be nonmonotonic in character; that is, conclusions are often revised when additional evidence is presented. Various alternative logics have hence been proposed and used in some artificial intelligence systems called truth maintenance systems.

Knowledge representation mechanisms such as frames and semantic networks have been widely used in applications such as natural language processing. The task in this context is for machines to process input sentences in a language like English, represent it in a way that captures the intended meaning of the sentence rather than merely the grammatical form, and answer questions about the knowledge contained in those sentences. Another related task is machine translation: The representation of an English sentence, for example, should be used together with the grammar rules for French, in order to obtain a French translation of the original sentence.

A semantic network is a graphlike structure, consisting of concept nodes whose interrelations are explicitly indicated using pointers or network connections. This mechanism combines the representations of concepts and their associations into one overall structure. Each concept or object has a unique representation, and disambiguation is performed when reference is made to the same object in different parts of a text or conversation, possibly using different words or phrases, as in "John left. *He* lives in Rome." Here, the italicized pronoun refers to "John," and understanding the latter assertion involves making the appropriate connection between the nodes representing the concepts "John" and "Rome." When a question "Where does John live?" is asked, such a system would start searching the network at the node representing "John," and produce the appropriate answer when a connection representing residence is located to the node representing "Rome."

Structured objects called relations modified by logical operations have been used successfully in large databases (stores of structured information). Each relation is essentially a table or two-dimensional array, in which each column is a separate field. Each row in a relation contains an entry from the domain (set of eligible values or elements) of each field. Each row is considered a member of the relation in which it occurs. Relations are generally manipulated using three special operations called selection, projection, and join. Selection extracts some rows of a relation, which satisfy some predicate. Projection results in a relation with fewer columns. Join is the operator that combines relations that have some common fields, combining pairs of rows with matching values in the common fields. The following tables illustrate these ideas. (EMP and MGR are two separate relations.)

EMP

Employee	*Salary*	*Age*
John	33	39
Peter	31	40
Mary	42	34
Lee	39	31

MGR

Employee	*Manager*	*Department*
John	Mary	Shoes
Peter	Mary	Shoes
Joe	Ed	Computers
Martha	Ed	Computers
Ella	David	Clothes

The following relations show the results of: selection on EMP using the predicate "salary > age" and projection of MGR onto "manager" and "department" fields.

Employee	*Salary*	*Age*
Mary	42	34
Lee	39	31

Manager	*Department*
Mary	Shoes
Ed	Computers
David	Clothes

The following relation is obtained by performing a join operation of EMP and MGR relations, which have a common "employee" field.

Employee	*Salary*	*Age*	*Manager*	*Department*
John	33	39	Mary	Shoes
Peter	31	40	Mary	Shoes

Context

Computing techniques can be broadly classified as numeric, symbolic, or connectionist. The first is the classical approach, which historically preceded the others and is still of great importance in fields such as scientific computing. A disproportionately large number of computer programs being used today have been built using programming languages and techniques dating back to the 1960's. In artificial intelligence applications, however, symbolic processing techniques constitute the dominant paradigm. The connectionist paradigm of computing with artificial neural networks has also been applied successfully to many tasks related to pattern recognition.

Unless properly represented, knowledge is useless. For example, although everything is stored in terms of 1's and 0's in a computer, that level of representation is completely incomprehensible to humans and therefore very little can be done with it. In computer programming, the lowest level of communication with a computer is via machine language programs, which are merely sequences of numbers. Assembly-language programs are a little higher in level: They consist of simple instructions, typically using mnemonic alphabet sequences. Programming languages such as FORTRAN are one level higher and are easier to work with than assembly-language programs. Programs that manipulate symbolic representations, logical formulas, and abstractions are at a higher level and are much easier to understand. The higher the level of representation, the easier it is to perform high-level reasoning with the represented knowledge.

The lofty goal of research in artificial intelligence is to develop intelligent systems that should be able to reason in ways analogous to people. Hence, logic and theorem proving and automated reasoning are important. Present formulations of logic date their development from the work of Friedrich Ludwig Gottlob Frege in the early part of the twentieth century. In the 1970's and 1980's, there was much progress in building theorem-proving computer systems that perform logical deductions efficiently. In addition to proving mathematical theorems, they have been applied to problems such as proving the correctness of computer hardware and software.

In the 1980's, commercial and industrial applications of artificial intelligence became increasingly popular, under the name of "expert systems." An important trend for future development of such successful artificial intelligence applications is the combination of various knowledge-representation and database techniques with powerful reasoning methodologies. A central problem in artificial intelligence is the task of simultaneously obtaining a good representation language, inference mechanism, and domain knowledge. Often, the requirement of expressive adequacy conflicts with that of reasoning efficiency, and suitable compromises have to be chosen.

Bibliography

Barr, Avron, Edward A. Feigenbaum, and Paul R. Cohen, eds. *Handbook of Artificial Intelligence*. 4 vols. Stanford, Calif.: HeurisTech Press, 1981-1989. These volumes constitute comprehensive reference material for various techniques and applications of artificial intelligence.

Brachman, Ronald J., and Hector J. Levesque, eds. *Readings in Knowledge Representation*. San Mateo, Calif.: Morgan Kauffmann, 1985. This collection contains several oft-quoted articles by leading researchers on various approaches to knowledge representation. Recommended for interested readers who have some familiarity with the basics of artificial intelligence and logic.

Chang, Chin-Liang, and Richard C. Lee. *Symbolic Logic and Mechanical Theorem-Proving*. New York: Academic Press, 1973. This book presents logic from the viewpoint of automating reasoning techniques. The main focus is on the development of resolution, a well-known theorem-proving method discovered by J. A. Robinson in 1965.

Horowitz, Ellis. *Programming Languages: A Grand Tour*. Rockville, Md.: Computer Science Press, 1983. Contains an introduction to various modern programming languages and is useful in obtaining a panoramic view of different styles of computing.

Liskov, Barbara, and John Guttag. *Abstraction and Specification in Program Development*. Cambridge, Mass.: MIT Press, 1986. This book presents an excellent approach to program design. Recommended for its discussion of systematic software development.

Chilukuri K. Mohan

Cross-References

Artificial Intelligence: Expert Systems, 139; Artificial Intelligence: An Overview, 146; Finite Element Methods, 886; Neural Networks, 1481.

SYNCHROTRON RADIATION

Type of physical science: Condensed matter physics
Field of study: Surfaces

Synchrotron radiation is the electromagnetic radiation emitted by charged particles when their paths are bent from a straight line. It provides a source of X rays and ultraviolet light, which are used for microcharacterization of materials.

Principal terms

ELECTROMAGNETIC RADIATION: a propagating wave consisting of an electric field and a perpendicular magnetic field, such as light or radio waves

ORBITAL FREQUENCY: the number of revolutions completed in a unit time

PHOTON: the basic quantum of electromagnetic radiation; it has zero rest mass and an energy proportional to its frequency

ULTRAVIOLET LIGHT: electromagnetic radiation in the wavelength range from 4×10^{-7} meters to 2×10^{-8} meters, occurring between visible light and X rays

X RAYS: electromagnetic radiation in the wavelength range between 2×10^{-8} meters and 4×10^{-12} meters

Overview

Synchrotron radiation is emitted by a charged particle when the path of that particle is bent from a straight line. James Clerk Maxwell, a nineteenth century English physicist, predicted that any charged particle experiencing accelerated motion would emit electromagnetic radiation. This radiation carries energy and momentum away from the charged particle and propagates outward from the particle. If the acceleration occurs along a straight path, such as the electrons moving in a linear antenna wire, the electromagnetic waves will propagate outward in a "dipole" radiation pattern. This principle is used for radio transmission. When a charged particle is forced to move around a circular path, it experiences a continuous acceleration, called the centripetal acceleration, which is directed toward the center of the circle. Because of this acceleration, the particle will radiate electromagnetic energy continuously as it moves around this circular path. If the speed of the particle is low, the radiation is emitted into a fan-shaped pattern directed tangentially to the circle of motion of the particle. If the speed of the particle approaches the speed of light, the emission pattern is distorted by relativistic effects and the radiation is concentrated into a cone facing forward along the instantaneous direction of the particle motion.

The frequency of the emitted radiation also varies with the speed of the particle. At low velocities, most of the radiation is concentrated at the orbital frequency, although a small amount of energy is emitted at frequencies that are integer multiples of the orbital frequency. As the particle velocity increases, more energy goes into these higher-frequency or shorter-wavelength harmonics. Thus, with increasing particle speed, the peak of the emitted radiation moves from the infrared to the visible ultraviolet, and then the X-ray region, of the electromagnetic spectrum. As the speed of the particle approaches the speed of light, the orbital frequency decreases rapidly, and the harmonics are so closely spaced that the frequency distribution can be considered to be continuous.

In 1912, G. A. Schott described the properties of the electrons orbiting a nucleus of an atom and established the basic features of synchrotron radiation. A complete analytical treatment of synchrotron radiation was developed by Julian Seymour Schwinger in 1945. Schwinger calculated the rate at which energy would be emitted, the detailed frequency spectrum, and polarization properties of the radiation. Using circular particle accelerators such as cyclotrons and synchrotrons, scientists attempted to detect synchrotron radiation. In these devices, particles are confined to circular or near-circular orbits by magnets that bend their paths. As the particle path is bent, the centripetal acceleration gives rise to synchrotron radiation. In 1946, John P. Blewett detected synchrotron radiation for the first time as emissions from the bending magnets of the General Electric Research Laboratories' electron accelerator. Synchrotron radiation was observed visually a year later by Glen Elder, also using the General Electric accelerator. In 1948, Elder verified Schwinger's predictions of the polarization and frequency spectrum of the emitted radiation.

Initially, synchrotron radiation was regarded as a hindrance by particle physicists who were interested in obtaining a beam of maximum energy from their accelerators. Synchrotron radiation was the mechanism by which particles being accelerated in these circular particle accelerators radiated away some of the energy being added by the machine. In 1956, however, Diron Tomboulian and Paul Hartmen demonstrated that the synchrotron radiation from the Cornell 320-megaelectronvolt electron synchrotron could be used as an intense source of ultraviolet light. By the early 1960's, researchers around the world had become interested in the use of synchrotron radiation as a bright source of ultraviolet and, from the higher-energy accelerators, X-ray photons. Synchrotron radiation from existing particle accelerators was made available to these researchers, and plans were made to convert some older electron synchrotrons, which had outlived their usefulness as particle accelerators, into synchrotron radiation light sources.

Synchrotron radiation has several unique properties that make it superior to other ultraviolet or X-ray sources for many experiments: It is emitted in a continuous spectrum from the infrared, through the visible, ultraviolet, and into the X-ray energy range; it is highly polarized, with its electric field vector lying in the plane of the electron orbit, and thus the primary synchrotron radiation can be distinguished by its polarization from radiation emitted by interaction with a sample; it is emitted in discrete bursts, since the electrons orbiting in the synchrotron are clustered; it is highly collimated in the vertical plane, making the beam quite intense; and it is emitted in a high-vacuum storage ring, so that it can be provided to an apparatus requiring a high vacuum. Each of these characteristics has been used to advantage in the applications of synchrotron radiation to the investigation of the physical properties of materials, including their geometric and electronic structures, elemental compositions, and optical characteristics.

As the demand for access to synchrotron radiation increased and the researchers recognized that many of the characteristics of the radiation could be tailored to particular experiments, the first accelerators and storage rings designed to be synchrotron radiation sources were constructed. In the United States, the major national facility is the National Synchrotron Light Source (NSLS) at Brookhaven National Laboratory. The NSLS includes two electron storage rings, an intermediate-energy ring optimized to emit ultraviolet radiation and a high-energy ring optimized to emit X rays.

Although the first source of synchrotron radiation was the bending magnets that confine electrons to circular paths in accelerators and storage rings, the need for even higher-energy X rays has resulted in the insertion of other magnets, called wigglers and undulators, into the electron beams. These wiggler and undulator magnets are inserted in the straight, magnetic-field-free regions of storage rings to force rapid deviations in the electron paths. The accelerators experienced by the electron beam in the wiggler or undulator magnets are much higher than in the bending magnets, so these magnets provide intense pulses of X rays of higher energy than can be achieved by the bending magnets.

Applications

The availability of intense beams of ultraviolet light and X rays from synchrotron radiation sources opened up a wide variety of applications in many disciplines, including materials science, physics, biology, geology, and planetary science. Experiments that had previously been difficult or impossible because of limited source intensity quickly became common.

In the late 1960's, synchrotron radiation sources were employed to study the interaction of atoms, molecules, and solids with ultraviolet and X rays. X rays are absorbed in matter primarily through the "photoelectric effect," in which a bound electron is ejected from an atom when hit by an incident X ray. In the Bohr model of the atom, electrons orbit the nucleus in a series of permissible energy levels, the lowest of which is called the K-shell, then the L-shell, and so on. Once an electron is ejected from a shell through the photoelectric effect, the atom is in an unstable state, having a vacancy in a low-energy shell and being charged as a result of the electron deficiency. The low-energy vacancy is filled by an electron from a higher-energy state; that electron drops down to the vacant lower-energy state. In the process, the electron must emit energy in the form of a photon. The energy of this photon corresponds to the energy difference between the two shells. This emission process is called fluorescence.

The emission of fluorescence photons, generally in the X-ray energy range, makes possible the determination of the elemental composition of a sample. Since the energy difference between the shells varies from element to element, the elemental composition of a sample can be determined by measuring the number and energy of the fluorescence X rays from that sample. This analysis technique, known as X-ray fluorescence, is routinely employed using an incoming beam of electrons to knock out the bound electrons in the sample. The incident electrons, however, decelerate in the sample and emit X rays of their own, a process called bremsstrahlung. This background of X rays from the decelerating electrons limits the sensitivity of this technique to elements present at the 0.1 percent level. If an X-ray beam is substituted for the incident electron beam, this background radiation is eliminated, since the X-ray beam is not charged. Using the X-ray beams available at synchrotron radiation sources, researchers have detected trace elements present at the 10-parts-per-billion concentration in particles as small as 10 billionths of a gram.

The photoelectric emission process also provides a tool to determine the structure of molecules using extended X-ray absorption fine structure (EXAFS). Originally demonstrated in the 1930's, EXAFS came into practical use only with the development of intense X-ray sources. As the energy of an incident X-ray beam increases, a sharp increase in the absorption of X rays by the sample is seen at the energy required to eject electrons from each shell. There is no structure in the absorption for single atoms such as krypton gas; when samples of diatomic molecules such as bromine gas are studied, however, sinusoidal oscillations in the quantity of X rays absorbed are observed near the shell ejection energies. Studying the detailed shape of the fine structure of the absorption as a function of energy, which is the subject of EXAFS analysis, provides information about how one atom is bound to another. The EXAFS technique also has been applied to solids, providing information on the distance between atoms in the structure.

Synchrotron radiation is used in the fabrication of integrated circuits. For many years, these circuits were fabricated by optical microlithography using ultraviolet light. The ultimate size, and thus the speed, of integrated circuits fabricated by this technique is limited by the wavelength of the ultraviolet radiation. Using X rays from synchrotron sources, industrial researchers have fabricated much smaller integrated circuits. Synchrotron radiation X-ray lithography is favored over other techniques from both a technical and an economic point of view.

Synchrotron radiation sources have been used in a variety of microscopy experiments, in which an X-ray beam is substituted for the optical beam in a traditional microscope. Because the wavelength of an X ray is much smaller than that of visible light, the resolution of an X-ray microscope can be made much higher than the optical microscope. When combined with either X-ray absorption or fluorescence techniques, an X-ray microscope makes it possible to map the distribution of a particular element or elements in the sample. Although the resolution of the synchrotron X-ray microscope is comparable to that obtained using the transmission electron microscope, samples in the synchrotron X-ray microscope can be analyzed in air, whereas the transmission electron microscope sample chamber must be held under high vacuum, which causes many biological specimens to dehydrate and alter in form.

Synchrotron radiation can also provide the intense, collimated source of monochromatic, single-energy X rays required in X-ray diffraction and scattering experiments, which are used to identify the geometrical structure and atomic spacing of crystalline samples.

Context

The radiation emitted by charged particles moving in a circular path is called synchrotron radiation, since it initially was studied in association with particle accelerators such as synchrotrons. In these machines, strong magnets are used to bend the path of high-energy electrons or nuclei into circular paths. As the particles pass through each bending magnet, they experience an acceleration and, thus, emit radiation. The synchrotron had to be designed to add energy to the particles, in order to replace the energy that was radiated away. Synchrotron radiation imposes an upper limit on the energy to which particles can be accelerated in any given particle accelerator, since an equilibrium is reached between the rate at which energy can be added by the accelerator and the rate at which it is radiated away by synchrotron radiation.

By the early 1960's, it was recognized that the unique properties of synchrotron radiation made it an ideal source of intense ultraviolet and X-ray photons. Synchrotron radiation was employed in a wide variety of experiments for the determination of the electronic and geometrical structures, chemical compositions, and optical properties of materials. The small beam size made synchrotron radiation particularly suitable for experiments with microscopic samples.

The first synchrotron radiation facilities were parasitic, in that they were grafted onto particle accelerators constructed principally to produce beams of high-energy electrons for particle physics experiments. The rapid rise in applications for the synchrotron light resulted in the design and construction of facilities optimized for their output of synchrotron radiation. By 1986, dedicated synchrotron radiation facilities had been constructed in China, Great Britain, France, Germany, Japan, Sweden, the United States, and the Soviet Union.

Commercial applications of synchrotron radiation have been examined, particularly in X-ray lithography for the fabrication of ultraminiature integrated circuits. The first dedicated industrial synchrotron light source is expected to be constructed before the end of the twentieth century.

Bibliography

Bienenstock, Arthur, and Herman Winick. "Synchrotron Radiation Research: An Overview." *Physics Today* 36 (June, 1983): 11-18. A review of how synchrotron radiation is generated and its uses as a research tool. Readable by nonspecialists and extensively annotated to reference original sources for the various research areas described.

Rowe, Ednor M., and John H. Weaver. "Synchrotron Radiation." *The Physics Teacher* 15 (May, 1977): 268-274. An elementary discussion of the process of synchrotron radiation and a description of its uses. Well illustrated with photographs and diagrams of the Synchrotron Radiation Center at the University of Wisconsin.

Sparks, Cullie J., Jr. "Research with X-Rays." *Physics Today* 34 (May, 1981): 40-49. A well-illustrated description, suitable for general audiences, of the variety of experiments on the geometric and electronic structure of matter that have been performed using synchrotron radiation.

Winick, Herman, George Brown, Klaus Halbach, and John Harris. "Wiggler and Undulator Magnets." *Physics Today* 34 (May, 1981): 50-63. A detailed account of the design and capabilities of the wiggler and undulator magnets used to extend the spectral range and increase the brightness of synchrotron radiation sources.

Winick, Herman, and Sebastian Doniach. *Synchrotron Radiation Research*. New York: Plenum Press, 1980. An extensive discussion of the techniques by which synchrotron radiation is generated and the applications to different areas of research.

George J. Flynn

Cross-References

Centrifugal/Centripetal and Coriolis Accelerations, 332; Electromagnetic Waves: An Overview, 735; Electron Emission from Surfaces, 764; Radiation: Interactions with Matter, 2026; Optical Properties of Solids, 2268; Special Relativity, 2333; Storage Rings and Colliders, 2388; Synchrotrons, 2464.

SYNCHROTRONS

Type of physical science: Elementary particle (high-energy) physics
Field of study: Techniques

A synchrotron is an accelerator of uniform radius which uses an oscillating magnetic field to move charged particles in a circular motion. Synchrotrons are used to produce elementary particles resulting from nuclear collisions.

Principal terms

ELECTRONVOLT: the energy acquired by an electron or unit charge as it moves through a potential difference of 1 volt

GAUSS: a unit measure of magnetic field strength

MAGNETIC FLUX: the passing of magnetic field lines through an area

PHASE STABILITY: the situation that exists when a particle is kept in synchronous phase with the accelerating voltage

RADIO FREQUENCY: the oscillating voltage applied to the accelerator

RELATIVISTIC: refers to moving at high speeds that are comparable to the speed of light

RESONANCE: a vibration in phase or unison with a reference signal

REST MASS: the normal mass of an object, measured in kilograms, when it is not moving at high speeds

SUPERCONDUCTING TEMPERATURE: the temperature at which a material loses its electrical resistance and becomes a perfect conductor

Overview

Synchrotrons are large accelerators that move charged particles in a fixed circular path. They were developed to solve the problem of how to gain higher energy without using a single large magnet; instead they use many smaller magnets that are located around the orbital path. Particles such as protons are first given a preliminary acceleration to high kinetic energies and then are injected into the synchrotron. Acceleration is provided on each revolution by a radio-frequency generator. This voltage is synchronized with the orbital frequency, thus increasing the particle's energy on each turn. To maintain a fixed orbit with increasing energy, the magnetic field must also be incremented. After reaching the desired speed and energy, the particles are deflected from their circular orbit by a magnet to an experimental area containing targets and detectors.

Prior to the synchrotron, other accelerators were used with varying degrees of success. The Cockcroft-Walton voltage multiplier and the Van de Graaff electrostatic generator had been used to accelerate electrons, but they were limited to energies of a few million electronvolts. The cyclotron experienced problems with electron acceleration because of a large relativistic mass increase at low energies, which caused electrons to get out of step with the accelerating voltage and quickly slow down. Synchrocyclotrons compensated for the mass increase by using a variable oscillating voltage which progressively slowed down with the mass increase. Particles were now permitted to get back into phase or step with the voltage and slowly build up their energies. Like cyclotrons, however, these machines were limited by the size and cost of their huge single magnets.

The theoretical principles of the synchrotron were first announced by Vladimir Veksler and Edwin M. McMillan. These principles maintain that, by increasing the magnetic field while reducing their frequency of the cyclotron voltage, one can increase the particle's orbit size and its energy. If the change is accomplished slowly, then phase stability is maintained during the acceleration. Ions or charged particles may also be accelerated by keeping the electrical field frequency constant and changing the magnetic field. The term "synchrotron" was proposed by McMillan because the machine behaves like a synchronous motor.

Electrons can be injected into a circular chamber at speeds high enough to remain in a stable orbit, which is accomplished by accelerating the electrons up to 1 or 2 million electronvolts by a betatron accelerator. The betatron relies on the same principle as the transformer, whereby an alternating current which is applied to a primary coil induces an oscillating secondary coil current. With the betatron acting as a transformer, a swarm of electrons inside a doughnut-shaped vacuum chamber serves as the secondary coil. If the chamber is placed between the poles of an electromagnet activated by a pulsed current, then a strong magnetic flux permeates the region. The electrons move in response to the changing magnetic field and gain energy by induction.

The betatron, like the cyclotron and synchrocyclotron, has the disadvantage of using a large magnet to supply the required variable magnetic flux for acceleration. By employing a radio-frequency voltage which creates a variable electric field, however, the size of the magnet can be reduced. When electrons reach energies of around 2 million electronvolts, a speed very close to that of light, then speeds and revolution frequency change very little, permitting an application of a nearly constant voltage. With increasing revolutions, the electron energy and mass become larger and the magnetic field must also be increased. All of this can be accomplished with a fixed-radius machine and a ring-shaped magnet. The accelerator that is designed in this manner is the electron synchrotron. Electron synchrotrons use flux bars inside the orbit as an aid for the preliminary acceleration. These small bars, made of a high-permeability metal, serve initially to block the magnetic field at low current values and reach a saturation point at high values, thus aiding the field.

The energy limitations for electrons that are accelerated in a betatron or synchrotron are dependent on the radiation losses. The electron, when accelerated, suffers a loss in energy which is proportional to the fourth power of its energy. This maximum is reached when essentially all the additional energy that is supplied to the electron is radiated away during each revolution. It is believed that the practical limit for energy reached in such machines is 1 billion electronvolts. More than twenty electron synchrotrons have been built, and energies of 300 million electronvolts have been reached.

The synchrotron located at the Massachusetts Institute of Technology has an orbital radius of 1.1 meters with a magnet of more than 45,000 kilograms. Electrons are initially injected at 80,000 electronvolts, then accelerated to 7 million electronvolts by betatron principles, and finally reach energies of 300 million electronvolts through electron synchrotron action. The proton synchrotron was designed to penetrate into the billion-electronvolt (gigaelectronvolt) range. In 1952, the proton synchrotron named the Cosmotron, located at the Brookhaven National Laboratory, reached an energy of 3 gigaelectronvolts. Two years later, the Bevatron at the University of California reached 6 gigaelectronvolts, and in 1957, a machine built in the Soviet Union reached 10 gigaelectronvolts. In the 1960's, the Argonne National Laboratory was able to achieve energies of 12.5 gigaelectronvolts.

The proton synchrotron differs from the electron synchrotron in two important areas. First, radiation problems that are associated with electron acceleration become significant at high energies but are negligible with protons. Second, protons do not achieve relativistic velocities until their energies reach the billion-electronvolt range. The rest mass of a proton is about two thousand times that of the electron, which would allow it to reach an energy of 10 gigaelectronvolts, radiating as much energy as a 5-megaelectronvolt electron. Energy loss by an electron in this range, however, is not high. This means that protons will not suffer large radiation losses until energies considerably beyond 10 gigaelectronvolts are reached. Other problems with the design of the multibillion-volt accelerators are encountered with their sheer size. To obtain higher magnetic field strength, the machines were built larger. For example, the radius of the proton orbit of the Cosmotron was 10 meters. Newer accelerators measure kilometers across.

The radial and vertical deviations of the protons inside the proton synchrotron's acceleration chamber are subject to more precise control than they were with the electron synchrotron. The oscillations experienced by the betatron were dependent on the initial injection conditions set up for the electrons. Injection in the Cosmotron was from a Van de Graaff generator, which delivered a well-focused beam with amplitude deviations less than 10 centimeters.

The Brookhaven Cosmostron magnet was built in sections consisting of 288 magnet blocks reaching 23 meters across. The magnet was capable of a maximum of 14,000 gauss but was constructed to allow straight sections of the vacuum chamber to be free from the magnetic fields for injection, acceleration, and finally ejection of the ions. The Van de Graaff electrostatic generator provided protons in pulses of 3.5 million electronvolts. Acceleration is increased in intervals of one second, with some 800 electronvolts of energy added at each revolution. Protons end up making three million revolutions and travel more than 160,000 kilometers while reaching their highest energies. The radio-frequency voltage is changed from 0.37 million cycles per second to 4 million cycles per second during the operation.

At the high operating power levels of these machines, accurate timing of the magnetic pulses is not possible. The magnetic field at this stage must be used to control the cycles of the radio-frequency voltage. A signal proportional to the rate of change of the magnetic field is integrated electronically and supplied as a control device for the radio-frequency oscillator. The oscillator, through a resonance circuit with a magnetic core, is able to sense the required voltage needed from variations in the magnetic field.

The increase in energy to which ions have been accelerated is followed by a corresponding increase in the complexity and cost of the required accelerators. The energies attainable by proton synchrotrons, although high in theory, were limited by their size and costs. For example, the Bevatron, which reached 6 gigaelectronvolts, used a magnet that weighed more than 9 million kilograms and an orbital radius of 24 meters, at a cost of $15 to $20 million. Using the same design, a 30-gigaelectronvolt synchrotron would require a magnet that weighed more than 90 million kilograms at a prohibitive cost.

Improving the synchrotron design has been aided by the realization that an improvement in the efficiency of focusing the ions could greatly reduce the size of the magnet. The large magnet of the Cosmotron was required to keep the ions in the desired orbit by use of corrective forces that act on these particles when they stray from their path. These deviations arise from collisions with air molecules in the tube or from fluctuations in the accelerating voltage. It was thought that if the straying of the particles could be more precisely controlled, then the circular pipe in which the particles move could be made much narrower, and only a small, thin magnet would be necessary to surround it.

A device called the alternating gradient synchrotron was built to compensate for ion path variations. This machine operates on the principle that alternate focusing and defocusing of the ion beam can have a net focusing effect. The principle is similar to the focusing of a beam of light by a combination of converging and diverging lenses. The design arranges a number of C-shaped magnets in a circle, allowing alternate magnets to face in opposite directions; the back of one magnet faces toward the center of the circle and the back of the next toward the outside. With the proper design, it is possible to make the focusing forces very strong.

In addition to providing a guide field for the particle path, the magnets of the alternating gradient synchrotron have another important purpose. The poles of each magnet are shaped to allow the magnetic field to increase and then decrease in an outward direction. This alternation of the magnetic field gradient permits the proton beam to focus and then defocus both vertically and horizontally, presenting a beam with greater focusing power than that of a conventional constant gradient machine. The very large magnetic gradients employed require that all the component magnets be precisely built and accurately aligned to avoid errors that could lead to ion path deviations and possible collisions with the vacuum chamber walls.

Since the ion beam can be contained in a much smaller pipe around the accelerator, the required magnets are much smaller, with a considerable reduction in the amount of steel and copper needed in construction. As an example, the quantity of steel in a 30-gigaelectronvolt accelerator of this type is 3.6 million kilograms, as contrasted with more than 32 million kilograms in the 10-gigaelectronvolt machine built in the Soviet Union.

The 30-gigaelectronvolt alternating gradient accelerator at Brookhaven National Laboratory accelerates protons with a succession of twelve radio-frequency oscillation stations. A high-frequency voltage is created across two gaps at each station. Protons that arrive at the gaps when the voltage and electric field are in the forward direction receive an acceleration. Provided that the applied radio frequency is correct, these protons will be accelerated at each gap. Very large energies can be reached after many cycles with small voltage increments. Each of the twelve stations provides a 7,500-volt increment each time the protons pass, gaining up to 90 kiloelectronvolts per revolution. Thirty-billion-electronvolt protons are achieved after about 325,000 revolutions in the machine. At this time, protons are traveling at more than 90 percent of the speed of light and the mass of the proton has increased more than thirty times. After each pulse and a two-second recovery period, the cycle is repeated. Twenty pulses are produced for each minute of operation, with each pulse containing about ten billion protons.

The 30-gigaelectronvolt machine utilizes a two-stage preliminary acceleration method involving both a Cockcroft-Walton generator and a linear accelerator operated in tandem. After emerging from the linear accelerator, protons travel with a velocity one-third that of light and an energy of 50 million electronvolts. The third stage of acceleration is performed in the main doughnut-shaped vacuum tube, which is 257 meters across. The magnet is divided into 240 units of mass equal to 14,500 kilograms. The actual vacuum tube measures only 18 by 7 centimeters and is serviced at a pressure equal to one hundred-millionth of an atmosphere.

Applications

Electromagnetic radiation is emitted by charged particles moving in a circular motion at relativistic energies. The emission of light by electrons as they spiral around the celestial magnetic field is an example of synchrotron radiation. The background light of the Crab nebula, as well as the pulsed beam of light from pulsars, is an example of this type of radiation. The construction of electron synchrotrons, and other types of synchrotrons for nuclear physics, has permitted the production of synchrotron radiation in the X-ray and ultraviolet regions of the spectrum.

Synchrotron radiation has such features as high intensity, small source size, high polarization, a pulse time structure, and a broad spectral bandwidth. Important research in such fields as biochemistry, materials science, surface science, and crystallography makes use of the techniques of X-ray absorption and scattering, as well as photoemission spectroscopy and X-ray microscopy.

Synchrotron radiation from protons is some thirteen orders of magnitude weaker than that from electrons of the same energy level and orbital radius. For multigigaelectronvolt accelerators, also known as storage rings, the rate of energy loss is so high that the electrons would lose all of their energy very quickly. In contrast, synchrotron radiation from the 28-gigaelectronvolt proton storage rings located in Geneva, Switzerland, is so small that protons could orbit for years without significant losses.

X-ray absorption is used to investigate atomic structure in complex solids, liquids, and gases. The intense, continuous synchrotron radiation from the high-energy storage rings provides the necessary photon energy for studies of amorphous and other noncrystalline materials, including catalysts and proteins that are not determinable by other research methods. The high intensity and strong collimation of synchrotron radiation provide an ideal source for the small-angle scattering and diffraction of X rays. Research in the timed contraction cycles of muscular tissue has been undertaken to resolve the associated molecular changes.

Photoemission spectra obtained from the surfaces of crystals or oxidized gallium arsenide show that oxygen attaches itself to the arsenic atom rather than the gallium. This was surprising because the reverse was expected from chemistry theory. The strength of the photoemission spectra appears to depend on the oxidation state as well as the photon energy.

Cosmic-ray research became centered on synchrotrons when it was discovered that protons which strike a stationary target at an energy of several hundred megaelectronvolts produce small particles called pions and muons, which were first observed in cloud chambers. The collision of a proton beam with target nuclei creates pions. As the beam continues a short distance, some of the pions decay into muons. A magnetic field deflects the positive particles of the beam in one direction and the negative particles in the opposing direction. The length of the track that each particle leaves is a function of its energy and momentum. Because the initial momentum is exactly known, the mass of each particle may be calculated by comparing the length and density of their tracks.

Context

The desire to accelerate particles to higher energies has led to the development of accelerators of increasing dimensions and advanced technology. The older accelerators are sometimes converted into facilities for preliminary acceleration. The cyclotrons of the 1930's and 1940's evolved into synchrocyclotrons and finally into synchrotrons, which were first built in the early 1950's. The sizes of the accelerators of the 1930's were no more than a few meters, but by the 1970's and 1980's, the dimensions had stretched to several kilometers in machines that were able to accelerate protons to thousands of gigaelectronvolts. Construction and operation costs have run into the billions of dollars, requiring both national and international efforts.

Energies into the gigaelectronvolt range were first reached in the 1950's with the construction of two large proton synchrotrons. The Cosmotron, built in 1952 at the Brookhaven National Laboratory in Long Island, New York, reached energies of 3 gigaelectronvolts. The second machine, activated in 1954 at the University of California in Berkeley, reached an energy of 6.4 gigaelectronvolts, sufficient to create proton-antiproton pairs. By the 1960's, a 28-gigaelectronvolt proton synchrotron was in operation at the Centre Européen de Recherche Nucléaire (CERN) near Geneva, Switzerland, and involved the cooperation of twelve nations. A superproton synchrotron with a circumference of 7 kilometers was on line at CERN in 1976 and placed in an underground tunnel. Breaking into the trillion-electronvolt range was accomplished in the 1980's with the Fermilab synchrotron, located near Batavia, Illinois. This accelerator—renamed the Tevatron—first produced 1,000-gigaelectronvolt protons in 1984.

The major breakthroughs of the 1980's were made possible by two important developments in accelerator technology. The first technique for increasing the energy of synchrotrons was to replace existing magnets with powerful superconducting ones. Superconducting magnets can generate much stronger magnetic fields, and they boosted the energy of the Illinois machine from 400 to 1,000 gigaelectronvolts. The second technique involves storing particles in intersecting storage rings. Protons are accelerated to 28 gigaelectronvolts at the CERN proton synchrotron and then alternately deflected by large magnets into two storage rings. The two beams of 28-gigaelectronvolt protons traveling in opposite directions are allowed to intersect each other in an experimental area. This process gains more energy than would be available from a single beam, and the increase in energy comes from the difference between the initial and final kinetic energies of the particles traveling at relativistic speeds. The final energy from two 28-gigaelectronvolt protons colliding is equivalent to the collision of a 1,500-gigaelectronvolt proton with a stationary proton. Accelerators that employ this technology are called colliders.

The largest accelerator ring will be built at the Superconducting Supercollider (SSC), south of Dallas, Texas. This collider is planned to produce proton-antiproton energies of up to 20 trillion electronvolts. The ring will be 87 kilometers in circumference and will accommodate ten thousand superconducting magnets.

Bibliography

Condon, E. U., and Hugh Odishaw. *Handbook of Physics.* New York: McGraw-Hill, 1967. A summary of the most important accelerator designs with a detailed and well-illustrated discussion of both the electron and proton synchrotron. Alternate gradient focusing is treated mathematically, and its applications are described.

Kaplan, Irving. *Nuclear Physics.* 2d ed. Reading, Mass.: Addison-Wesley, 1962. An entire chapter is devoted to the acceleration of charged particles, covering ion sources, cyclotrons, synchrocyclotrons, sychrotrons, and linear accelerators. Presents accelerator and component dimensions, magnet sizes and weights, cycling frequencies and energy levels.

Myers, Stephen, and Emilio Picasso. "The LEP Collider." *Scientific American* 263 (July, 1990): 54-61. A discussion of the building and results of the large electron-positron collider in Geneva, Switzerland. Includes full-color illustrations and photographs.

Neeman, Yuval, and Yoram Kirsh. *The Particle Hunters.* New York: Cambridge University Press, 1983. The physicist is compared to a detective in the hunt for particles created by the accelerators. Detecting methods are presented along with an extensive listing of the elementary particles already found and yet to be discovered.

Parker, Sybil P., ed. *McGraw-Hill Encyclopedia of Physics.* New York: McGraw-Hill, 1982. Diagrams and photographs adequately cover synchrotrons and their operating characteristics. Graphs illustrate phase stability and accelerating frequency. The total energy and electrical currents carried by the various accelerators are presented in table form.

Wehr, M. Russell, James A. Richards, Jr., and Thomas W. Adair III. *Physics of the Atom.* Reading, Mass.: Addison-Wesley, 1985. A good qualitative presentation on accelerators that is well illustrated. Methods of increasing the effective energy of these machines and limitations are included. Suitable for the general reader.

Michael L. Broyles

Cross-References

Bubble Chambers, 273; Cyclotrons, 635; Detectors on High-Energy Accelerators, 643; Linear Accelerators, 1250; Nuclear Forces, 1580; Nuclear Reactions and Scattering, 1605; Radiation: Interactions with Matter, 2026; Storage Rings and Colliders, 2388.

THERMAL PROPERTIES OF MATTER

Type of physical science: Classical physics
Field of study: Thermodynamics

When heat is added to a substance, it will respond by changes in temperature and volume. Substances will also conduct heat through themselves. An understanding of the thermal properties of matter has affected the development of areas of physics, including thermodynamics, statistical mechanics, quantum theory, and solid-state physics, and has both scientific and commercial applications.

Principal terms

CONDUCTOR: a substance, such as a metal, through which heat can flow at a rapid rate

EQUIPARTITION THEOREM: a classical theorem that predicts the value of the specific heat for substances, which is obeyed for most substances in the limit of high temperature

HEAT: the energy that flows between substances because of their difference in temperature

HEAT CAPACITY: the ratio of the heat added to a system divided by the resulting temperature change of the system

INSULATOR: a substance that is a poor conductor of heat, such as an ionic solid

MOLE: a unit for quantity equal to the number of atoms in a 12-gram sample of isotopically pure carbon, approximately 6×10^{23}

SPECIFIC HEAT: the heat capacity for a specified amount of substance, either 1 gram of substance (specific heat per gram) or 1 mole of substance (specific heat per mole)

TEMPERATURE: a measure of the magnitude of thermal energy contained in a quantity of matter

THERMAL CONDUCTIVITY: the flow of heat through a substance, caused by a difference of temperature

Overview

When two objects at different temperatures are placed in contact, there will be a net flow of energy from the object at higher temperature to the object at lower temperature. This flow of energy will continue until thermal equilibrium is reached and the two objects acquire the same temperature. The energy that moves between two objects as a result of their difference in temperature is defined as heat, or thermal energy. From a microscopic point of view, this thermal energy can be pictured as the energy resulting from random motion of the molecules making up a substance. The changes in a substance that occur when heat is added or removed, or when the temperature of a substance is changed, and the rate of flow of heat through a substance are examples of the thermal properties of matter.

When heat is added to a system, the system will respond by an increase in temperature. The heat capacity of the system (C) is defined as the ratio of the amount of heat added to the system (q) divided by the change in the temperature of the system (T), that is, $C = q/(T)$. One drawback in this definition is that heat capacity is proportional to the mass of the system. It is convenient to define a new term, the specific heat (c), as the heat capacity for a fixed quantity of substance. Two different types of specific heats are usually used. The specific heat per gram (c_g) is equal to the heat capacity of 1 gram of the substance. Alternatively, the specific heat can be defined for a sample containing a specific number of molecules of a substance. The number of molecules is usually given in units of moles, where 1 mole of molecules is a number equal to the number of atoms in 12 grams of an isotopically pure sample of carbon, approximately 6×10^{23}. The specific heat per mole (C_m), sometimes called the molar heat capacity, is equal to the heat capacity of a sample containing 1 mole of molecules. For a sample of substance with a mass of m grams, and containing n moles of molecules, the specific heat per gram of substance is $c_g = q/m(T)$, while the specific heat per mole of substance is $C_m = q/n\ (T)$.

The specific heat of a substance depends both on the properties of the substance and on the conditions used to add heat to the system. Heat can be added to a system under conditions where the volume of the system is held constant or under conditions where the external pressure applied to the system is held constant. When heat is added to a substance at constant volume, all the heat goes into raising the temperature of the substance. When heat is added to a substance at constant pressure, however, some of the heat will be converted into work associated with the change in volume of the substance. A general relationship for the difference between the specific heat measured at constant pressure and at constant volume can be found from thermodynamics. For substances that expand when heated, as is usually the case, the specific heat measured for conditions of constant pressure will be larger than that measured under conditions of constant volume. This results from the fact that some of the heat is used to expand the substance against the applied pressure. For substances where the volume decreases as temperature increases, the specific heat measured at constant volume will be larger than the value found at constant pressure, since the work performed on the substance resulting from the decrease in volume will be converted into an equivalent quantity of heat.

Heat added to a gas can be used to increase the average translational kinetic energy of the gas particles. For gases composed of molecules, some of the thermal energy will also be used to increase the total energy of the molecule contained in vibrational motion of the atoms making up the molecule and in rotation of the molecule. Since the number of molecular vibrations increases as the number of atoms making up the molecule increases, the specific heat per mole of gas molecules in general increases as the size of the molecules increases. The equipartition theorem, a result from statistical mechanics, predicts that there will be a contribution to the specific heat of a gas from translational, rotational, and vibrational motion of the molecules, which can be calculated from the theorem. The values observed for specific heats for gases are in general agreement with the predictions made from the equipartition theorem at high temperatures, but are less than predicted by the theorem at low temperatures. The discrepancy is caused by the failure to take into account the fact that vibrational and rotational energy in molecules is quantized. When quantum theory is used to model the vibrational and rotational motion of gas molecules, the values calculated for specific heat are in general agreement with experimental results at all temperatures. For ideal gases, the difference between constant pressure and constant volume specific heats per mole of substance is equal to R, the gas constant, independent of the type of gas involved.

For solids, added heat is used to increase the amount of vibrational energy of the particles making up the solid. The equipartition theorem predicts that the specific heat per mole of solid, measured at constant volume, should be equal to three times the gas constant. As with gases, the equipartition theorem correctly predicts the specific heat for solids at high temperatures but fails for low temperatures. When quantum theory is used to model the vibrational motion of the particles making up a solid, agreement is again achieved between theory and experiment at all temperatures. Constant pressure and constant volume specific heats for solids are usually close in value, but can in some cases differ from each other by as much as 50 percent. The difference between the constant pressure and constant volume specific heat can be calculated from general thermodynamic relationships; the difference depends on the compressibility of the solid and its change in volume, or thermal expansion, with temperature.

The change in the specific heat of a substance with temperature is also used as a means of classifying different types of phase transitions, or transformations to different forms of matter. For a first-order phase transition, such as occurs when a substance is converted from the solid to liquid phase, the specific heat of the substance measured at constant pressure changes discontinuously, and is infinitely large at the temperature at which the phase transition occurs. This results from the fact that all the heat being added is used to convert the substance from the initial to the final phase. Other types of phase transitions, such as the conversion of the normal liquid phase of helium to the superfluid liquid phase, show different behavior for the specific heat at the point where the phase transition occurs.

A second thermal property of substances is the change in the volume of a substance that occurs when heat is added. This volume change is usually expressed in terms of the coefficient of thermal expansion for the substance, defined as the relative change in the volume of the substance with temperature, measured under conditions of constant applied pressure. For an ideal gas, the coefficient of thermal expansion is equal to $1/T$, where T is the temperature of the gas measured in Kelvins. Real gases can show deviations from this simple behavior. For solids and liquids, the coefficient of thermal expansion tends to be much smaller than for gases. Occasionally, substances will over some ranges of temperature decrease in volume as the temperature of the substance increases, as is the case for liquid water below 4 degrees Celsius.

Substances also differ from one another in the rate that they conduct heat. For a solid, the rate of heat flow can be determined by measuring the movement of heat through a rod of the substance whose two ends are held at different temperatures. The conduction of heat through such a rod is observed to be directly proportional to the temperature difference between the two ends of the rod, and to depend also on the physical properties of the substance composing the rod. Conduction of heat by liquids and gases is measured by observing the rate at which heat passes through a sample of the fluid when placed between two walls maintained at different temperatures. The ability of a substance to conduct heat is expressed in terms of the coefficient of thermal conductivity for the substance, which is the constant of proportionality that appears in the equation describing the heat flow through the substance.

Solids differ widely in their ability to conduct heat. Solids can be qualitatively divided into two categories: thermal insulators (which have low values for the coefficient of thermal conductivity and conduct heat slowly) and thermal conductors (which have large values for the coefficient of thermal conductivity and are efficient conductors of heat). The presence of trace amounts of impurities in an otherwise pure solid can have a large effect on the ability of the solid to conduct heat, particularly at low temperatures. Thermal conduction in solids takes place by two mechanisms: conduction by electrons in the solid and conduction by vibrational motion of the particles making up the substance. For crystals, these vibrations are called lattice vibrations. Since thermal conduction by electrons is in general more efficient than conduction through lattice vibrations, metals—which possess a large number of free electrons—tend to be good conductors of heat, while solids without large numbers of free electrons, such as salts, are thermal insulators. The theory used to describe electrical conduction in metals can be modified to describe thermal conduction by electrons as well.

For gases, thermal conductivity is observed to be independent of gas pressure except at extremely low pressures, when the thermal conductivity is directly proportional to the pressure of the gas. This surprising result can be understood by consideration of the mechanism by which heat flow occurs. The rate of heat flow through a gas depends both on the number of gas molecules available to transport heat through the gas and the average distance a gas molecule moves through the gas before undergoing a collision, and giving up part of its energy to another molecule. As the pressure of a gas increases, the number of molecules available to carry thermal energy also increases. Nevertheless, because the density of a gas increases with pressure, the average distance a molecule in a gas can travel before colliding with another molecule decreases with increasing pressure. For most pressures, the two effects cancel, resulting in no net change in the thermal conductivity of the gas. At very low pressures, however, the probability of a gas molecule colliding with another molecule becomes negligible, and the rate of heat conduction then becomes proportional to the density of molecules in the gas. Vacuum represents an extremely efficient thermal insulator, since heat can move through a vacuum only by radiative processes.

Applications

Substances respond in different ways to addition of heat and changes in temperature. Substances also differ in their ability to conduct heat. These differences can be exploited for practical benefits. Substances with a large value for specific heat can be used as coolants to remove heat from systems. One common coolant is water, which has an unusually large specific heat on a per gram basis, approximately ten times that of iron. This makes water an excellent substance for the removal of heat, and it is used as a coolant in everything from automobile engines to some types of nuclear power reactors. The large quantity of water found in the environment has a moderating effect on the temperature of the earth, tending to minimize variations of temperature with latitude or time of day. Other liquids with a large value for specific heat, such as sodium (which melts at 97 degrees Celsius) and ammonia (a gas that can be liquefied by the application of pressure), are also used as coolants in specialized applications.

Knowledge of the specific heat of materials is used in scientific applications as well. A common method for determining the heat evolved or taken up during a chemical reaction is calorimetry. In bomb calorimetry, for example, a sample of a compound is burned in a closed container in the presence of several atmospheres of pure oxygen. If the specific heat of the system is known, the change in temperature observed when combustion takes place can be used to calculate the energy of combustion for the reaction. Calorimetry represents the most common method for obtaining precise information on the thermodynamic properties of matter.

The conducting and insulating properties of substances have a variety of practical applications. In many cases, such as in electronic chips and cutting tools, it is important to remove heat as quickly as possible. While metals are in general good thermal conductors, other substances are even better conductors of heat. Diamond, for example, is four times more efficient at conducting heat at room temperature than copper and has been used in cases where rapid removal of heat is of critical importance. In other cases, the desire is to prevent the flow of heat into or out of a system. Materials such as Styrofoam are used to hold hot and cold beverages, or to keep foods warm, while foam insulation is used to minimize heat flow between a building and its surroundings. The insulating properties of vacuum are also used to minimize heat flow, as in a Thermos bottle. Specially designed insulating tiles are used in the space shuttle to shield it from the intense heat generated upon reentry into Earth's atmosphere.

The change in the volume of a substance with temperature can also be put to practical use. A mercury thermometer makes use of the fact that the change in volume for a fixed quantity of mercury is, to a first approximation, proportional to the difference between the temperature of the mercury and some arbitrary reference temperature. By confining the mercury to a narrow tube, small changes in the volume occupied by the mercury can be detected easily. Mercury thermometers were among the first devices developed to measure temperature and are still the most common devices for measuring temperature. Since solid metals also expand with increasing temperature, metal coils are often used in thermostats for regulating the temperature of a room or building. The change in the length of the coil with temperature can be used as a switch to turn a heater or air conditioner on and off.

Context

The study of the thermal properties of matter has played a central role in the development of thermodynamics, statistical mechanics, kinetic theory of gases, quantum theory, and solid-state physics. In case after case, the failure of theory to explain experimental observations of thermal properties has resulted in advances in the understanding of the laws of nature.

Qualitative observations of the relationship between heat, temperature, and volume date back to ancient times. In the seventeenth century, Galileo developed the first thermometer, using the change in the volume of a gas to measure temperature. Improvements in the thermometer occurred when liquids replaced gases as the working fluid. In 1714, Daniel Gabriel Fahrenheit devised the first mercury thermometer.

In the early nineteenth century, two French scientists, Pierre-Louis Dulong and Alexis-Thérèse Petit, developed a procedure for the determination of the molecular weight of an element based on measurements of specific heat. Dulong and Petit observed that for elemental solids with known molecular weights, the specific heat per mole of element was always approximately 25 joules per Kelvin. Assuming that this number was a universal constant, the molecular weight for a newly discovered element could be determined by measurement of its specific heat. This method was in fact used to determine the molecular weight of a number of elements.

At about the same period of time, consideration of the thermal properties of matter was leading to the general relationships summarized in the laws of thermodynamics. One consequence of this development was the derivation of exact expressions for the difference between the specific heat of a substance measured at constant volume and that measured at constant external pressure. When statistical thermodynamics was developed by James Clerk Maxwell, Ludwig Boltzmann, and others in the late 1800's, one of the results was the equipartition theorem, which made it possible to predict the value for the specific heat of a substance from theory.

While some substances were found to have the value of heat capacity predicted from statistical mechanics, discrepancies between prediction and experiment soon appeared. The specific heat for diatomic and polyatomic gases was far below the predicted value, although agreement between theory and experiment improved at high temperatures. For solids, it was found that at low temperatures, the specific heat was also smaller than calculated from theory. Also, as noted by the American chemist Josiah Willard Gibbs, there seemed to be no contribution to the heat capacity of a metal from its free electrons.

In 1907, Albert Einstein recalculated the specific heat for a solid, assuming that the vibrational energy for the particles making up the solid was quantized, that is, that it could only take on particular values. The quantization of energy had previously been used by Max Planck to explain black body radiation and by Einstein to explain the photoelectric effect. Using the assumption of quantized values for vibrational energy, Einstein was able to predict the specific heat of a solid even at low temperatures. The success of Einstein's model in predicting the specific heats of solids was one factor leading to the development of quantum mechanics. Modification of Einstein's model by Peter J. W. Debye and others soon led to better agreement between theory and experiment.

Following the development of quantum mechanics in the 1920's, the combination of quantum mechanics and statistical mechanics was used to advance the understanding of thermal properties of matter. At temperatures approaching absolute zero, the spin properties of the particles making up a substance become an important factor in determining the specific heat and thermal conductivity. In the 1940's, Lars Onsager and others developed general methods for describing heat flow through matter. These methods have made it possible to understand heat flow in different types of matter under a variety of conditions of temperature and pressure.

Bibliography

Hemminger, Wolfgang, and Gunther Hoehne. *Calorimetry.* Translated by Y. Goldman. Deerfield Beach, Fla.: Verlag Chemie, 1984. A detailed survey of the theory and application of calorimetry to problems in physics and chemistry. There is a good discussion of how heat capacity is related to calorimetry.

McClintock, P. V. E., D. J. Meredith, and J. K. Wigmore. *Matter at Low Temperatures.* New York: Wiley, 1984. The introductory chapters of this book give the basic theoretical background for the heat capacity and thermal conductivity of matter. The discussion is well organized, although at a fairly sophisticated level. The remainder of the book is a survey of low-temperature properties of matter, with emphasis on the unusual behavior of liquid helium at low temperatures.

Mott-Smith, Morton. *Heat and Its Workings.* New York: Dover Press, 1963. A general introduction to thermodynamics for nonscientists. There are a large number of illustrations, and the discussion is supplemented with examples from everyday experience. Chapter 4 focuses on heat and heat capacity.

Yates, Bernard. *Thermal Expansion.* New York: Plenum Press, 1972. A discussion of the thermal properties of solids. The emphasis is on the change in volume of solids with temperature, and includes sections on the properties of metals, salts, glasses, polymers, and semiconductors. There is also a brief discussion of specific heat and its connection to thermal expansion.

Zemansky, Mark W. *Heat and Thermodynamics.* New York: McGraw-Hill, 1968. One of the best introductory texts for thermodynamics. The book presents both classical and statistical thermodynamic theory. Chapter 4 is devoted to heat, and includes a discussion of heat capacity and its measurement. Heat capacity and other thermal properties of matter as applied to thermodynamics also appear in other sections of the book.

Jeffrey A. Joens

Cross-References

The Absolute Zero Temperature, 1; Calorimetry, 290; The Behavior of Gases, 935; Phase Changes, 1777; Thermal Properties of Solids, 2276; Statistical Mechanics, 2367; Thermodynamics: An Overview, 2493.

THERMOCOUPLES

Type of physical science: Condensed matter physics
Field of study: Solids

Thermocouples determine temperature differences, relative to a reference junction, by monitoring electromotive forces in systems consisting of junctions between dissimilar, electrically conducting materials. This useful temperature measurement technique involves low thermal masses and responds quickly to temperature changes.

Principal terms

CURRENT DENSITY: current flow per unit cross-sectional area in an electrically conducting material

JOULE HEAT: irreversible heat dissipated within an electrical conductor by virtue of current flow

REFERENCE JUNCTION: a connection of dissimilar metals in good thermal contact with a thermal reservoir at a known temperature

THERMAL ELECTROMOTIVE FORCE: a potential difference generated solely by temperature differences

THERMAL GRADIENT: a temperature difference spread over a spatial extent

THERMOCOUPLE: a thermometer that uses thermoelectrical properties of dissimilar metals to monitor temperature differences relative to a selected reference

THERMOELECTRIC POWER: for a junction between dissimilar materials, the ratio of thermoelectric potential difference to the temperature difference between test and reference junctions

THERMOMETER: a device that uses a particular practical temperature scale to measure temperature

THOMSON HEAT: reversible heat given off or absorbed by an electrical conductor that is subjected to a thermal gradient

Overview

Thermoelectricity involves the study of energy transfer by charged particles flowing through electrical conductors as a result of thermal gradients. Different electrically conducting materials promote charge carrier diffusion throughout their volume at rates determined by atomic structure. Consider a pair of thermal reservoirs at temperatures T_1 and T_2 separated physically from each other. The reservoir at temperature T_1 is referred to as the reference reservoir, and the one at temperature T_2 is the test reservoir. These reservoirs are connected by a pair of dissimilar metallic conductors. Both conductors are joined at a junction where they are in thermal contact with each thermal reservoir. At no other point in the space between the two reservoirs are the dissimilar metallic conductors in physical or thermal contact. A number of thermoelectric effects can be observed simultaneously in this system.

Charge carrier density is determined by atomic structure; in other words, the number of free electrons per unit volume and the mobility of those electrons are different in the two metals. Thus, diffusion of charge carriers in each metal connected between the pair of thermal reservoirs will not be equivalent. As a result, there will be a thermal emf (electromotive force) across the combination of connected dissimilar metals. The production of such a thermal emf is called the Seebeck effect. It is the fundamental operating principle of a practical device called a thermocouple, which is often used for temperature measurement in the laboratory.

The net motion of charge carriers in the dissimilar metals could be described as driven by an electric field of nonelectrostatic origin. Using the relationship defining electric potential in terms of electric field, the thermal emf can be calculated by evaluating the line integral of this nonelectrostatic field around the two dissimilar metals joined at each of the two thermal reservoirs. This Seebeck potential difference is a function of the temperature difference between the two reservoirs. The exact value of the Seebeck effect is determined by the given value of the reference junction temperature and the temperature difference between the two reservoirs. The rate of change of thermal emf with reservoir temperature difference is independent of the value of the reference temperature T_1. This rate is referred to as the thermoelectric power or Seebeck coefficient of the junction of dissimilar metals. For many common pairs of materials used to construct thermocouples, the thermoelectric power is in the range of several microvolts per unit degree Celsius. Thus, for many thermocouples, the Seebeck emf is a small effect, significantly observable only when there is a reservoir temperature difference of more than 100 degrees Celsius.

Because the Seebeck emf is not externally compensated by external sources, it will result in a current, or a flow of free charge carriers, through the thermocouple. When current passes through a normal conductor (possessing resistance to the flow of electrical charge under the influence of an emf), heat is generated and dissipated within the conductor. The power dissipated in this Joule heating effect is given by I^2R, where I represents the current and R is the resistance of the conducting material.

Suppose that the temperatures T_1 and T_2 of the two reservoirs joined by the thermocouple are identical. If an external emf (not of thermal origin) is established across the thermocouple, a current will flow in a direction determined by the sense of the external emf. This current will alter the temperature of the thermocouple junctions by conduction of heat in excess of that produced by Joule heating alone. The heat conduction is determined by the current flow direction; whereas Joule heating can raise only the temperature of a junction, this effect, known as Peltier heating, can either raise or lower the junction temperature. Peltier heating is reversible, whereas Joule heating is irreversible. The Peltier heat production rate is proportional to the current flow. The constant of proportionality is called the Peltier coefficient or Peltier power. Its value for a given junction of dissimilar materials depends only on the temperature of that junction and the composition of the junction materials.

Along the different materials that make up the thermocouple attached to the two reservoirs, there exists a thermal gradient. The distribution of temperature along the length of each side of the thermocouple is uniform if the material contains no defects or impurities. When a current passes through each side of the thermocouple, excess heat beyond the Joule heating effect will be produced. This effect is known as Thomson heat. The current flow direction determines whether that heat is absorbed or reflected. Thus, like the Peltier effect, the Thomson effect is reversible. The power of Thomson heat transfer is proportional to the product of the current and thermal gradient. The constant of proportionality, called the Thomson coefficient, is determined by the composition of the material and the average temperature. Unlike the Seebeck and Peltier effects, which involve two dissimilar materials, the Thomson effect occurs in a single material. The Seebeck, Peltier, and Thomson coefficients are not independent of one another. Thermodynamic formalism relates the three coefficients using a pair of equations, one of a differential nature and one of an algebraic form, known as Kelvin relations.

Because the Peltier and Thomson effects are reversible, it is possible to use these effects with a suitably directed current flow to lower temperature. This simple process forms the basis of the process of thermoelectric refrigeration. Thermoelectric effects also can be used to generate electricity directly from thermal gradients.

Applications

The primary application of thermoelectric properties of current-carrying materials is the thermocouple. The basic thermocouple design consists of a pair of dissimilar conductors in the form of wires joined together to construct two junctions of the dissimilar conductors in series. The two unconnected wires must be made of the same conducting material. The wires of the other conducting material connect the two junctions together in series. The reference junction is inserted into a thermal reservoir of known temperature. The test junction is in good thermal contact with the object whose temperature is being measured via the Seebeck effect.

To be useful, a thermocouple must be calibrated accurately. The thermocouple must be placed in thermal equilibrium with materials at precisely known temperatures. Depending upon the size of the temperature range for which the thermocouple is meant to be useful, the number of calibration points varies, as does the degree of the polynomial fit between measured thermal emf and temperature. Usually, a quartic fit is sufficient over a temperature range of several hundred degrees Celsius. Once calibrated against a sufficient number of known temperatures, the thermocouple can be used to ascertain unknown temperatures by measuring experimentally the thermal emf generated across the two junctions. Thermocouples are particularly useful when determining the temperature of small amounts of a substance. Because the junction has such small thermal mass, thermal equilibrium between the test junction and the sample is rapidly reached. As a result, relatively rapid changes in temperature of the sample can be monitored. One practical difficulty in using a thermocouple involves keeping the wires that are connected to a digital voltmeter or potentiometer at constant temperature and vibration-free. Temperature and vibration changes can alter the emf read by the meter beyond the generated thermal emfs that are indicative of the sample's temperature.

The choice of thermocouple depends on the temperature range for which it will be used. An upper limit of availability is the lower melting point of the two materials used to construct the thermocouple. High-melting-point metals, such as platinum and platinum-rhodium, are used for high-temperature measurement (above 1,000 degrees Celsius). Commonly used thermocouple materials and alloys include iron, copper, constantan, alumel, chromel, platinum-rhodium, and platinum. Such materials are chosen because of their high melting points and (when used in pairs) thermoelectric powers. Some of these materials have thermoelectric powers as large as 40 microvolts per degree Celsius. Occasionally, semiconducting materials with thermoelectric powers as high as about 1 millivolt per degree Celsius will be used to construct a thermocouple. Voltmeter accuracy determines which materials can be used for thermocouples in some applications.

A thermopile is useful in measuring very small temperature differences for which the thermal emf that is produced will be small. It consists of a number of identical thermocouples placed in series with each other. One junction of each thermocouple in the thermopile is in thermal equilibrium with the reference reservoir. The other junctions are in thermal contact with the sample. Each thermocouple generates the same thermal emf; and by being connected in series, the total thermal emf across the entire thermopile is the linear sum of the individual thermal emfs. This larger voltage can be read more easily and be corrected to determine the sample temperature.

Powders of specially heat-treated Y-Ba-Cu-O can be pressed under high pressure into the shape of thin disks to make high-temperature superconductors. Superconductivity is a unique state of electrical conduction without resistance. This property is lost in a phase transition from the superconducting state to normal conductivity at a critical temperature. Measurement of the superconducting property can be accomplished using a technique referred to as a four-point probe. Four wires are connected to the sample disk. Two wires carry applied current and the other two measure the voltage between two points on the disk. Resistance, which is the voltage divided by the current, is monitored as a function of temperature as the sample is either cooled or warmed. Below the critical temperature, the resistance of the superconductor vanishes; above the critical temperature, the sample resistance rises abruptly. A thermocouple can be included in the preparation of the sample disk so that the test junction of the thermocouple is a part of the sample. In such good thermal contact, the test junction of the thermocouple can respond quickly to temperature changes in the sample, which permits an accurate determination of the superconductor's critical temperature.

Context

The history of thermoelectrical understanding is intimately linked to the developments of electromagnetism and thermodynamics. As with most scientific endeavors, practical use of an identified phenomenon occurs well after the initial discovery. Thermocouples remained an experimental curiosity until the thermoelectric properties of many pairs of dissimilar conducting or semiconducting metals were thoroughly investigated and categorized. Measurement of temperature by thermocouple has become a standard experimental technique for temperatures ranging from near absolute zero to the melting points of many common metals and is used in pure physics research laboratories and in the industrial applied research community.

In 1821, Thomas Johann Seebeck demonstrated that an electric current flows between different conductive materials that are kept at different temperatures—the Seebeck effect. Thermoelectric power, the constant of proportionality between thermal emf and junction temperature difference, is also referred to as the Seebeck coefficient of a thermocouple. This latter designation is perhaps more appropriate, since the dimensionality of thermoelectric power is not energy per unit time as is a normal power rating. Jean-Charles-Athanase Peltier first demonstrated the Peltier effect in 1834. The constant of proportionality between the heat production in excess of Joule heating and the current density flowing through the junction of a thermocouple bears Peltier's name as well. William Thomson, Lord Kelvin, first predicted the Thomson effect in 1854. Lord Kelvin also predicted this phenomenon on theoretical grounds. Several years later, he experimentally demonstrated excess heat generation in a current-carrying conductor exposed to a thermal gradient. The coefficient relating the excess heat thus generated to the product of current density and thermal gradient bears Thomson's name. Thermodynamic theory eventually related the Seebeck, Peltier, and Thomson coefficients to one another through standard Kelvin relations.

From these basic thermoelectric properties, a host of applications followed, such as the thermocouple, thermopile, thermoelectric power generation, and thermoelectric refrigeration.

Bibliography

Besancon, Robert M., ed. *The Encyclopedia of Physics*. 2d ed. New York: Van Nostrand Reinhold, 1974. A thorough compendium of all aspects of physics. Graphs and illustrations.

Morse, Philip M. *Thermal Physics*. New York: W. A. Benjamin, 1969. Thorough discussion of thermodynamics. Thermoelectric effects are presented from a theoretical standpoint.

Reif, Frederick. *Fundamentals of Statistical and Thermal Physics*. New York: McGraw-Hill, 1965. Excellent treatise on all aspects of thermodynamics. Describes thermoelectric effects that are pertinent to the operation of a thermocouple.

Wedlock, Bruce D., and James K. Roberge. *Electronic Components and Measurements*. Englewood Cliffs, N.J.: Prentice-Hall, 1969. Practical examples of thermocouple use in experimentation. Basic electronics principles presented in a manner accessible to the hobbyist.

Zemansky, Mark W. *Heat and Thermodynamics*. New York: McGraw-Hill, 1968. A classic intermediate textbook for undergraduate instruction in thermodynamics. Excellent treatment of thermoelectric effects. Includes a theoretical description of thermocouple principles.

David G. Fisher

Cross-References

Infrared Astronomy, 1118; Thermal Properties of Solids, 2276; Superconductors, 2417; Thermal Properties of Matter, 2472; Thermodynamics: An Overview, 2493; Thermometers, 2501.

LAWS OF THERMODYNAMICS

Type of physical science: Classical physics
Field of study: Thermodynamics

Thermodynamics deals with the macroscopic properties of bulk matter, which involve thermal effects in an essential way. The results of an immense body of experimental knowledge can be formulated in terms of a small number of generalized abstract principles known as the laws of thermodynamics.

Principal terms

ENERGY: a fundamental conserved physical attribute that a system possesses by virtue of its state of motion or interaction with other systems; operationally, the capacity to perform work or generate heat

ENTROPY: a measure of molecular disorder in a system; entropy increases in spontaneous processes

EQUILIBRIUM: a condition of stability whereby the state of a system does not perceptibly change over some period of time

HEAT: thermal energy transferred into or out of a system, commonly as a flow from a warmer to a cooler body

STATE: the condition of a thermodynamic system, as described by the appropriate physical variables, such as mass, volume, pressure, and temperature

SYSTEM: an explicitly defined part of the world being observed; the system may be isolated from its surroundings or it may be interacting in some specified way

TEMPERATURE: a measure of the thermal energy of a system

THERMAL ENERGY: energy possessed by a system by virtue of the random motions of its constituent molecules

WORK: energy transferred into or out of a system by means of some mechanical or electrical interaction with the surroundings

Overview

The science of thermodynamics is concerned with transformations of matter and energy in physical and chemical processes. In particular, it seeks to describe states and processes in which thermal effects—those involving transfer of heat and changes in temperature—play a significant role. In this regard, mechanics and electromagnetic theory in their purest forms can be considered to treat only the behavior of matter and energy at the absolute zero of temperature. The name thermodynamics originates from the Greek words *thermos*, meaning heat, and *dynamis*, meaning power.

The branch of the subject that concerns the equilibrium macroscopic properties of matter is called "classical thermodynamics," to distinguish it from "irreversible thermodynamics"—whose subject matter includes nonequilibrium properties and rate processes—and "statistical thermodynamics"—which treats the subject from a microscopic or molecular point of view. This article will be principally concerned with classical thermodynamics, which has long been a central topic of physical chemistry and also plays a significant role in various branches of physics, engineering, biology, and geology.

Thermodynamics rests on an experimental foundation of immense breadth and depth, encompassing a vast and diverse array of accumulated experience in chemistry, physics, biology, and engineering. In common with other branches of science, the specific details and individual peculiarities of a large number of experimental facts are abstracted and, through inductive reasoning, expressed as a compact set of fundamental laws. The abstract terminology of such inductive principles tends to obscure their origin in specific phenomena, but in consequence their universality and range of application becomes significantly enhanced. A set of inductively derived fundamental laws can subsequently be regarded as a system of postulates. From these there should follow, as logical consequences, all the experimental results that originally led to their formulation, as well as others not previously known.

The modern formulation of classical thermodynamics is based on four fundamental principles, designed as the zeroth, first, second, and third laws. The zeroth law of thermodynamics was enunciated by Sir Ralph Fowler in 1931. It stands most logically *before* the first and second laws in the theoretical framework of thermodynamics. Since the designations "first" and "second" laws were by then too well established to change, however, the new principle was dubbed the "zeroth law." The zeroth law, also called the law of thermal equilibrium, can be stated: Two systems in thermal equilibrium with a third system are in thermal equilibrium with each other. Two systems are said to be in thermal equilibrium when there is no net heat exchanged when they are brought into thermal contact. Although this principle may appear at first sight to be rather obvious and trivial, it is well to point out that the equivalent statement about "chemical" equilibrium between substances is not true. Thus, two systems unreactive with a third system are not necessarily unreactive with each other. (Consider, for example, gaseous ammonia, hydrogen chloride, and helium; neither of the first two reacts with helium, but they do react with each other to produce a white mist of ammonium chloride.) The zeroth law leads to an operational definition of temperature: If two systems coexist in thermal equilibrium, they have the same temperature.

The temperature scale used in most scientific work is the Celsius scale, defined such that 0 degrees Celsius is the freezing point of water and 100 degrees Celsius is the boiling point (both under a pressure of 1 atmosphere). Of more fundamental significance is the absolute, or Kelvin, temperature scale. Zero on the Kelvin scale represents the lowest conceivable temperature, known as absolute zero. This corresponds to -273.15 degrees Celsius. The degree has the same size as on the Celsius scale but is now designated as 1 Kelvin. Thus, the freezing and boiling points of water are approximately 273 Kelvins and 373 Kelvins, respectively. Room temperature is about 300 Kelvins.

Friction in mechanical systems results in the evolution of heat, with an associated rise in temperature. This would be especially evident in the operation of a grindstone or in the boring of cannon barrels. In the same category is the heating effect of an electric current, in which electrical work is transformed into heat. The interconversion of heat and work was studied in the first half of the nineteenth century, with major contributions by Benjamin Thompson (later to become Count von Rumford), Sir Humphry Davy, and Julius Robert von Mayer, culminating in the definitive experiments of James Prescott Joule in 1840. These measurements determined the mechanical equivalent of heat, the modern value being 4.184 calories of heat per joule of work. This led to the understanding of heat as a form of energy transfer. In earlier times, heat had been thought to be a material substance, variously called "phlogiston," or "caloric." The first law of thermodynamics can now be stated: The energy of the universe is constant. A thermodynamic system can indeed either gain or lose energy by heat transfer or performance of work, but this energy change must be exactly compensated by that of the surroundings.

The law of conservation of energy plays a central role in classical mechanics, electromagnetism, and quantum theory. According to this fundamental principle of science, energy can neither be created nor destroyed, but only converted from one form to another. (The theory of relativity encompasses, in addition, the interconvertibility of matter and energy, but this need not be considered in ordinary physical and chemical phenomena.) The first law of thermodynamics augments the concept of energy by taking into account "thermal energy," that concept associated with the randomized motions of individual molecules, in addition to mechanical and electrical energy. More precisely, heat represents a flow of thermal energy, analogous to work, which is a transfer of mechanical or electrical energy. Whereas work usually involves perceptible motion of part of a system, heat does not, occurring on two small a scale to be macroscopically detectable.

The first law for a thermodynamic process can be expressed in the following mathematical form: $U = q + w$. Here, q represents the heat transferred *into* the system and w, the work done *by* the system; U is the energy of the system. As defined, a negative value of q would mean that the system loses heat, and a negative w, that the system does work on its surroundings. The quantity U introduced by the first law of thermodynamics is a new function of state; that is, it depends only on the variables defining the state of the thermodynamic system, independent of how the system got there. By contrast, heat and work depend critically on the path taken by the system.

The second law of thermodynamics is one of the most profound and intriguing principles in all of science, with implications going far beyond its original subject. It deals with the direction of spontaneous change. It is a well-documented observation that a drop of ink will diffuse uniformly throughout a beaker of water. The reverse situation will never occur: that the drop of ink will reassemble out of tinted water. More generally, all changes in the universe appear to have a natural direction, with the reverse processes usually being regarded as impossible. The second law of thermodynamics can be understood most readily from a molecular point of view. To every thermodynamic state of a macroscopic system, as described by a small number of variables, there must correspond a huge number of microscopic or molecular arrangements that behave like the very same state. Thus, exchanging two molecules in a mole of a substance (approximately 6×10^{23} molecules) will not alter any of the observable properties of the system. The essence of spontaneous change is that a transition is made to a final state, which has many more possible molecular arrangements than the initial state. The result of a spontaneous change can therefore be regarded as an increase in the "disorder" or the "randomness" of the system. In 1865, Rudolf Clausius and William Thomson (later to become Lord Kelvin) defined the thermodynamic function that precisely describes this randomness: the "entropy," symbolized by S. The original definition was based on a formula involving heat and absolute temperature ($dS = dq/T$). Later, in 1896, Ludwig Boltzmann postulated the relation, $S = k \ln W$, in which W represents the number of different molecular arrangements that correspond to a given thermodynamic state: This is an immensely large number, on the order of 10^{23}; "*ln*" stands for natural logarithm; $\ln W$ will be reduced to a number of the order of 10^{23}; finally, k stands for Boltzmann's constant, equal to 1.38×10^{-23} joules per Kelvin. To illustrate Boltzmann's idea, consider the possible arrangements of a deck of 52 playing cards. (There is a total of 8×10^{67} possible ways to order the cards.) Obviously, the number of well-shuffled arrangements of the deck will far exceed those that have a highly ordered pattern, for example, the sequence in a brand new deck: ace of spades, king of spades, deuce of clubs. Thus, a shuffled deck exhibits a higher value of entropy than an ordered one. The numbers involved in a macroscopic thermodynamic system are immensely larger, with 10^{23} molecules rather than 52 cards.

The second law of thermodynamics can be stated: Every spontaneous process increases the entropy of the universe. Since the "universe" includes both a system and its surroundings, there is no requirement that the entropy of the system alone has to increase. For example, in the freezing of water, the resulting ice has a decreased entropy. This is more than compensated by an increase in the entropy of the surroundings.

Some classic examples of natural processes, all associated with increasing entropy, are: the flow of heat from a warmer to a cooler body; the conversion of work into heat by friction; the expansion of a gas to fill its container; the freezing of supercooled liquid water (meaning water at a temperature below 0 degrees Celsius), and the reaction of gaseous ammonia and hydrogen chloride to form ammonium chloride. Any of the preceding processes can in fact be reversed, but to do so would require some external intervention. For example, heat *can* be made to flow from a cooler to a warmer body (this is precisely what a refrigerator does) by means of compressors, and the like, which are part of a larger system. The second law of thermodynamics still holds for the composite system, even though there is an entropy decrease within one of its component parts.

The second law of thermodynamics is sometimes stated as the "law of degradation of energy," meaning that energy in some low-entropy form is spontaneously converted to energy in a higher-entropy form. The classic example is the degradation of work into heat by friction. Modern civilization is crucially dependent on the conversion of heat into work, by means of heat engines. This reversal of energy degradation entails the rejection of heat to a cold reservoir to compensate for its conversion into work at a hot reservoir. The theory of the heat engine, based on the second law of thermodynamics, was first developed by the French engineer Nicolas-Léonard-Sadi Carnot in 1824.

A useful corollary to the second law of thermodynamics focuses entirely on the behavior of the system, independent of its surroundings. This entails definition of a function known as the "free energy," in terms of which the second law can be restated: Every spontaneous process decreases the free energy of a system. Free energy was first introduced by the American physicist and mathematician Josiah Willard Gibbs. Actually, there are two variants of free energy: for systems in which the pressure and temperature are controlled, the Gibbs energy function G is used; for systems for which the natural variables are volume and temperature, the Helmholtz energy function A is used.

The third law of thermodynamics was discovered by Walther Hermann Nernst in 1912. Later reformulations allow the statement of the third law in either of the equivalent forms: The absolute zero of temperature is unattainable in a finite number of operations; or, as the absolute temperature is reduced toward zero, the entropy of a perfect crystalline solid approaches zero. The state of a thermodynamic system at absolute zero can be characterized as being perfectly ordered on a molecular level. Thus, for this state alone, the microscopic arrangement is unique, corresponding to $W = 1$ in Boltzmann's formula. Also, since $ln\ 1 = 0$, the entropy S equals zero. According to classical mechanics, molecular motion would cease entirely at absolute zero. Quantum mechanics, the more correct theory for molecules, predicts that at absolute zero, the amplitudes of molecular motion are reduced to their lowest possible levels, but not zero.

In discussing the laws of thermodynamics, extensive use of molecular concepts have been made. It should be noted that, in its purest form, classical thermodynamics is actually independent of molecular details, or indeed of the very existence of molecules. Historically, this has been an advantage, allowing the science of thermodynamics to develop in advance of the discoveries of modern theories of molecular structure.

Applications

The laws of thermodynamics find application in many branches of science and technology. Applications to chemistry constitute the subfield called chemical thermodynamics. From extensive tabulations of free energies of the elements and their compounds, it is possible to make predictions about the course of chemical reactions involving these substances. A pioneer in this field was the American physical chemist Gilbert Newton Lewis. In the commercially vital synthesis of ammonia by the Fritz Haber process, the temperature and pressure can be optimized in accordance with thermodynamic considerations so as to achieve maximum product yield.

Ammonia is used as a fertilizer and as a precursor for other essential chemicals.

It was realized in the latter part of the nineteenth century that the free energy change in a chemical reaction could be directly harnessed in an electrochemical cell, now commonly called a battery. Many of the advances in thermodynamics during this period were, in fact, developments in electrochemistry. Electrochemical technology has given scientists efficient and portable sources of energy. The continuing development of fuel cells promises a clean, efficient method of converting fuels directly into electricity, circumventing the thermodynamic restrictions on heat engines. Electrolysis is the reverse of the above process, whereby electrical energy is used to carry out chemical change. Electrolytic methods are extensively used in electroplating, material purification, and chemical synthesis. Electrochemical phenomena are known to play an important role in a number of essential biological processes, including conduction of nerve impulses and transport across cell membranes.

By the methods of statistical mechanics, it has become possible to make theoretical predictions of thermodynamic properties of substances, given the appropriate molecular parameters, as obtained usually from spectroscopic measurements. This has enabled the extension of thermodynamic methods to reactions involving highly unstable or exotic molecules that cannot be measured directly. Such applications have been made in atmospheric chemistry, astrophysics, and in the study of reactions involving short-lived intermediates.

In 1875, Gibbs published his definitive work on the equilibrium of heterogeneous substances. From this work comes one of the simplest, yet most powerful, results in all of science: the "phase rule," which states that $F = C + P - 2$, in which P represents the number of phases in a system (that is, solids, liquids, gases), C the number of components (distinct chemical species), and F the number of degrees of freedom (the number of variables such as pressure, temperature, and composition needed to specify the state of the system). The phase rule has played a key role in the development of material science, leading to the design of new materials with desired physical and chemical properties.

Context

In the overall scheme of modern science, thermodynamics has played the role of a censor somewhat. Thermodynamics has seen to it that energy is strictly conserved in all physical and chemical processes, that entropy always increases in the direction of spontaneous change, and that absolute zero is never attained. These rules are to be obeyed unconditionally whatever the detailed forms of the underlying atomic and molecular mechanics. In a sense, classical thermodynamics has now been superseded by statistical mechanics. With the development of the quantum theory of matter, augmented by statistical laws governing the immense numbers of molecules in a macroscopic sample of matter, it has become possible to "derive" the laws of thermodynamics. Still, the elegant simplicity of thermodynamic formulas, coupled with their independence of the complex details of molecular mechanics, guarantees that this subject will remain useful.

Historically, thermodynamics arose from very practical concerns on how to optimize the performance of steam engines. This was the focus of Carnot's research. The law of conservation of energy is usually credited to Hermann Ludwig Ferdinand von Helmholtz in 1847. In 1859, Clausius formulated the first and second laws of thermodynamics. Gibbs is responsible for much of the subsequent mathematical development of thermodynamics, featuring the free energy, chemical potential, and phase rule. Boltzmann and James Clerk Maxwell, in the latter part of the nineteenth century, developed the kinetic molecular theory of heat, which evolved into statistical mechanics. The twentieth century brought the flowering of chemical thermodynamics.

Irreversible, or nonequilibrium, thermodynamics remains a developing area of research. This field considers nonequilibrium states of systems, which are beyond the scope of classical thermodynamics. Much of this subject was pioneered by the Belgian physical chemist Ilya Prigogine. Irreversible thermodynamics has important implications for the behavior of biological systems and might contain the secret of life itself.

Bibliography

Atkins, Peter W. *The Second Law*. New York: Scientific American Library, 1984. A popular exposition on thermodynamics. Profusely illustrated, with appended BASIC computer programs.

Bent, Henry A. *The Second Law*. New York: Oxford University Press, 1965. A breezy discussion of the second law of thermodynamics with applications to chemistry. Contains a large number of problems with answers.

Blinder, S. M. *Advanced Physical Chemistry*. New York: Macmillan, 1969. A treatise emphasizing the conceptual foundations of thermodynamics and statistical mechanics.

Moore, Walter J. *Physical Chemistry*. 4th ed. Englewood Cliffs, N.J.: Prentice-Hall, 1972. A very lucidly written undergraduate text. About one-third of the text concerns thermodynamics.

Nash, Leonard K. *Elementary Chemical Thermodynamics* and *Elementary Statistical Thermodynamics*. Reading, Mass.: Addison-Wesley, 1962. Elementary expositions of thermodynamics and statistical mechanics with chemical applications. On the level of a general chemistry course.

S. M. Blinder

Cross-References

Batteries, 202; Carnot Cycles, 310; Electrochemistry, 721; Entropy, 800; The Behavior of Gases, 935; Liquefaction of Gases, 1256; Perpetual Motion, 1771; Phase Changes, 1777; Statistical Mechanics, 2367; Thermal Properties of Matter, 2472; Thermometers, 2501.

THERMODYNAMICS: AN OVERVIEW

Type of physical science: Classical physics
Field of study: Thermodynamics

Added heat can alter the temperature, pressure, entropy, and other properties of a system, thus being stored in the system, and it can cause the system to do work on the surroundings of the system. The laws of thermodynamics relate the heat to the stored energy and work done.

Principal terms

ADIABATIC PROCESS: a process undergone by a system without addition or extraction of heat

COORDINATE: any of the quantities temperature, pressure, volume, or entropy

ENTROPY: the amount of heat transferred to or from a system (during a small interval of time) divided by the temperature at which it was transferred

HEAT CAPACITY: the amount of heat an object can absorb divided by the temperature change in that absorption

INTERNAL ENERGY: energy stored by a system, indicated by a change in temperature and volume

POTENTIAL: any of the measures of stored energies measured as internal energy, enthalpy, Gibbs function, or Helmholtz function

PRESSURE: the force per unit area exerted by the system on its surroundings

SYSTEM: an amount of matter or radiation contained by a boundary

WORK: the pressure exerted on the surroundings (container) of the system multiplied by the change in volume of the system (measured in joules)

Overview

Thermodynamics deals mainly with the energy content and production of systems. The "systems" are most often enclosed quantities of matter (solids, liquids, gases, plasmas, and mixtures of these components). In these cases, the laws of thermodynamics are developed to describe changes in properties of these systems as different forms of energy flow into or out of the system. The two principal kinds of energy of most concern are "heat" added to or taken from the system, and "work" done by the system on the confining environment. Theoretical thermodynamics can accommodate more abstract systems, such as confined (massless) radiation or fields or more subtle phases of matter, and even has extraordinary power in nonphysical systems, such as economic models (where economic analogies to thermodynamic terms and laws can be constructed).

The "properties" of a system can be virtually any measurable attribute, such as degree of magnetization or concentration of a dissolved chemical, but are usually illustrated by a basic discussion of a few typical measurables. There are two groups: the "coordinates" (pressure of the system on the environment, volume of a fixed mass of the system, temperature at points in the system, and entropy of the system) and the "potentials" for the system (internal energy, enthalpy, Gibbs function, and Helmholtz function). These potentials represent various forms in which energy is "stored" or "released" by the system as the coordinates change. For example, if a gas is compressed such that no heat is added from the environment, the temperature will rise, and the internal energy is said to rise to represent the energy stored in the gas. This example represents the main application of the "first law of thermodynamics," a thermodynamic version of the law of conservation of energy. Generalizing from mechanics, it includes energy dissipated by "friction" in considerations of energy balance (a bouncing ball eventually settles down and loses its energy of motion and position to heat energy). One advantage of thermodynamics is that it accounts for energy conversions when more than one coordinate changes at a time (for example, simultaneous change in pressure and temperature). The resulting change in the other properties (density, internal energy, and the like) can be computed from the thermodynamic laws.

The properties of a system are usually described in terms of a few basic familiar properties. "Temperature" is defined as the reading on a standard thermometer, although the second law of thermodynamics allows a more abstract definition. "Pressure" is defined as the mechanical force exerted by the system on its environment (or container) divided by the area over which this force acts. This definition allows the "work" done by the system on its environment to be computed as simply the product of the pressure and the volume change. This "work" only includes the "mechanical" energy delivered by changing the system, but it is of the most interest, since it represents the "useful" output of a system, while the heat added represents the "expense" of energy input. In some arrangements, such as refrigeration, the useful energy is the heat transferred and the expense is the work done (say by compression by a motor). In either case, the energy balance of heat, work, and (occasionally) other kinds of energy transfer is the focus of concern.

Heat, the major activating energy of systems, is described practically in terms of its effects on standard measuring systems. For example, one standard "calorie" is defined as the amount of heat required to raise the temperature of a standard substance (water) by 1 Kelvin. Another substance exposed to this same standard heat will have a temperature rise characteristic of its composition. The "heat capacity" of any substance is then its heat absorbed per degree of temperature rise. Experiments (originated by James Prescott Joule) can relate the standard calorie to the unit of energy used in mechanics. By experiment, 1 calorie equals about 4.2 joules. The more abstract use of the laws of thermodynamics allows a more abstract definition of heat, and this definition is at the heart of the theoretical structure of the subject.

There are three central laws of thermodynamics. The first is the conservation of energy (often used to "define" the concept of internal energy). It asserts that the heat added to a system must reveal itself in the sum of work being done by that system and a rise in internal energy of the system. The system can respond to an input of heat in many ways under this rule, changing its coordinates to conserve energy. In an "isochoric" (constant volume) process, for example, no work is done by the system (since work done is defined through a change in volume), and all the heat goes into raising the internal energy. Energy relations in an "isobaric" (constant pressure) process are often theoretically revealing, since many chemical processes are isobaric, particularly if they occur on a tabletop where the atmospheric pressure is constant. "Adiabatic" processes, where no heat is added, are the most useful, for they describe how the internal evolution of a system might proceed. This latter process becomes more interesting from the point of view of the second law of thermodynamics, the centerpiece of most discussions of the subject.

There are several statements of the second law of thermodynamics, which are all equivalent. The simplest statement asserts that heat cannot be entirely converted to work by a system. For example, a bouncing ball can eventually come to rest, thus converting all of its mechanical energy (or work) into heat; it heats up as it settles. Nevertheless, one does not expect to heat a resting ball and have it start bouncing; the heat cannot be entirely transformed into work. An equivalent statement asserts that heat will not flow spontaneously from a cold object to a hot object; an ice cube on a hot pavement is expected to warm, not cool. Another version asserts that when systems are out of "equilibrium"—that is, when parts of a system are at different temperatures—the temperature changes to reduce the differences (toward equilibrium). In the case of the ice cube, one expects the pavement to cool a little while the ice cube heats up, not vice versa. The justification for asserting these laws can be either a statement about the observed properties of natural systems or a more abstract imposition of the second law as a condition of intelligibility of a system. In either case, the second law has fascinating and profound consequences.

A particularly convenient view of the second law of thermodynamics is provided by defining a new property of the system, a coordinate called entropy. Returning to the ice-cube example, the entropy is defined by the amount of heat entering the cube (in a short instant) divided by the temperature (at that instant). If one claims that the second law requires the heat to flow from the pavement to the cube (not in reverse), then one notices that the entropy received by the cube is greater than that given up by the pavement. This is simply by definition, for the heat given and taken was the same (by the first law of thermodynamics), but the temperatures were different. The lower-temperature ice cube got more entropy (heat divided by temperature) than the hotter pavement gave up. In general, the second law of thermodynamics asserts that the entropy of a "universe" (system plus its surroundings) must always increase as the system tends toward equilibrium with its surroundings. This is a much-discussed aspect of the second law, as it provides an "arrow of time" for physical systems—that is, a natural direction of evolution—the reverse of which is prohibited. Quantitatively, the second law allows the prediction of the particular equilibrium state toward which chemical and other interactions will tend. This is immensely practical in studying such diverse systems as chemical reactions, heat flow in low-temperature solids, nuclear reactions, and even information flow in the analog nonthermodynamic systems of "information theory."

Another abstract view of the first two laws of thermodynamics is that they prohibit certain processes in nature. Perpetual motion machines that deliver more work out than they take heat in, or that cycle work and heat without "running down," violate the laws of thermodynamics. There are extensions of the basic arguments. Some natural systems, particularly biochemical ones, do seem to "cycle" from low- to high-entropy states, and this has led to an extension of the laws of thermodynamics to "nonequilibrium" systems, where self-organizing systems are permitted to tend toward ultimate equilibrium in surprising ways, sometimes lowering local entropy, as in the case of spontaneous organization ("de-evolution") of a genetic system from a state of molecular disorder to one of symmetric genetic material.

There are other forms of system-energy considered besides internal energy, such as enthalpy, a combination of internal energy and the pressure and volume coordinates. Its use for isobaric (constant pressure) processes is similar to internal energy for adiabatic processes. The variants of internal energy called Helmholtz and Gibbs functions are similar, the latter being particularly useful in isothermal processes such as phase changes, where the system maintains its fixed temperature while using energy to change its physical structure, as when ice melts.

There is also a third law of thermodynamics, asserting that the rates of change in entropy will always diminish as the temperature of a system diminishes. This amplifies another of the consequences of the second law, which states that there must be a minimum (theoretical) temperature (0 Kelvin) for all equilibrium systems. The third law of thermodynamics is thus particularly useful in experimental low-temperature systems.

Applications

Thermodynamics is very general in structure. It describes systems of a wide range of compositions and a very wide variety of properties. Examples of a system include the classical enclosed vapors of a steam engine, the mixtures of sand, lime, and chemicals that make up cement, and a volume of atmosphere that is the model for a storm at sea. Any knowledge of pairs (or more) of attributes of a system (such as the temperature and density of the elements mixed to make cement) allows the specification of all the other attributes (pressure, entropy, enthalpy, Gibbs function, and the like). Applying the laws of thermodynamics predicts how the variables of the system will evolve (toward equilibrium). Thus, the density (composition) and the "hardness" of the cement is calculable, or the speed and moisture content of the storm, or the "efficiency" of the steam engine.

Historically, the applications of thermodynamics are most visible in the study of engines, and particularly in the comparisons between the efficiency of steam engines and other types of such devices using chemical or electrical systems. The competition between steam, diesel, and electrical engines for railroads was an example of a thermodynamics-driven technology. Periodically, new kinds of engines are proposed for transportation or energy production, and their proposed merits must be analyzed. New combinations of processes for internal combustion (such as in the Wankel engine of the mid-twentieth century) or new working systems (such as the use of fuels such as methane, ethane, or hydrogen) are periodically proposed and require thermodynamic description.

Studies of weather patterns provide excellent opportunities for the thermodynamicist. A mass of wet air moving up a mountainside seems like an obvious "system," its pressure and temperature and the density of its moisture being some of its thermodynamic coordinates. The work done against gravity and the heat added by sunlight and the surrounding atmosphere raise its internal energy (and its other "potentials," such as Gibbs function), and the heat exhausted in the form of rain or snow represent calculable quantities in a model of the mass of air. In practice, the atmosphere is so complicated that the most advanced computers are required to keep track of all the details of a weather front, but the principles governing the computer's computations of detail are the simple principles of thermodynamics.

Low-temperature science and technology, from the early liquefaction of gases to the modern exploitation of the electrical peculiarities of very-low-temperature systems, have required extensive use of the principles of thermodynamics. In particular, the so-called superfluidity of helium at ultra-low temperatures (below 3 Kelvins) has provided both a theoretical challenge and a technical opportunity to investigate allied properties of matter—"superconductivity" (electrical conduction) and heat conduction at low temperatures. Hydrodynamics (fluid flow), particularly highly frictional turbulent flow, can be organized by thermal principles, though much is still unknown about the transition from smooth to erratic (even "chaotic") patterns of flow. Analogies to eddies and whirlpools in fluid flow can be mathematically constructed in other pictures, such as field-theoretical models of the flow of light or particles in the subatomic regime, and so current algebraic models of radiation fields carry the models of thermodynamics into fundamental particle theory.

Practical thermodynamic engineering is more directly illustrative of the basic principles of thermodynamics. One can easily see the problems of the power output of electric or steam-generating plants as directly predictable from principle. Nonsteam engines and refrigerators are equally obvious candidates for analysis, as are ideas to generate work and energy in novel ways, such as solar energy, wind, and tidal power.

The characterization of forms or states of matter itself requires considerable study. The old classifications of matter—solid, liquid, and gas—have been augmented by another major class (plasma) and many subclasses. A plasma is an electrically charged gas, and has considerable utility in anything from lighting to energy generation. The subclasses of matter (phases) caused by different molecular crystalline patterns suggest ever more new kinds of matter, such as spin-glass and superfluid liquids.

Chemistry and biochemistry are perhaps the main arena of classical thermodynamics, for there the evolution and combination of diverse chemical systems are particularly sensitive to the energy balances of their successive states. Understanding the transfer of chemical energy across membrane boundaries is part of the explanation of muscle action, itself a user of chemical energy derived from nutrition and respiration (two fields of study prominent in the prehistory of thermodynamics). Rhythmic mechanisms in controlling the daily and monthly cycles of chemical changes in organisms yield secrets to application of nonequilibrium thermodynamic models.

Astrophysics and cosmology are perhaps the most dramatic arena of thermodynamic theory, for the evolution of stars and galaxies, the origins of the cosmic expansion, the explosions of stars, and the collapse of systems into black holes are dramatic cosmic examples of systems exchanging energy and matter with their surroundings—the central thematic topic of the theory. Like its companion subjects, mechanics and electrodynamics, thermodynamics is a central pillar of the kind of thinking known as physics.

Context

Modern thermodynamics arose in the middle of the nineteenth century, particularly from the work of Rudolph Clausius and Hermann Von Helmholtz. Before that time, there were many scattered investigations into phenomena of heat and temperature, but no clear theoretical central framework. In the eighteenth century, Sir Isaac Newton focused attention on thermodynamic problems (in addition to his better-known concerns), and experimental developments soon followed in measuring temperature and heat involved in chemical and digestive processes.

In the early nineteenth century, Nicolas-Léonard-Sadi Carnot formulated an early version of the heat-energy relations in his investigations into "the motive power of fire," or the nature of the "power" of steam engines. Carnot still thought of heat as a material, fluidlike substance (caloric) and did not have the advantage of the later notions of mechanical energy (developed by Helmholtz). Nevertheless, his consideration of the power and action of a steam engine (molded after his father Lazare-Nicolas-Marguerite Carnot's ideas of making more efficient waterwheels by minimizing the splash) contained the germs of the first and second laws of thermodynamics.

Around the middle of the nineteenth century, Clausius combined the new notion of heat as a "kind of motion" with Carnot's "heat" and Helmholtz's "energy." He produced the first modern statement of the first and second laws of thermodynamics. Physicists rapidly adapted the newest advances in mathematics to generalize and refine the theory. Engineers and mechanics began to investigate thermal devices and phenomena. For example, Rudolf Diesel, one of Clausius' students, developed the engine that now bears his name. Also around the middle of the nineteenth century, Ludwig Boltzmann, James Clerk Maxwell, and Clausius developed a related discourse on heat (known as the kinetic theory), comparing notions of temperature, energy, and heat with the mechanical features of mixtures of atoms and molecules in motion. This later kinetic theory eventually grew into the science of statistical mechanics (particularly in the hands of Gibbs) by the early twentieth century.

Until the success of the theory of atomic-molecular motions of Albert Einstein and Jean-Baptiste Perrin with the atomic-molecular motions known as heat, most early-century thermal inquiries were conducted with the classical thermodynamics of Clausius and Helmholtz. These investigations involved such diverse topics as the source of volcanoes, the cooling of the earth (too fast to allow geologists their evolutionary epochs), heating of the upper atmosphere, and chemical reactions requiring or delivering heat. The latter study led to a new field of physical chemistry sponsored by Wilhelm Ostwald, himself a skeptic about the new idea of atoms. Such investigations spawned interest in liquefying gases at low temperatures (Lord Rayleigh discovered argon in this manner) and the achievement of ever lower temperatures. Walther Hermann Nernst was led to formulate the third law of thermodynamics during this quest early in the twentieth century.

The twentieth century theory became ever more mathematically sophisticated and general, while the practical applications multiplied still further. Fluid flow at low temperature and solid resistance to high pressure won Nobel Prizes for their investigators. Numerous chemists and biochemists adapted the laws for their use. By the late twentieth century, new areas of nonequilibrium thermodynamics were pioneered by Ilya Prigogine.

Engineering thermodynamics also kept pace with rapid industrialization and the new age of science. New internal combustion and electric engines and generators, and Linde and Carrier's systems of air-conditioning, were precursors to twentieth century expansion of low- and high-temperature technology, energy production, polymer and biochemistry, weather studies, and space-engineering.

In a sense, thermodynamics is incomplete, for its laws can be applied to ever subtler and more abstract systems. One can extend, for example, the number of independent coordinates (such as temperature and pressure) from two to three, or even to infinity, requiring perhaps new properties and potentials to be defined. It is clear that thermodynamics has even more development to come.

Bibliography

Angrist, Stanley, and Loren Hepler. *Order and Chaos*. New York: Basic Books, 1967. This excellent work, intended for young people, surveys the concepts of heat, energy, entropy, and the laws of thermodynamics with clear (though occasionally dated) examples from everyday life.

Atkins, P. W. *The Second Law*. New York: W. H. Freeman, 1984. From the publishers of *Scientific American*, this well-illustrated book focuses on the kinetic-theoretical and statistical-mechanical view of entropy, including excellent sections of thermodynamic background (particularly chapter 5) as well. No mathematics background is required.

Fermi, Enrico. *Thermodynamics*. New York: Dover, 1936. Fermi, a Nobel laureate in physics, gave introductory lectures in college-level thermodynamics in the 1930's, and these have become a classic model for a clear, intuitive survey of the subject. The Dover reprint is beloved by teachers of the subject, and the book is highly recommended to readers with a command of elementary calculus.

Harman, P. M. *Energy, Force, and Matter: The Conceptual Problems of Nineteenth Century Physics*. Cambridge, England: Cambridge University Press, 1982. A distinguished historian of ideas, Harman is especially adept at describing nineteenth century German scientific thought, here especially appropriate to the thermodynamics of Clausius and Helmholtz. This may be somewhat difficult reading for the general audience, but well worth the effort.

Mendelssohn, Kurt. *The Quest for Absolute Zero*. London: Taylor and Francis, 1979. Addressed to the general reader, this volume introduces most of the important thermodynamic concepts (such as heat capacity) in the context of the rise of low-temperature physics. Very readable, and more comprehensive on thermodynamics than his biographical work (listed below).

____________. *The World of Walther Nernst*. Pittsburgh: Pittsburgh University Press, 1973. This is a historical survey of the life and times of the author of the third law of thermodynamics and describes both his work and the Germany of his time (from the late nineteenth century to World War II). Mendelssohn is a distinguished successor to Nernst in low-temperature physics.

Van Ness, H. C. *Understanding Thermodynamics*. New York: Dover, 1985. College instructors have adopted this short survey as a classic commentary on the more detailed texts used in introductory college-level thermodynamics courses. Goes especially well with the Dover reprint of Fermi's classic (listed above).

Zemansky, Mark, and Richard Dittman. *Heat and Thermodynamics*. New York: McGraw-Hill, 1979. A survey of American college courses in thermodynamics shows this text as one of the most-often adopted, and it is recommended by many authors as a standard text. Somewhat difficult for the introductory college audience (with calculus), but some topics (such as cyclic engines) are accessible to those with a precollege mathematics capability and are worth looking up in this text.

Peter D. Skiff

Cross-References

The Absolute Zero Temperature, 1; Carnot Cycles, 310; Chemical Reactions and Collisions, 407; Entropy, 800; The Behavior of Gases, 935; Liquefaction of Gases, 1256; Perpetual Motion, 1771; Phase Changes, 1777; Superfluids, 2425; Thermal Properties of Matter, 2472; Laws of Thermodynamics, 2486; Thermometers, 2501; The Physics of Weather, 2735.

THERMOMETERS

Type of physical science: Classical physics
Field of study: Thermodynamics

A thermometer is a device used to measure temperature and thus assign a numerical value to the degree of hotness or coldness of matter. Temperature is an essential parameter used to understand a wide variety of scientific phenomena and to characterize chemical and physical properties of matter.

Principal terms

ABSOLUTE ZERO: the lowest possible temperature (assigned a value of 0 on the Kelvin scale) and the temperature at which atomic motion and randomness of atomic arrangement are at a minimum

HEAT: the energy resulting from random motion of molecules, atoms, or subatomic particles

INTERNATIONAL PRACTICAL TEMPERATURE SCALE: the standard temperature scale based on a platinum resistance thermometer or optical pyrometer and one to which other thermometers are referenced

PYROMETER: a device used to determine temperatures above 900 Kelvins by measuring the intensity of light emitted by a hot object at a known wavelength

RESISTANCE THERMOMETER: a device used to measure temperature by measuring the change in resistance to electrical flow with temperature

TEMPERATURE: the average energy resulting from random motion of molecules, atoms, or subatomic particles

THERMOCOUPLE: a device used to measure temperature by monitoring the potential between two different metals

TRIPLE POINT: the unique temperature where a substance's solid, liquid, and vapor phases all exist together in equilibrium

ZEROTH LAW OF THERMODYNAMICS: the observation that two bodies, each in thermal equilibrium with a third body, are in thermal equilibrium with each other

Overview

Temperature is a quantitative measurement of the degree of hotness or coldness of a collection of matter. Thermometers are used to determine temperature by measuring some physical property that varies with temperature. One important property that varies with temperature is the direction of heat flow. The natural direction of heat flow between two collections of matter or bodies is from the warmer to the cooler. When two bodies are in contact, heat will flow from the hotter to the cooler until the two bodies are in thermal equilibrium. At this time, no additional heat will flow, and the two bodies are said to have an identical temperature.

The observation that two bodies, each in thermal equilibrium with a third body, are in thermal equilibrium with each other is sometimes referred to as the "zeroth law of thermodynamics." This universal observation is not dependent on the composition of the body or the state of matter—that is, whether it is a solid, liquid, or gas. The zeroth law observation is important because it allows scientists to define a universal temperature scale. One can imagine the third body of matter to be a device called a thermometer. The thermometer can be used not only to establish that two different collections of matter are at the same temperature but also to establish numerical values of temperature that are independent of the state of matter or its composition. In this sense, temperature is a universal property.

The problem involved in measuring temperature is to establish a specific temperature scale and to have a reproducible method of determining the numerical values of temperature. Yet, to understand what a thermometer represents when a particular temperature is reported, it is necessary to understand something about the connection between atomic motion and temperature.

The temperature of a collection of matter is dependent on the collective atomic motion. An increase in atomic motion is associated with an increase in temperature, and a slowing of atomic motion is associated with a decrease in temperature. This atomic motion may be from vibration of atoms or from the rotation or translation (moving from place to place) of molecules. The total energy caused by the constant random motion of a collection of molecules is the heat, whereas the temperature is the average energy of molecular motion. Thus, a large collection of matter may have a greater heat content than a smaller collection of matter, even though both have the same temperature. In solids, the vibrational motion of atoms stores heat energy, and the vibrational energy can be increased or decreased as the temperature is increased or decreased. The average speed of motion of the molecules in a gas is proportional to the temperature of the collection of molecules. The average speed of the molecules increases as the temperature is increased, and the average speed of molecules decreases as the temperature is decreased.

With this connection between atomic motion and temperature made clear, it becomes obvious that there is a lowest possible temperature. At absolute zero, atomic motion is at the lowest possible value and the randomness or disorder in an ordered solid is 0. Entropy is a measure of this disorder, and the entropy of a perfect crystalline solid is 0 at absolute zero. The thermodynamic temperature scale is based on a heat flow ratio proportional to temperature ratio and the realization that there is a lowest possible temperature. In the thermodynamic temperature scale, there are no negative values. The low end of the scale is fixed at absolute zero, where the temperature is exactly 0 Kelvin.

The Kelvin scale is one temperature scale, but there are other scales that are used to quantify the hotness or coldness of matter. The three most important temperature scales are Fahrenheit, Celsius, and Kelvin. The Celsius and Kelvin scales are the most widely used in science. A unit on the Fahrenheit and Celsius scales is referred to as a degree; a unit on the Kelvin scale is referred to as a Kelvin.

It is most convenient to compare these temperature units by comparing the values each scale assigns to the melting point and boiling point of water under a pressure of exactly 1 atmosphere. On the Fahrenheit scale, water freezes at 32 degrees and boils at 212 degrees. On the Celsius scale, water freezes at 0 degrees and boils at 100 degrees. On the Kelvin scale, the freezing point of water is 273.15 Kelvins and the boiling point of water is 373.15 Kelvins. To convert Celsius degrees to Kelvins, one simply adds 273.15. For example, 20.00 degrees Celsius is equivalent to 293.15 Kelvins.

The thermodynamic temperature scale is based on defining the triple point of water (where liquid, solid, and gas all coexist) at exactly 273.16 Kelvins, or 0.01 degree Celsius. The Kelvin scale is the most useful temperature scale in science because there are no negative temperatures. Absolute zero has a value of exactly 0 on the Kelvin scale. On the Celsius and Fahrenheit scales, absolute zero has a value of −273.15 and −459.67, respectively.

One method to determine absolute temperature values is through the use of a gas thermometer, which is based on how temperature affects the pressure of a gas. The gas is considered to be an ideal gas, meaning that there are no interactions between gas molecules. Since all gases have some attractions between molecules, it is necessary to correct for nonideality.

Helium gas behaves most like an ideal gas and so it is the best choice for a gas thermometer. For a fixed volume and a fixed amount of helium, the pressure is measured with a mercury manometer at various temperatures. A manometer is a device used to measure pressure by measuring the height of a column of mercury supported by that pressure. Pressure can be used to measure the temperature because for an ideal gas the height of the column of mercury is proportional to the pressure, and according to the ideal gas law, the pressure is proportional to the temperature. As the temperature is changed, one must correct for gas imperfections and the expansion and the contraction of the gas container.

Because the use of a gas thermometer to obtain accurate temperature values is an extremely tedious process, it is normally used only to determine a few fixed point temperatures. Fixed point temperatures do not vary, because they are based on equilibrium between two or three different phases or states of matter. A triple point is based on an equilibrium between solid, liquid, and vapor. A melting point is based on an equilibrium between solid and liquid, and a boiling point is based on an equilibrium between a liquid and its vapor.

The International Practical Temperature Scale is selected to correspond to the thermodynamic temperature scale by using a few fixed points as reference values to define the scale. Useful fixed points under a pressure of exactly 1 atmosphere include the triple points of hydrogen (13.81 Kelvins), oxygen (54.361 Kelvins), and water (273.16 Kelvins); the boiling points of hydrogen (20.28 Kelvins), neon (27.102 Kelvins), oxygen (90.188 Kelvins), and water (373.15 Kelvins); and the melting points of zinc (692.73 Kelvins), silver (1,233.23 Kelvins), and gold (1,337.58 Kelvins). A platinum resistance thermometer is used from −259.34 degrees Celsius (13.81 Kelvins) to 630.74 degrees Celsius (903.89 Kelvins), the melting point of the metal antimony, to measure values on the International Practical Temperature Scale. The temperature scales of most other thermometers are fixed by comparing them to a standard platinum resistance thermometer.

A variety of thermometers have been developed, including mercury, Bourdon tube, bimetallic, thermocouple, resistance, and thermistor. Liquid expansion thermometers are the most common type of thermometer and normally use mercury as the working liquid. (Alcohol can be used in regions where the temperature goes below the freezing point of mercury, 234.3 Kelvins, or −38.85 degrees Celsius.) A bulb containing mercury is connected to a tube partially filled with mercury. As the temperature increases, the liquid mercury expands to fill a greater volume and rises in the tube. As the temperature decreases, the liquid mercury contracts to fill a smaller volume and goes down in the tube. Conveniently, the expansion and contraction of mercury with increasing and decreasing temperature are fairly linear over a broad range of temperature, so that the height of the mercury column in the tube is directly related to the temperature of the environment in which the thermometer is placed. A scale marked on the glass tube is used to determine the numerical value of the temperature.

Another type of liquid expansion thermometer is a Bourdon tube thermometer. A Bourdon tube is a curved tube made of a flexible metal that is filled with a liquid. As the temperature is increased, the liquid expands and the curved tube straightens out slightly. The temperature is indicated by a pointer or recorded by a marker attached to the end of the tube.

A bimetallic thermometer is based on the expansion of solids. Two metals that have different coefficients of expansion, such as brass and iron, are connected together to make a thin, spiral bar. Temperature changes cause the spiral bar to bend into a tighter spiral or relax into a looser spiral. A pointer attached to the end of the bar indicates the temperature along a scale marked above the pointer.

The two main types of electrical thermometers depend on monitoring either changes of voltage potential or changes in electrical resistance. A thermocouple is used to measure temperature by monitoring potential. In a thermocouple, two different wires, such as copper and constantan (a copper and nickel alloy), are connected together to make two junctions. One junction is kept at a known temperature, such as in melting ice, which is exactly 0 degrees Celsius. This junction is known as the reference junction, and the temperature of the other junction is measured relative to it. When the second junction is at a temperature different from that of the reference junction, then there is a voltage potential across the two thermocouples. As the temperature difference increases, the potential difference between the two thermocouple junctions also increases. This potential is measured with a voltmeter capable of measuring millivolts or better. Either standard tables are used to convert millivolt potential difference to a temperature difference or a conversion to a temperature scale is done by an integrated circuit and the temperature is read directly on a digital scale or on a computer monitor.

Resistance thermometers, the other type of electrical thermometer, measure electrical resistance. Thermistors (a contraction from "thermally sensitive resistors") are metal oxides that have an extremely large decrease in resistance to electrical current flow as the temperature is increased. Thermistors such as nickel oxide and manganese oxide are about ten times as sensitive as metal resistors to temperature change. Numerous complications in measuring accurate resistances are eliminated by switching from metallic to metallic oxide resistors. Thermistors are convenient to use, but the resistance characteristics tend to change with use and therefore need to be recalibrated periodically.

Applications

The accurate measurement of temperature is of vital interest in a wide array of scientific and engineering endeavors. Temperature is important in understanding weather patterns and the extent to which the planet is being subjected to global warming as a result of the greenhouse effect. Accurate temperature measurements are a routine part of determining whether a patient has a fever that could indicate infection or illness. Clinical thermometers are used to measure the internal temperature of the human body. Normal body temperature is 98.6 degrees Fahrenheit, or 37 degrees Celsius.

Temperature is important in determining the utility of new superconducting materials, in monitoring energy production with calorimetric experiments such as the early cold fusion work, in determining properties such as boiling point and melting point of new molecules and compounds, in following reaction conditions in chemical and industrial manufacturing, and in maintaining comfortable but energy-efficient living conditions in buildings and vehicles.

Since many different types of temperature measurements are routinely used, many different methods of measuring temperature have been developed. All these different types of thermometers accomplish their goals, not by measuring temperature directly but rather by measuring a physical property that changes in a known way with temperature. Temperature affects a wide variety of physical, electrical, magnetic, and optical properties.

Mercury thermometers are the most common type of thermometer. Most household thermometers are mercury thermometers. These devices are used to measure indoor and outdoor temperature. Knowledge of outdoor temperature is important in determining how to dress properly for either cold or warm days. On very cold days, there is the danger of hypothermia if a person is exposed to cold temperatures outside for an extended period and his or her internal body temperature drops. On hot days, it is important to drink fluids so as not to become dehydrated. Measuring the temperature accurately helps one to be prepared for whatever weather is encountered.

Measuring indoor temperature is important in maintaining comfortable surroundings while being as efficient in energy usage as possible. In a bimetallic thermometer, temperature changes cause the spiral bar made of two kinds of metal to bend into a tighter spiral or relax into a looser spiral. These bimetallic thermometers are used commonly in thermostats to control indoor temperatures. The thermostat is set on the desired temperature; as the bimetallic bar in the thermostat changes position, it causes an electrical contact to be made or broken. In this manner, heating or cooling devices are turned on or off, and a constant indoor temperature is maintained.

Thermocouples and resistance thermometers are useful for remote monitoring, since the temperature probe holding the thermocouple junction or resistor can be far from the site where the potential or resistance is measured. In many industrial plants, it is important to be able to monitor the temperature continuously at remote locations throughout the facility. In chemical engineering complexes where large-scale chemical reactions are carried out, it is necessary to be able to follow the temperature of the reaction to make sure that the reaction is not going too rapidly and that not too much heat is being produced. In nuclear power plants, it is also important to follow the temperature closely to ensure that the reactor is operating at the proper conditions.

Cooking is another activity in which thermocouple and resistance thermometers prove to be essential. In any type of cooking, it is important to maintain careful control over the temperature of the food. Electrical monitoring and control of oven temperature allows one to ensure that food is cooked to a proper temperature and for the correct length of time.

Optical pyrometers are used to determine temperatures above 900 Kelvins, where the wavelength or color of a hot object changes in a known way with temperature. The electromagnetic radiation or light of a predictable wavelength given off by heated objects is known as black body radiation. Optical pyrometers operate by using a photometer to measure the intensity of light given off at a particular wavelength. Optical pyrometers are useful in iron and steel manufacture, where molten metals must be maintained and poured from one container to another. It is essential to maintain the hot liquid metal at the correct temperature, and hand-held pyrometers allow the operator to check the temperature of the molten metal without touching it directly.

There are various other specialized devices for measuring temperature, including, but not limited to, quartz thermometers, in which the resonance frequency of a vibrating quartz crystal is changed with temperature; integrated-circuit transducers, which provide a linear output with temperature change; and vapor pressure thermometers, in which the vapor pressure of a liquid or solid that is dependent on temperature is measured.

Context

A thermoscope is a device that can measure changes in temperature without accurately measuring actual temperatures. The first thermoscope was made by Galileo in

1603 by observing changes in the height of water in a glass tube inserted into a container of water. Galileo's thermoscope responded to temperature changes but was also affected by changes in air pressure. About fifty years later, Ferdinand II, the Grand Duke of Tuscany, and Robert Boyle independently realized that effects resulting from air pressure could be eliminated by sealing the water into a straight glass tube. With air pressure effects removed, the first true thermometers were made.

Early thermometers used water or alcohol as the working liquid. The first modern mercury thermometer was made in 1714 by Gabriel Daniel Fahrenheit, who used the expansion of mercury in a sealed glass tube as a measure of temperature. His zero point was obtained by mixing equal weights of snow and sodium chloride salt. On the Fahrenheit scale, water freezes at 32 degrees Fahrenheit and boils exactly 180 degrees higher, at 212 degrees Fahrenheit. In 1742, Anders Celsius developed a scale with one hundred divisions between the freezing and boiling points of water. It was Sir William Thomson (Lord Kelvin) who first suggested a scale using degrees with the same spacing as Celsius but starting at absolute zero.

In 1858, a German physician, Karl August Wunderlich, introduced the practice of determining body temperature as an indicator of illness with a mercury thermometer. The clinical thermometer was invented about ten years later by Sir Thomas Clifford Allbutt. In a clinical thermometer, the liquid mercury rises in the glass tube at elevated temperatures but does not return to lower values, because of a constriction in the glass tube that causes the mercury in the bottom bulb to separate from the mercury in the tube as the temperature is lowered. The bulb and tube mercury is rejoined by vigorously shaking the tube downward.

While liquid expansion thermometers continue in widespread use, modern measurements of temperature can also utilize bimetallic, Bourdon tube, thermocouple, resistance, and pyrometric thermometers, depending on the temperature range and application of interest. In addition to providing information about the daily weather, the measurement of temperature remains important in medicine, engineering, and science. As new frontiers of low and high temperature and properties of matter are explored, the careful and exact measurement of temperature will continue to play a central role in scientific observations.

Bibliography

Asimov, Isaac. *Asimov's New Guide to Science*. New York: Basic Books, 1984. Although this book presents an overview of all the biological and physical sciences, it has great detail for historical development and contributions of individual scientists. It has good sections on history of low-temperature developments and history of developments in measuring temperature.

Atkins, P. W. *The Second Law*. New York: W. H. Freeman, 1984. This book presents a lively discussion of the second law of thermodynamics and the intellectual origins and implications of the tendency toward randomness and disorder. Includes discussion of connections between temperature, heat, and absolute zero. Excellent color drawings aid significantly in giving a clear understanding of these phenomena.

Fenn, John B. *Engines, Entropy, and Energy*. New York: W. H. Freeman, 1982. A cartoon caveman appears throughout chapters to help explain with wit and humor energy, heat, and entropy, and how these ideas affect real processes occurring in machines or in nature. Although equations are used throughout, the discussions make these concepts accessible to those without science and mathematics background.

Spielberg, Nathan, and Bryon D. Anderson. *Seven Ideas That Shook the Universe*. New York: John Wiley & Sons, 1987. This is an examination of the key ideas in physics and includes, in a section on energy, discussions on how heat, motion, and temperature are related. The section on entropy discusses the second and third laws of thermodynamics, temperature, and absolute zero.

Weber, Robert L. *Heat and Temperature Measurement*. Englewood Cliffs, N.J.: Prentice-Hall, 1950. This book includes tremendous detail on all manner of methods used to measure temperature and heat flow. Although parts are more technical, it is worth the read to learn something of how much is involved in careful, classic measurements of temperature.

Thomas R. Rybolt

Cross-References

The Absolute Zero Temperature, 1; Calorimetry, 290; Carnot Cycles, 310; Entropy, 800; The Behavior of Gases, 935; Liquefaction of Gases, 1256; Phase Changes, 1777; Thermal Properties of Solids, 2276; Thermodynamics: An Overview, 2493; Thermal Properties of Matter, 2472; Laws of Thermodynamics, 2486.

THERMONUCLEAR REACTIONS IN STARS

Type of physical science: Nuclear physics
Field of study: Thermonuclear reactions

The study of thermonuclear reactions is only decades old, but it enables scientists to understand how the sun has continued to shine for the first 5 billion years of its lifetime. This information can be used to explain the birth, life, and death of various types of stars.

Principal terms

ENERGY: the ability to do work

NEUTRINO: an elementary particle with zero mass

NEUTRON STAR: a star that has undergone gravitational collapse

POSITRON: the antiparticle of an electron

STRONG NUCLEAR FORCE: the strongest of the four known fundamental forces in the universe; it acts over a very short range

THERMONUCLEAR REACTION: the fusing of nuclei into other nuclei of higher atomic number, such as two deuterons colliding to form a helium-4 nucleus

WORK: the product of a force on an object and the component of distance traveled that is parallel to that force

Overview

The nature of the sun's energy source has baffled scientists for centuries. In 1905, however, while Albert Einstein was making calculations for his special theory of relativity, an equation unexpectedly emerged: $E = mc^2$, where E is energy, m is mass, and c is the speed of light. This equation indicated that matter was merely another form of energy, albeit a very compact form. Another implication was that a small amount of mass could be transformed into a large amount of energy. The c^2 term is a conversion factor for converting mass units into energy units. The equation also states that a 1-kilogram mass, if converted into energy, would yield 90 million billion joules. A joule, the basic unit of energy, is the energy needed to lift a large can of peas from waist height to a high shelf. The idea that mass could be converted into energy became the basis for thermonuclear weapons and for fission reactors that convert a small amount of a uranium atom's mass into electrical power. It also determined the source of the sun's energy.

Scientists have calculated that the sun's output is 383 trillion trillion joules per second. One second of the sun's energy output is equivalent to 10 million times the annual electricity consumed in the United States. This permits the calculation of how much matter the sun converts into energy every second and leads to insights about the birth, life, and death of the stars.

Astronomers believe that the solar system, with its sun, planets, satellites, and other associated objects, developed from a cloud of gas and dust several light-years in diameter. The slowly spinning cloud started to contract under its gravitational attraction. As it shrank, the cloud rotated faster and became disk-shaped. At the center was a glowing ball of hot gas called a protosun. It was not a star, since it was not yet converting mass into energy. Its high temperature was caused by the compression of the gas and the conversion of the potential energy of the gas into thermal energy. As the contraction continued, the temperature, density, and pressure at the interior of the protosun rose. When these were high enough, nuclear fusion began, and the protosun became a star.

Since one cannot physically measure conditions in the center of the star, astronomers model the sun's interior by assuming that it is composed of many concentric shells of matter. Starting with known conditions at the surface and the sun's total mass, diameter, and energy output, computers step through the various layers from the surface into the interior. The model's mass and energy output must equal those of the sun. These calculations suggest that the core temperature is 16 million Kelvins, at 200 billion atmospheres of pressure, with a density of 160 grams per cubic centimeter. It is under these conditions of high temperature, high pressure, and high particle density that the fusion of four hydrogen nuclei (protons) into helium 4 occurs. It is this series of thermonuclear reactions, known as the proton-proton cycle, in the 350,000-kilometer-diameter core that fuels the sun.

The reactions start with the collision of two high-speed protons. Normally, as two protons approach each other, their positive electrical charge produces a mutual repulsion, and they slow down, stop, and then move away in an elastic collision. If they are moving fast enough, however, they move sufficiently close for the short-range nuclear force to cause an attraction between the protons. At this point, one of the protons is converted into a neutron, and a deuterium nucleus, an isotope of hydrogen with one proton and one neutron, results. Since electrical charge must be conserved during a nuclear reaction, the change of a proton into a neutron results in the formation of a positron, a positively charged electron, which is the antimatter form of the electron. Electrons, positive or negative, cannot exist in the confined space of the nucleus, and therefore the positron is emitted from the nucleus. When a positive electron and a negative electron collide, they are annihilated, and their mass is completely converted into 0.16 billionth of a joule, as prescribed by Einstein's mass-energy equation. The energy is released as gamma rays, a high-frequency, high-energy form of electromagnetic radiation. A neutrino, a neutral particle with a negligible or zero mass, is also released.

Next, a fast-moving proton collides with a deuterium nucleus, yielding a helium-3 nucleus and a gamma ray. The helium-3 nucleus has two protons and one neutron. Finally, two helium-3 nuclei collide to produce one helium-4 nucleus (with two protons and two neutrons) and two protons. The net result of the interactions is that four protons fuse to yield one helium-4 nucleus, two positrons that are annihilated to form gamma rays, two neutrinos, gamma rays, and energy in various forms.

The mass of four protons is 6.6942 billionth, billionth, billionth of a kilogram, while the mass of a helium-4 nucleus is 6.6466 billionth, billionth, billionth of a kilogram. That mass difference of 0.0476 billionth, billionth, billionth of a kilogram is converted into energy in agreement with Einstein's mass-energy equation. The amount of hydrogen converted per second into other forms of energy is equivalent to 4 billion kilograms, or the mass equivalent of 1 million automobiles. One hundred million, trillion, trillion nuclear reactions in one second transform 610 billion kilograms of hydrogen into 606 billion kilograms of helium.

The gamma rays are absorbed and reradiated many times during their journey to the surface. This energy requires from 100,000 to 1 million years to make the 700,000-kilometer trip from the core of the sun to its surface, where most of the energy departs as ultraviolet, visible, and infrared electromagnetic radiation.

The released energy also prevents the sun from collapsing. The gravitational attraction of the sun's mass produces an inward pull, but the thermal energy of the sun heats the interior gases. This produces an outward pressure to counter the inward pull of gravity. This results in a sun that has been stable for 5 billion years. With the remaining hydrogen in the core, the sun should last for another 5 billion years. At the end of that time, the core's hydrogen supply will be depleted but not exhausted. The increased amount of helium "ash" will start to interfere seriously with the thermonuclear reactions. The energy conversion process will stop, and the sun will start to collapse. That will be the beginning of its death.

This collapse will raise the internal temperature to the point at which the hydrogen in a shell around the core fuses to helium 4. Later, as the core temperature increases to more than 100 million Kelvins, two of the core's helium-4 nuclei will fuse into a beryllium-8 nucleus. Another helium-4 nucleus will fuse with the beryllium to produce carbon 12. In each case, a gamma ray is emitted. The sun's outer layers will expand because of this increased energy conversion, and the sun will become a red giant that will engulf the inner planets of the solar system. The core reactions will cease as the helium 4 is depleted, but the shell fusion will continue. This will lead to the expulsion of the sun's outer layers to form a planetary nebula.

The sun will start to shrink as the shell's thermonuclear reactions cease and the thermal induced pressure diminishes. The sun will now begin its white dwarf stage. The size of the earth, the white dwarf will radiate only by its residual heat, since no thermonuclear reactions will now occur within it. It will slowly cool to the final, black dwarf phase of a star with the initial mass of the sun.

The collapse of a nebula also may result in a star with a mass of more than 1.4 solar masses. The thermonuclear reactions, known as the CNO cycle, will start as the star's core temperature, pressure, and particle density increase to levels higher than those of a 1-solar-mass star. These conditions are caused by the larger mass of the star producing a greater gravitational pull inward. Because of these higher levels, the reactions will proceed at a higher rate. The star now has a larger mass but consumes it more quickly. This results in a shorter lifetime for the star. As the core depletes its hydrogen, it collapses, and helium fusion commences. The hydrogen to helium fusion is initiated in the hydrogen shell around the core.

When the core helium is expended, the core collapses again, with a resulting temperature and pressure increase. This cycle of fusion, depletion, and collapse continues until the nuclei in the core are fused into iron nuclei and the core is surrounded by several concentric shells of fusing nuclei. The outer layer consists of hydrogen fusing to helium. Iron is the core's end product, since further fusion requires the input of additional energy rather than its release. With no source of thermal energy to prevent the collapse of the star, it implodes, the material bounces off the core, and the star explodes into a supernova. Energy is so abundant during the explosion that the lower atomic number nuclei are fused into elements with atomic numbers higher than iron, such as uranium and thorium. The remnant of the star collapses to a mass of 10 kilometers in diameter—a neutron star. Because of the high pressures, the electrons and protons of the star are fused into neutrons.

When the collapse of a nebula produces a star with more than 3 solar masses, an even more dramatic death occurs. The very high temperatures and pressures in the core lead to a very high rate of thermonuclear reaction that depletes the core hydrogen very rapidly. This results in a lifetime that is measured in millions of years. As the core fuels are depleted, the star's high mass and gravitational attraction will continue the collapse beyond the neutron-star stage into a black hole. Nothing can stop the gravitational attraction that collapses the star to a size from which not even light can escape.

Applications

The study of thermonuclear reactions concerns humanity's most urgent need—energy. The standard of living is governed by the availability of easily obtained energy, and humanity's present control of energy permits people to perform feats that previously were thought impossible. Human and animal muscle power, wind and water, coal, natural gas, petroleum, nuclear fission: Each of these sources of energy has been drained. Each has proved to be inadequate in one way or another, either because it is insufficient in amount or intensity, or because it produces harmful waste. Humanity needs a reliable source of energy, and thermonuclear reactions, either directly or indirectly, will be that source.

Solar energy is the product of a natural thermonuclear reactor 150 million kilometers from Earth, and humans use solar energy in many forms without knowing it. In fact, except for tidal and geothermal energy, all energy sources are solar. Coal, oil, and natural gas, for example, are forms of fossilized solar energy. The direct application of solar energy is difficult, since solar energy is a diffuse energy resource. Large arrays of solar panels are needed to supply the required energy, and they operate only when the sun shines. One method of avoiding the Earth-bound problems of solar power is to construct, in geostationary orbit around the earth, large solar-cell panels called solar power satellites. One station, 5 kilometers by 15 kilometers, would supply the needs of New York City, with energy to spare. Its solar cells would convert light into electricity, which would be converted into microwaves. These would be beamed to Earth, where a receiving antenna would convert them into electrical power. One of these satellites would provide six to ten times the energy that an array of the same size on Earth would provide. Several hundred of these satellites would supply a large portion of humanity's energy needs.

Another possibility is to design machines that would permit the control of fusion reactions on Earth. For decades, engineers and scientists have struggled with the problem of obtaining sufficiently high particle densities, temperatures, and pressures. The main problem is the design of a vessel that will hold the deuterium long enough for the fusion to occur. Magnetic fields and lasers have been employed to initiate the fusion process, but the goal of a net energy output on an economic scale is still elusive. When practical fusion reactors are developed, hydrogen from the water of the earth's oceans will provide an abundant source of fuel.

Scientists who construct models of the thermonuclear reactions and other processes that occur in the cores of stars use the sun as their test case. Certain predictions about the sun can be made from the model. One of these is the number of neutrinos emitted by the reactions in the sun's core. Neutrinos, which are particles with little or no mass and no electrical charge, travel at close to the speed of light. As a result, they do not interact with matter to a large extent. A neutrino can easily pass through a light-year-thick layer of lead shielding.

The model predicts the number of solar neutrinos that should be counted by experiments designed to detect them. The number of detected neutrinos is, however, one-third of the expected number. Scientists question if the models of the sun's interior processes are incorrect, if there is something unknown happening in the sun's interior that the models have not taken into account, and if the neutrinos change into other particles during the trip. Since neutrinos make the trip from the core to the detectors at the speed of light, they provide information about what is occurring now in the sun's core. The electromagnetic radiation discloses what occurred in the core 100,000 or more years ago. Further theoretical and experimental work should produce an answer on determining if the sun's energy output varies substantially over time.

Context

In the eighteenth century, Immanuel Kant estimated that if the sun were composed of coal, it would burn only for several thousand years. This posed a problem even for those who based their belief that the earth was only six thousand years old on the Bible. In the nineteenth century, Hermann von Helmholtz reasoned that the gravitational contraction of the sun could provide its energy by converting gravitational potential energy into thermal energy. This theory increased the sun's age to 20 million years, which satisfied the biblical literalists but not the geologists and the biologists, who saw that billions of years would be necessary for the evolution of life and the development of the earth.

In the nineteenth century, physicists conducted experiments to determine how the speed of light changed as the speed of the medium through which it traveled changed. The conclusion was that there is no medium through which light moves. It was also observed that the speed of light is constant, no matter how fast its source moves. This is similar to observing a plane that moves at 500 kilometers per hour, no matter who observes it. A person on the ground, someone in a car moving 100 kilometers per hour, and a person in a rocket moving 50,000 kilometers per hour would all measure the plane's speed as 500 kilometers per hour.

Physicists were unable to accept this nonsensical result for the speed of light. Einstein, however, stated that the speed of light is a constant in any frame of reference, and went on to investigate the consequences of this result. While he was making these calculations, his mass-energy equation emerged. In 1939, Hans Albrecht Bethe and Carl von Weizsäcker determined the nuclear reactions that generate the sun's energy. They found that the four protons could fuse into helium 4 and that the mass difference was converted into energy.

It was at that time that nuclear fission was discovered. This led Einstein to write to President Franklin D. Roosevelt that a superbomb, an atomic bomb, could be developed that could help win World War II. An atomic bomb contains several pieces of uranium 235 that are rapidly pushed together by a chemical explosion. A rapid chain reaction of fissioning uranium atoms occurs, and mass is rapidly converted into energy, which causes an explosion of massive proportions. The United States launched the Manhattan Project to perfect the bomb, which was dropped on Japan in August of 1945. This led to the nuclear arms race with the Soviet Union.

In the early 1950's, the thermonuclear bomb was developed. A thermonuclear bomb is an atomic bomb surrounded by deuterium. The ignition of the atomic bomb produces high temperatures and pressure that lead to the fusion of the deuterium and the release of even more energy. This produces the hydrogen bomb's greater destructive power.

Bibliography

Friedman, Herbert. *The Amazing Universe.* Washington, D.C.: National Geographic Society, 1980. Produced in the style that is typical of National Geographic publications, this book covers the universe from the solar system to the cosmos. It is rich in colorful photographs and figures. Chapter 3 discusses the sun's interior and its surface features, which are the expression of the processes occurring in the interior.

Jastrow, Robert. *Red Giants and White Dwarfs.* New York: Warner Books, 1979. A wide range of subjects are covered: the life and death of stars, the origin of the solar system, the beginning of life on Earth, and UFOs. There are many black-and-white photographs and diagrams, and a section of color photographs of nebulas and the planets.

Kaufmann, William III. *Black Holes and Warped Spacetime.* San Francisco: W. H. Freeman, 1979. Kaufmann discusses the evolution of stars and in particular their deaths as white dwarfs, neutron stars, or black holes. There are many black-and-white photographs and diagrams. There is a section of color photographs of nebulas and another of galaxies.

Kuhn, Karl. *In Quest of the Universe.* St. Paul, Minn.: West, 1991. This college-level textbook provides an overview of astronomy, from the solar system to the origin of the universe. The appendices contain useful astronomical data, while the main body of the text has many tables, diagrams, figures, and black-and-white and color photographs. There are chapters on the sun and on the births, lives, and deaths of stars.

Moore, Patrick, and Garry Hunt. *The Atlas of the Solar System.* New York: Crescent Books, 1990. This updated version of the authors' 1983 *Atlas of the Solar System* covers the various objects of the solar system and the history of their observation, such as Viking's coverage of Mars. Sixty pages are devoted to the discussion of the sun. There are many figures, diagrams, tables, maps, a bibliography, glossary, and black-and-white and color photographs.

Wheeler, J. Craig, and Robert P. Harkness. "Helium-Rich Supernovas." *Scientific American* 257 (November, 1987): 50-58. This article, which contains tables, photographs, figures, and diagrams, relates the last days of a star's life, when its fuel is depleted and it collapses under its own mass. This implosion leads to the catastrophic supernova explosion.

Williams, George E. "The Solar Cycle in Precambrian Time." *Scientific American* 255 (August, 1986): 88-95. Some sedimentary rocks, such as varves in glacial lakes, record seasonal changes in the weather. Some changes in the weather are caused by changes in the sun's activity. The study of these Cambrian rocks can lead to a better understanding of the sun's activity over long periods of time. There are many diagrams and figures typical of *Scientific American* articles.

Stephen J. Shulik

Cross-References

Models of the Atomic Nucleus, 176; The Structure of the Atomic Nucleus, 182; Black Body Radiation, 231; Black Holes, 238; Nuclear Forces, 1580; Neutron Stars, 1503; Nuclear Synthesis in Stars, 1621; Special Relativity, 2333; Star Formation, 2360; Supernovas, 2434; White Dwarf Stars, 2742.

CONTROLLED THERMONUCLEAR REACTORS

Type of physical science: Nuclear physics
Field of study: Thermonuclear reactions

Light nuclei, such as isotopes of hydrogen, release energy when they combine to form heavier nuclei. In order to overcome the electrical repulsion between the light nuclei, which carry positive charges, and to bring them near enough for fusion to take place, they must be heated to temperatures of hundreds of millions of Kelvins or brought together by other means.

Principal terms

BREAKEVEN: the condition in which a plasma generates as much energy through fusion reactions as is put into the plasma to heat it; at breakeven, the plasma only replaces the heat that it loses naturally and does not produce net energy

DEUTERIUM: an isotope of hydrogen whose nucleus contains one proton and one neutron

FUSION: the process of combining light nuclei to form heavier ones, with a resulting release of energy

IGNITION: the point at which a plasma generates enough energy to sustain fusion reactions without external heating

INERTIAL CONFINEMENT: the family of techniques for holding a plasma while heating it; these methods use a laser beam or a beam of charged particles to compress a fuel pellet

LAWSON CRITERION: the product of the density of nuclei in a plasma and the confinement time for the plasma that will allow a positive output of energy

MAGNETIC CONFINEMENT: the family of techniques for holding a plasma by using the forces that a magnetic field exerts on the moving charged particles that compose the plasma

PLASMA: a state of matter composed of free (detached) electrons and nuclei without electrons

TRITIUM: a radioactive isotope of hydrogen whose nucleus contains one proton and two neutrons

Overview

Nuclear fusion is the process of combining light nuclei to form heavier ones. Because nuclei are so tightly bound, fusion releases a measurable fraction of the mass of the fusing nuclei as pure energy. In particular, combining a tritium nucleus, which contains two neutrons and one proton, with a deuteron, which consists of one proton and one neutron, produces the nucleus of a helium atom with two protons and two neutrons, an extra neutron, and 17.6 million electronvolts of energy. In practical terms, the complete fusion of a kilogram of tritium and deuterium fuel would produce 364 million kilowatt hours of energy. For comparison, burning a kilogram of gasoline produces about 13.2 kilowatt hours. Most experimenters believe that the reaction between tritium and deuterium, rather than fusion reactions between deuterium and deuterium or other possible fusion reactions, provides the best chance of achieving a fusion reactor for practical electricity generation.

The nucleus of one of every sixty-seven hundred hydrogen atoms on Earth is a deuteron, and tritium can be formed when lithium absorbs neutrons from nuclear reactors or the extra neutrons from the fusion reaction itself. Thus, existing lithium and the deuterium in the oceans seem adequate to provide global energy needs into the distant future. Unlike nuclear fission reactions, the waste products of nuclear fusion reactions, helium and a neutron, are not inherently radioactive. The neutrons induce radioactivity in the reactor walls, however, and tritium itself beta-decays with a half-life of twelve years and three months. Thus, nuclear fusion will produce some radioactivity that might be released into the environment, but the radionuclides have relatively short half-lives and pose considerably less of a long-term threat than the by-products of nuclear fission reactors.

In view of these advantages, nuclear fusion would seem to offer large energy supplies with minimum environmental damage. Unfortunately, nuclear fusion can take place only when tritium and deuterium nuclei approach one another closely enough that the nuclear force can overcome their mutual electrical repulsion. The range of the strong nuclear force is about 10^{-15} meter, beyond which it drops abruptly to zero. The repulsive electrical force between the two positively charged nuclei extends to infinity, and its strength increases as the two nuclei move closer together. Most schemes for producing power from nuclear fusion impart enough kinetic energy to the nuclei that they can approach within 10^{-15} meter of each other and let the strong force fuse them. This kinetic energy corresponds to temperatures of more than 100 million Kelvins, with temperature in this context taken as a measure of the kinetic energy of the nuclei.

At such temperatures, electrons are ripped off atoms and the detached nuclei and electrons form a new state of matter called a plasma. In order to sustain a fusion reaction, a plasma must be produced and held at a temperature of hundreds of millions of Kelvins at sufficient density and for a long enough time to generate useful power. Because materials vaporize long before they reach the required temperatures, the technology necessary to generate energy from nuclear fusion has not yet been developed. Nuclear fusion energy provides most of the explosive energy in thermonuclear weapons, in which the plasma is compressed and heated by the radiation from a nuclear fission warhead, but a more controlled release of nuclear fusion power has eluded researchers.

A fusion power scheme must reach breakeven; that is, the power put into the plasma to heat it must be less than the power produced in the plasma by fusion. Increasing the density of the plasma increases the numbers of collisions between nuclei and the output of power from fusion. The plasma cools down when it develops instabilities that let it escape from the magnetic fields that hold it. A measure of the time during which the plasma cools down by a certain fraction unless more heat is added is called the confinement time. A plasma is said to reach ignition when it produces enough energy from fusion reactions to compensate for energy losses from the plasma and to produce a self-sustaining fusion reaction. In order for a plasma to reach ignition, it must not only reach ignition temperatures above 100 million Kelvins, but the product of its density measured in particles per cubic centimeter and its confinement time measured in seconds must be greater than 10^{14} seconds per cubic centimeter as well. This is known as the Lawson criterion, and it is the goal toward which the designers of nuclear fusion reactors strive.

In addition to schemes that rely on the kinetic energy of nuclei to overcome the electrical repulsion between their positive charges, a fundamentally different technique for triggering nuclear fusion relies on other mechanisms for bringing nuclei close enough together that the nuclear force can bind them to one another with a resulting release of energy. A negative particle called a muon behaves like an electron with 207 times the electron mass. The muon can enter orbit around a deuterium or tritium nucleus and bind it to a second nucleus in a molecule just as electrons bind ordinary hydrogen molecules. Because of the greater mass of the muon, the nuclei in the muonic molecule are bound closely enough together that the nuclei have a relatively high probability of fusing. Muon-catalyzed fusion has been demonstrated in the laboratory, where it is a subject of great interest to students of both nuclear and molecular structure. As yet, however, practical fusion reactor designs have not employed it. Reports by two groups in Utah of using absorption by a platinum matrix to force deuterium nuclei close enough to start fusion reactions were controversial, and the effect was not demonstrated to the satisfaction of the majority of the scientific community. The fusion techniques that do not involve a hot plasma are known collectively as cold fusion.

Applications

Attempts to construct fusion reactors fall into two basic categories: magnetic confinement and inertial confinement. Magnetic confinement schemes use the forces of magnetic fields on moving charged particles to hold these particles of the plasma while they are heated. The magnetic fields generated by electrical currents of the charged particles in the plasma itself play an important role in magnetic confinement. In principle, the hot plasma eventually reaches ignition temperatures and the energy from fusion reactions is released steadily. Inertial confinement schemes use a laser beam or a beam of charged particles to condense and heat a plasma simultaneously, producing a burst of fusion reactions. In effect, inertial confinement schemes generate a series of tiny thermonuclear explosions.

The first efforts at magnetic confinement relied on the pinch effect, which takes advantage of the fact that moving charged particles constitute electric currents. Electric currents flowing in the same direction attract one another and tend to compress the plasma into a narrow tube. Compressing the plasma heats it because plasmas are effectively gases. Pinch confinement is limited by the interactions of the plasma with the ends of the tube that contains it and by instabilities in the plasma tube, which cause the hot plasma to bend into contact with the walls of the tube or break into a series of lumps rather than a continuous path. The theory of hot plasmas relies both on the behavior of charged particles in electromagnetic fields and on the dynamics of the plasma considered as a fluid. Although modern computers have helped to model plasma behavior, the detailed theory of hot plasmas remains an unsolved problem in theoretical physics.

Plasma machines contain very little matter and thus very little heat, although they operate at extremely high temperatures. Any contact with the walls of the tube quickly cools the plasma. All proposed designs for nuclear fusion reactors contain a limited amount of fuel at any one time. In case of an accident, fusion reactors would automatically shut themselves down as the limited supply of plasma would quickly cool in contact with construction materials. This automatic shutdown is one of the major advantages of nuclear fusion over nuclear fission reactors.

The interaction with the ends of the tube can be eliminated by bending the plasma current into a circular path. In order to reduce instabilities within the plasma, a magnetic field is applied along the line of the circular current loop. The most successful magnetic confinement scheme has been the tokamak, in which external magnets provide a carefully adjusted magnetic field in the direction of the plasma currents which is combined with the magnetic fields generated by the plasma currents to produce a corkscrew magnetic field containing the plasma. This device is particularly effective in preventing the loss of energetic electrons, which carry off much of the energy contained in the plasma. Modern modifications of the original tokamak design are among the experiments approaching an effective breakeven process.

Other methods for exploiting magnetic confinement use magnetic mirrors (shaped magnetic fields) to prevent contact between the plasma and the ends of a linear tube. Reactors using magnetic mirrors are popularly known as magnetic bottles. One modification of this scheme, which is believed to show promise for providing a sustained fusion reaction, is the dense z-pinch. In this scheme, a solid fiber of deuterium-tritium fuel suddenly carries a huge pulse of current. The fiber vaporizes and is compressed by the magnetic field of the current, which heats the resulting plasma. The development of solid fibers and better power supplies has increased experimental interest in the dense z-pinch.

Problems with magnetic confinement schemes arise when the plasma is heated to a higher temperature. The theory of hot plasmas is imperfectly understood, and unexpected instabilities tend to plague reactors as the temperatures of the plasma are increased. Experimenters attempt to remove these instabilities and increase the temperature of the plasma by modifications in the detailed design of the magnetic fields confining the plasma. The experiments divide themselves into those oriented toward attempts to understand the basic physics of the hot plasma's behavior and those oriented toward reaching breakeven and ignition. All the experiments involve constructions of large, complex devices and thus are extremely expensive. Efforts to promote international cooperation in the study of fusion have been encouraging because of both the flow of scientific knowledge and the sharing of the expense of the experiments among nations. The detailed design and evaluation of a reactor from which useful heat can be extracted and used to generate electric power await the choice of a scheme for producing the self-sustaining fusion reactions in the plasma.

Inertial confinement schemes inject a pellet of deuterium-tritium fuel into the point at which several laser or charged-particle beams meet. The beams rapidly compress and heat the pellet. As it is compressed, the pellet vaporizes, forming a hot plasma in which, it is hoped, fusion reactions will occur. Energy is released as the pellet explodes. The debris—consisting of deuterium, tritium, helium nuclei, and neutrons—then strikes and heats a surrounding medium from which useful power can be extracted.

In one scheme, the medium is liquid lithium, which is swirled to provide a cavity into which the fuel pellet is dropped. The neutrons from the fusion breed new supplies of tritium in their interaction with the lithium, which is drawn off the bottom of the container to be chemically cleaned for salvaging tritium, deuterium, and helium. Heat deposited in the lithium by the fusion interactions can be used to heat water for the generation of power before the lithium reenters the fusion reactor. The force of the small fusion explosion could possibly be absorbed by bubbles in the lithium.

There are several problems plaguing inertial confinement schemes. Such technologies demand very powerful laser or particle beams that must be sharply focused on the pellet so that the beams strike it from several directions at precisely the same time. If the timing is off, then the pellet will not be uniformly compressed and heated. These requirements challenge state-of-the-art technology for both generating and handling the beams. Design of the fuel pellets themselves involves new technologies. In some experiments, the pellets surround layers of tritium and deuterium with thin layers of gold and other metals. The entire multilayer pellet must be a perfect sphere, and hundreds of them can fit on the head of a pin.

Finally, inertial confinement schemes work by generating a series of small fusion explosions. These explosions can be used to mimic the effects of nuclear weapons on materials and electronic circuits. Much of the research on inertial confinement is used by the military and is classified. While there has been less discussion of inertial confinement in electric power generation, research programs have been proceeding vigorously in the classified world.

Context

Experimental designs for nuclear fusion reactors have concentrated on reaching first breakeven and then ignition. They are approaching these conditions, but achieving ignition will prove only that nuclear fusion can provide a self-sustaining energy source in a laboratory environment.

Such experiments provide new basic information on the structure of nuclei and the behavior of plasmas. Since nuclear fusion reactions are believed to provide the power of stars, these results also interest the astrophysics community. In addition, fusion reactor designs use such new technologies as superconducting magnets and provide laboratories for developing them. The need to provide high currents in the plasma challenges the technologies used in power supplies and has spurred developments in this area.

Before nuclear fusion can provide a useful energy source, experimenters must demonstrate engineering feasibility by incorporating the laboratory device in a pilot plant that can generate electric power. Developers must examine problems that may arise in the design of larger plasma machines, as well as such issues as the response of reactor materials to the high flux of high-energy neutrons produced during nuclear fusion. Systems must be developed to breed tritium for the reactor fuel and to prevent tritium leaks to the environment. Other health and safety issues will certainly include mechanisms for shutting down the reactor without damaging any of its components.

In inertial confinement schemes, for example, there will be concern over the interface between the lithium, which carries heat from the fusion reactions, and the steam used to drive electric generators. Lithium reacts strongly with water, and mechanisms must be developed to guard against explosions. These problems have little to do with the nuclear fusion process itself, but they are critical to the operation of a practical system for generating electrical power. Other issues will include exploring the degree to which fusion reactors will produce radioactive wastes from the materials of which the reactor is constructed. Finally, the high-energy neutrons produced in the fusion reaction will require very heavy shielding to protect the operating crew of the power plant.

Once engineering feasibility has been established, nuclear fusion reactors must next demonstrate commercial feasibility. The costs of an electric generating plant based on nuclear fusion cannot yet be estimated. Completed design of a pilot plant will permit assessment of costs per kilowatt of electric power in the terms of both operating expenses and capital costs. Pilot plants will define the environmental effects of the plants and permit authorities to establish licensing procedures for them. Finally, the costs for constructing a fusion plant must be comparable to the costs of constructing equivalent nuclear fission or fossil fuel plants and must demonstrate superior environmental effects.

Bibliography

Bromberg, Joan Lisa. *Fusion: Science, Politics, and the Invention of a New Energy Source.* Cambridge, Mass.: MIT Press, 1982. This volume describes the history and politics of the development of controlled fusion reactors in the United States as a serious effort in the history of science. The discussion is useful in explaining why some technical decisions were made at the time and the reason that only a few of the various approaches to producing fusion energy have been vigorously pursued.

Epstein, Gerald L. "Fusion Technology for Energy." In *The Energy Sourcebook: A Guide to Technology, Resources, and Policy*, edited by Ruth Howes and Anthony

Fainberg. New York: American Institute of Physics, 1991. This chapter provides a summary of developments in fusion technology, including a discussion of muon catalysis and other less prominent techniques. It is written for a general audience and is accessible to the educated lay reader.

Hunt, S. E. *Fission, Fusion, and the Energy Crisis.* 2d ed. Elmsford, N.Y.: Pergamon Press, 1980. This discussion of the British approach to fusion power provides a good analysis of the criteria that fusion must meet in order to provide a useful energy source. The discussion of some of the problems with various magnetic confinement schemes is particularly readable and interesting.

U.S. Congress. Office of Technology Assessment. *Starpower: The U.S. and the International Quest for Fusion Energy.* OTA-E-338. Washington, D.C.: Government Printing Office, 1987. This volume provides a clear and readable discussion of the technologies and economics of nuclear fusion power. It is written initially for the lay reader, but it provides a summary of technology for the reader who is more interested in the details of the technology.

York, Herbert F., ed. *Arms Control.* San Francisco: W. H. Freeman, 1973. Contains three articles on fusion. The articles are: Richard F. Post, "Fusion Power" (originally published in December, 1957); Moshe H. Lubin and Arthur P. Fraas, "Fusion by Laser" (June, 1961); and William C. Gough and Bernard J. Eastlund, "The Prospects of Fusion Power" (February, 1971). These articles provide an excellent general introduction to nuclear fusion power. The article by Post presents a very elementary discussion of the fusion process and the problems it faces. The article by Lubin and Fraas is an equally readable introduction to inertial confinement problems and potentials. Finally, the article by Gough and Eastlund introduces more detail on various types of reactors.

Ruth H. Howes

Cross-References

The Structure of the Atomic Nucleus, 182; Forces on Charges and Currents, 907; Lasers, 1227; Nuclear Forces, 1580; Quantum Mechanical (Gamow) Tunneling, 1979; Thermonuclear Reactions in Stars, 2509; Thermonuclear Weapons, 2523.

THERMONUCLEAR WEAPONS

Type of physical science: Nuclear physics
Field of study: Thermonuclear reactions

Cold War fears pushed scientists to harness nuclear fusion in a bomb they called the "super," or the hydrogen bomb.

Principal terms

CHAIN REACTION: a reaction sequence in which one event causes the next event, which causes the next event, and so forth

CRITICAL MASS: the minimum mass that can sustain a chain reaction

FISSION: the splitting of a heavy atomic nucleus into lighter atomic nuclei

FUSION: the joining of light atomic nuclei to form a heavier atomic nucleus

KILOTON: 1,000 tons

MEGATON: 1 million tons

NEUTRON: an electrically neutral particle and one of the two constituents of an atomic nucleus

PROTON: a positively charged particle having almost as much mass as the neutron; the other constituent of an atomic nucleus

TNT EQUIVALENT: the standard measure of energy released by a nuclear weapon is expressed as that mass of TNT (trinitrotoluene) that would produce the same amount of energy when exploded

Overview

The story of nuclear fusion is the story of man's efforts to capture the power of the sun. During the 1920's and 1930's British astronomer Robert Atkinson, Austrian physicist Fritz Houtermans, and German physicist Hans Albrecht Bethe solved the riddle of the sun's source of energy. It was nuclear alchemy. For generation upon generation, medieval alchemists had sought to transmute the base element lead into the regal element gold. Although they never succeeded, they did eventually learn much about chemistry. One lesson they learned was that no chemical reaction can change one element into a different element.

With the discoveries of Atkinson, Houtermans, and Bethe, however, it was revealed that nuclear reactions can do precisely that. The identity of an element is fixed by the number of protons in the atomic nucleus. Hydrogen nuclei have one proton each, helium nuclei have two, lithium nuclei have three, and so on. Nuclear reactions may add or subtract protons from the nucleus and thereby change one element into another. Such nuclear alchemy can even change lead into gold, but it would be prohibitively expensive to do so.

The simplest atomic nucleus is a single proton. Provided with an electron to orbit the proton, the combination becomes a hydrogen atom. The second simplest atomic nucleus consists of a proton and a neutron bound together by the strong nuclear force. Since it still has but a single proton, it is still a nucleus of hydrogen. To distinguish between the two kinds of hydrogen, they can be designated as hydrogen 1 and hydrogen 2. The number following the name denotes the total number of protons and neutrons in that nucleus. Nuclei with the same number of protons, but different numbers of neutrons are called isotopes of one another. A third isotope of hydrogen, hydrogen 3 (one proton plus two neutrons), can be formed, but it is unstable.

To oversimplify a little, one of the neutrons in hydrogen 3 sometimes changes into a proton and an electron. The electron is then ejected from the nucleus, leaving the nucleus with two protons and one neutron. The new nucleus is helium 3. The half-life for this process is 12.26 years; that is, in 12.26 years one-half of the original amount of hydrogen 3 will have decayed into helium 3. After another 12.26 years, one-half of the remaining hydrogen 3 will have decayed, and so forth. Nuclear weapons that use hydrogen 3 lose about 5.5 percent of their remaining potency each year.

In nature, 99.985 percent of hydrogen is hydrogen 1 and 0.015 percent is hydrogen 2. Hydrogen 3 must be produced artificially and is usually made in a nuclear reactor. Because of their importance, the hydrogen isotopes are given special names: deuterium for hydrogen 2 and tritium for hydrogen 3.

The sun gains the energy by which it shines from the fusion of hydrogen 1 into helium 4, a process that takes several steps. The helium-4 nucleus weighs a tiny amount, less than the sum of the constituents that went into its formation. The ultimate source of fusion power is the conversion of this mass difference into energy. It is estimated that in the sun's core where these reactions occur, the temperature reaches 16 million Kelvins and the density reaches 160 times that of water.

The high temperatures are required to give the protons enough energy of motion to overcome their mutual repulsion as positive charges. They must be nearly close enough to touch one another before nuclear reactions can occur. Even then, there is only a small probability that any given close encounter will result in fusion. The high density of the sun's core guarantees that there will be many close encounters.

In a hydrogen bomb, a fission bomb is used as a trigger to start fusion reactions. Even so, hydrogen-1 fusion is too improbable; however, deuterium and tritium fusion do occur. One likely reaction is deuterium fusing with deuterium to produce either tritium plus a proton, or helium 3 plus a neutron. An even more likely reaction is deuterium fusing with tritium to produce helium 4, plus a high-energy neutron. Tritium fusing with tritium produces helium 4 plus two high-energy neutrons. All of these reactions also produce considerable energy.

In order to be efficient, the fusion fuel must be dense. This dictates that one begin with a solid, a liquid, or a very high-pressure gas. The first successful test of a hydrogen "bomb" (Mike shot, 1952) used liquid deuterium, which must be kept at 20 degrees Celsius above absolute zero and required tons of refrigeration equipment. The first deliverable hydrogen bomb tested (Bravo shot, 1954) used a solid fusion fuel, lithium-6 deuteride (deuterium bonded to lithium 6). An incident neutron causes lithium 6 to fission to helium 4 plus tritium. The tritium and deuterium then fuse, as described.

Since fusion will not occur unless temperatures reach millions of degrees, any amount of fusion fuel can be assembled together without fear of a nuclear explosion. Thus, the only limit to the size of a hydrogen bomb is how the bomb is to be delivered.

Most of the ideas about fusion discussed predate the explosion of the first atomic bomb (Trinity shot, 1945). In fact, the first to propose the hydrogen bomb was the Japanese physicist Tokutaro Hagiwara of the University of Kyoto. He discussed it in a lecture entitled "Super-explosive U235" presented in May, 1941. The next oldest record is of a conversation between Edward Teller and Enrico Fermi in the fall of 1941. Teller followed up with a calculation that showed that, at least for the configuration he was considering, it would not work.

Applications

The first test of fusion in a nuclear weapon was to place some fusion material near the core of a fission device. The neutrons produced by fusion significantly increased the efficiency of the fission and thereby increased the yield. (The additional energy contributed by the fusion itself was small.) The process is referred to as "boosted fission" and is widely used because it allows the manufacture of smaller and more efficient fission bombs. A further advantage of a boosted fission bomb is that the amount of high-pressure tritium in the core can be adjusted to change the bomb's yield.

Yet, the obvious next step to the "super," as the hydrogen bomb was first named, did not work. That is, if a large amount of deuterium were assembled around a fission core, the deuterium was scattered before much fusion could occur. Teller, along with Stanislaw Ulam, eventually conceived of a solution. In the Teller-Ulam configuration, the fusion material is separated from the fission trigger.

Consider the elements of a hydrogen bomb with a yield equivalent to 300 kilotons of TNT. (Since the exact details are classified secret, it cannot be stated with certainty that the description is correct. The description is based on information in the public domain.) The bomb will have some kind of outer packaging that will protect the nuclear device and aid in its delivery. There will be circuits and devices to protect against accidental explosion and unauthorized detonation. Other circuits will arm the bomb and instruct it when it is to detonate. The nuclear device itself is a cylindrical casing about 1 meter long and 0.5 meter in diameter. For a fission-fusion-fission bomb, the casing will probably be made of uranium 238. The casing's first purpose is to contain the radiant energy momentarily from the trigger explosion and direct it on the fusion stage. Inside the casing and placed at opposite ends are the trigger (first stage) and the fusion device (second stage). The trigger is a boosted fission device about the size of a soccer ball. It consists of seven concentric spheres, one on top of the other like the layers of an onion. The innermost sphere is the booster. It contains a mixture of high-pressure deuterium and tritium gases. The second sphere consists of a few kilograms of plutonium; the third sphere consists of a few kilograms of uranium 235. The fourth sphere is air crossed by supports to hold the fission core in position. This is called a levitated core. The fifth sphere is a rather thick sphere of uranium 238 named the tamper. It is surrounded by a sphere of beryllium, which serves as a neutron reflector. It increases the efficiency of the device by reflecting neutrons, which would otherwise be lost, back into the reaction. The tamper reflects neutrons and its inertia also briefly restrains the expansion of the fireball. The outermost sphere of the trigger consists of a high explosive such as triamino trinitro benzene.

A neutron source initiator is mounted on the casing and aimed at the trigger's core. The initiators for the first atom bombs were neutron-emitting mixtures of radium and beryllium. The present-day initiator is a thumb-sized vacuum tube containing tritium and deuterium. The tritium is probably present as a low-pressure gas, or perhaps embedded in the tube's filament. The deuterium is probably embedded in the tube's target or else chemically bound to it. In action, the filament is heated to produce tritium ions. A high voltage (probably less than 1,000 volts) is applied between the target and filament so that the tritium ions are accelerated toward the target. The tritium-deuterium reaction will then produce high-energy neutrons.

As the initiator is pulsed on, the high explosives in the trigger are detonated. These explosives must be carefully formulated, shaped, and detonated so that the resulting explosion is very uniform. Then, the explosion will drive the beryllium reflector and uranium tamper inward. The air gap allows the tamper to gain some momentum before it smashes into the uranium-235 shell. The implosion will compress both the uranium and the plutonium to more than twice their original densities so that the conditions for critical mass are exceeded. Fission generations take about twice as long in uranium 235 as they do in plutonium. Since the tamper compresses the slower-fissioning uranium-235 shell first, a rampant chain reaction will begin there first, but that will soon be joined by fissioning in the plutonium shell.

The resulting heat and pressure will cause the booster deuterium and tritium in the central core to undergo fusion. Fusion neutrons will flood outward and significantly increase the fissioning of the plutonium and uranium-235 shells as they vaporize. The trigger is now an expanding fireball. The pressure is a billion atmospheres, and the temperature is 100 million Kelvins. The time since detonation is a few tenths of a millionth of a second at most.

The expanding fireball emits intense X rays that outdistance it, since X rays travel at the speed of light. The second stage, or fusion stage, is mounted at the end of the casing opposite from the trigger. The fusion fuel is lithium-6 deuteride, which is formed into a hollow rod. The rod is 30 to 60 centimeters long and perhaps 10 centimeters in diameter. A plutonium rod about 3 centimeters in diameter, called the spark plug, fills the hollow center of the fusion fuel rod, and a uranium-238 tamper covers the outside of the fusion fuel rod. The whole assembly is mounted along the axis of the casing. The end nearest the trigger is capped with a thick piece of uranium 238 called the pusher. The pusher first prevents premature detonation of the fusion fuel by shielding it from neutrons produced by the trigger. When the fusion fuel is compressed, the inertia of the pusher will keep the fuel from squirting out of the end of the rod assembly.

Dense polystyrene foam fills the remaining space inside the casing except around the trigger. The foam may contain bubbles of the explosive gas pentane and probably also contains small metallic particles to absorb high-energy X rays and reemit X rays at energies more easily absorbed by the foam. The insides of the casing and possibly some other interior fixtures are designed to receive X rays from the trigger and convey them to the foam.

Almost instantaneously, the X rays turn the foam into a high-pressure plasma that compresses and heats the fusion fuel. More heat and pressure will be added as the spark plug fissions. The pusher and fusion fuel rod tamper will now fission with neutrons released by fusion. If the casing is made of uranium 238, some of it will also fission. The whole bomb is now vaporized by the expanding fireball. Typical yields would be 40 kilotons of TNT for the trigger, 130 kilotons of TNT for the fusion component, and 130 kilotons of TNT for the uranium-238 components, for a total of 300 kilotons of TNT.

An enhanced radiation weapon, or neutron bomb, as it is called, is made by selecting a very low-yield fusion device, about 1 kiloton of TNT, and by using a non-neutron absorbing material in place of the uranium 238. The fusion fuel is tritium and lithium-6 deuteride; however, the temperatures are held down to enhance tritium fusion, which produces more high-energy neutrons than does deuterium fusion. Depending upon the yield, lethal damage from a blast might be limited to a circular area a few hundred meters in radius, with lethal and quickly incapacitating doses of neutron radiation extending outward a few hundred meters beyond that. The proposed use of the neutron bomb is against a massive tank assault. Nevertheless, it seems that maintaining a greater distance between tanks and lining them with lightweight neutron absorbers can counter the neutron bomb and make it little more effective than conventional explosives. The neutron bomb seems to be of limited use, and the United States has retired some of its enhanced radiation weapons from service.

Context

In the winter of 1948, the Czechoslovak communists staged a coup and took over that country's government. The next summer, communists blockaded Berlin in an effort to drive the Western allies out. Hungary enjoyed freedom under the Allied Control Commission, but through manipulation and purges, the Communist Party seized power. Upon their adoption of a Soviet-style constitution in August, 1949, contact with the Western world was restricted. The Communists took over China that same year. Then, on September 23, 1949, the Soviet Union shocked the West by detonating their first nuclear device, Joe I. (The Joe series of tests was named for Joseph Stalin.)

It seemed to Americans in general, and to Teller in particular, that with communism rampant, the most powerful response possible to Joe I needed to be made. Even so, an advisory committee headed by J. Robert Oppenheimer recommended against seeking to develop the hydrogen bomb at that time. It did recommend further improvements in fission weapons, including fusion boosting, but it had little faith that the proposed design for the "super" would work. Furthermore, some of the committee's members could see little use for a weapon of such mass destruction that its main potential targets seemed to be the civilian populations of cities.

Other advisers saw things differently. On January 31, 1950, President Harry S. Truman called for the development of the hydrogen bomb. In February, 1951, the Teller-Ulam configuration was conceived; any negotiating advantage that might have come from not developing the super was now lost. The first test of the Teller-Ulam configuration exploded with the force of 10.4 megatons of TNT (Mike shot, October 31, 1952). Two years later, a deliverable bomb was tested; in 1955, the Soviet Union tested the first deliverable superbomb. Bombs could now be made as big as could be delivered. The largest bomb ever tested was detonated by the Soviet Union in October 30, 1961; it had a monstrous yield equivalent to 58 megatons of TNT.

The yield of the United States' nuclear stockpile reached a maximum in 1960 of 19,000 megatons of TNT. There was a heavy reliance then on bombers and relatively inaccurate missiles, which carried high-yield devices of 10 to 20 megatons of TNT. As missile accuracy improved, such powerful weapons were not needed to ensure destruction of the target, and yield was reduced toward the range of 50 to 350 kilotons of TNT. By 1976, the total yield of the United States' nuclear stockpile was reduced to only one-fourth of its 1960 value. The yield of the Soviet stockpile peaked about 1975. By 1985, the stockpile had dropped to 70 percent of its maximum, but this is still three times the United States' value for that year.

The doctrine of Mutual Assured Destruction (MAD) has given these stockpiles a moral reason to exist, for it is probable that the fear of the consequences of nuclear war has restrained the nuclear powers from directly attacking one another. The major nuclear powers have pledged to work toward reducing the number of nuclear weapons, believing this to be in the world's best interest.

Bibliography

Blumberg, Stanley A., and Gwen Owens. *Energy and Conflict: The Life and Times of Edward Teller*. New York: G. P. Putnam's Sons, 1976. This is a biography of Teller, "the father of the H-bomb." Although he was a scientist of unquestioned brilliance, many find Teller's personality difficult. The authors' treatment of Teller is very charitable. The book is recommended because Teller's version of events deserves a hearing. A further caution: The authors confuse the Soviet test shot Joe 4 (1953) with a true "super" and place great emphasis on this point. Joe 4 is generally classed only as a boosted fission bomb.

Cochran, Tomas B., W. M. Arkin, and M. M. Hoenig. *Nuclear Weapons Databook*. Vol. 1, *U.S. Nuclear Forces and Capabilities*. Cambridge, Mass.: Ballinger, 1984.

An authoritative reference book. Includes a brief history of nuclear weapons and summarizes their operational principles. Describes in detail the capabilities of known nuclear devices and their delivery systems.

Davis, Nuel Pharr. *Lawrence and Oppenheimer*. New York: Simon & Schuster, 1968. This is a biographical work. While it emphasizes people, personalities, and politics, it does so against a background of the concepts and techniques for the development of both the atomic bomb and the hydrogen bomb. Treats Oppenheimer's security clearance hearing and the eventual efforts of the government to make amends.

Ehrlich, Robert. *Waging Nuclear Peace: The Technology and Politics of Nuclear Weapons*. Albany: State University of New York Press, 1985. While this book does discuss the makeup of nuclear weapons, the emphasis is on both the physical and political effects of these weapons. Contains a good treatment of various weapons systems and strategies for using them. It is well written and engrossing. Has the virtue of being a more balanced treatment than most.

Morland, Howard. *The Secret That Exploded*. New York: Random House, 1981. This book is Morland's story of how he learned details of the hydrogen bomb and successfully defeated government attempts to silence him. Contains one of the most detailed descriptions available to the public of the inside of a hydrogen bomb. While all details may not be correct, they do generally agree with information in the public domain.

Rhodes, Richard. *The Making of the Atomic Bomb*. New York: Simon & Schuster, 1986. This well-written book is probably the most comprehensive of the many books on this subject. It takes a historical approach and focuses on people, while not slighting the science. A lengthy epilogue describes the development of the hydrogen bomb. Extensive bibliography. Highly recommended.

Tsipis, Kosta. *Arsenal: Understanding Weapons in the Nuclear Age*. New York: Simon & Schuster, 1983. One of the most detailed works available on the workings of nuclear weapons and weapon systems. Suitable for the layperson, but has appendices with mathematical treatments for scientists.

Charles W. Rogers

Cross-References

Breeder Reactors, 259; Fission and Thermonuclear Weapons, 892; Nuclear Forces, 1580; Nuclear Reactions and Scattering, 1605; Nuclear Synthesis in Stars, 1621; The Effects of Nuclear Weapons, 1628; Radioactive Nuclear Decay and Nuclear Excited States, 2057; Controlled Thermonuclear Reactors, 2516.

THE NATURE OF TIME

Type of physical science: Relativity
Field of study: General relativity

Science has proceeded for centuries on the supposition that time is an absolute entity in the physical world. The theory of relativity demonstrated that time is intertwined with the space and materials in the universe. This relation of time with space and matter becomes significant in high-speed space travel and cosmology.

Principal terms

ATOMIC CLOCK: an extremely precise clock that operates on the basis of natural vibrations within the atomic structure

BLACK HOLE: an object of incredibly high gravity, formed when a star with a large mass collapses into itself under its own gravity

COSMOLOGY: the study of the origin and structure of the universe

EVENT HORIZON: a zone around a black hole within which time stops, and from which nothing, not even light, can escape

INERTIAL OBSERVER: a stationary observer or an observer with a uniform (unaccelerated) motion

LIGHT SPECTRUM: the combination of component frequencies or colors in a ray of light (as in a rainbow)

SPACE-TIME: a four-dimensional system of coordinates whose points represent events in the physical world; three coordinates represent the location of the event and one represents the time of the event

WHITE DWARF: a massive star approximately the size of the earth whose gravitation is tens of thousands of times greater than that of the earth

Overview

Time is a mysterious aspect of the physical world. The human concept of time is deeply rooted in everyday experience of the world and in common sense. It is such a fundamental concept that humans seldom inquire into its nature. To elucidate its nature, it is necessary to turn to other concepts, which are themselves closely related to time.

Two time-related entities that are used to characterize time are space and motion. Time is measured by motion in space—one measures the time of day by the motion of the earth about its axis and the time of year by the rotation of the earth around the sun. Time is also measured by using clocks and watches based on the motion of wheels and pendulums. Motion itself, however, is measured by time and space: To describe the motion of a moving object, one must specify the space in which the object moves and the time it takes for that object to move through that space. Finally, space can be measured by time and motion. Ancient Persians used the "parsang," the distance on level ground that an average person could walk in one hour. Even today, some distances are described in terms of space and time. The "light-year," a common measure of distance used in astronomy, is the distance that light travels in one year.

The use of this time-space-motion triangle for characterizing time has created much debate among scientists and philosophers for centuries. In classical physics (before the theory of relativity was described by Albert Einstein), time and space were both treated as absolute and independent entities. Specifically, time was assumed to be a universal concept, and the time between two events was considered independent of the state of motion of the body of reference. With this concept of time and space, all physical objects could be located in absolute space, and all events could be assigned positions objectively in a fixed three-dimensional space and on an ever-flowing stream of absolute time. Concerning the nature of time, Sir Isaac Newton states in his *Philosophiae Naturalis Principia Mathematica* (1687; principles of natural philosophy): "Absolute, true, and mathematical time, of itself, and from its own nature flows equably without regard to anything external." Newton believed that the rate at which time moments succeed each other is independent of all events and processes in the physical world.

The idea of absolute and uniformly flowing time has been criticized by a number of philosophers and scientists. Gottfried Wilhelm Leibniz, a philosopher and mathematician who was a contemporary of Newton, rejected the idea of absolute time and argued that events are more fundamental than moments (this is known as the relational theory of time). He defined time as the order in which events occur. In other words, events do not occur in "time"; time is derived from the occurrence of events.

Despite opposing views such as that of Leibniz, the classical view of absolute time prevailed until 1905, when Einstein published his work on the special theory of relativity. This work was based on the notion that laws of nature must remain the same for all inertial observers. As a result of this theory, the concepts of absolute time and absolute space became abolished. Time (as well as space) was found to be an aspect of the relationship between the universe and the observer. Accordingly, time and space must be regarded as interchangeable; only jointly in the form of a four-dimensional entity called space-time can they be used to characterize the physical world. Hermann Minkowski (a mathematician and contemporary of Einstein) was the first person to formulate the properties of space-time mathematically. In Minkowski's words: "Henceforth space by itself, and time by itself, are doomed to fade away into mere shadows, and only a kind of union of the two will preserve an independent reality."

One consequence of the theory of relativity is that clocks run more slowly at high speeds; that is, there is a longer wait (according to a stationary clock) between ticks of a moving clock. This is usually called "time dilation." A thought experiment, referred to as the twin paradox, has been given by Einstein to illustrate this concept. It involves two young twins; one stays on Earth, and the other goes for a long space journey at high speeds. When the traveling twin returns to Earth, still young, he finds his Earth twin quite old. Time in the spaceship has passed at a slower rate than time on Earth, and as a result, the twin in the spaceship has aged less. For long distances and at speeds close to the speed of light, time dilation becomes significant. The time dilation effect was verified by clocks flown on jet planes around the world in 1971.

In the general theory of relativity, Einstein included the effects of gravity. The general theory (or the theory of gravity) is based on the notion that laws of nature must remain the same for all observers regardless of their motion. Space-time, which in the special theory is the (objective) measure of time and space, in the general theory becomes dependent on matter. In the general theory of relativity, only the oneness of space, time, and matter has an independent existence.

According to general relativity, gravity has a slowing effect on time. This means that, in a stronger gravitational field, any cyclical motion, such as the ticking of a clock or the vibration of the nucleus of an atom, would take place more slowly.

The effect of gravity on time has been investigated and quantitatively verified since the 1950's. On Earth, because of small gravitational changes, such effects are too small to be measured by conventional timekeeping devices. In 1958, Rudolf Ludwig Mössbauer, a German scientist, discovered a phenomenon that opened the way to high-precision time measurement. Mössbauer timekeeping is based on the concept that, upon excitation, nuclei of certain atoms vibrate and emit electromagnetic waves with frequencies on the order of a million million million oscillations per second. Atomic clocks that utilize the Mössbauer effect are accurate to a fraction of a second in millennia.

Direct verification of the effect of gravity on time was made in a series of experiments on U.S. Navy aircraft in the Chesapeake Bay between September, 1975, and January, 1976. In these experiments, atomic clocks were carried on aircraft and flown at an altitude of more than 9,000 meters. The clocks aboard the aircraft (where gravitation was weaker as a result of the high altitude) gained about three-billionths of a second every hour, which was within 1 percent of the prediction of the general theory of relativity. Later, in June, 1976, an atomic clock was run at an altitude of approximately 9,600 kilometers. At that altitude, the clock ran faster than similar clocks on the earth's surface by one-billionth of a second every second.

The slowing effect on time of gravitational fields has also been investigated indirectly by studying the spectra of light emitted by massive bodies. This is based on the concept that light, in trying to escape from a massive object, loses energy in much the same way as an object moving away from the earth, and in doing so, its frequency decreases and its color becomes "redder." This is referred to as the "gravitational redshift." In a series of experiments at Harvard in 1959 and 1965, Robert V. Pound and his colleagues measured the effects of the earth's gravity on the frequency of light. In their tests, these researchers used the Mössbauer effect to measure the change in the frequency of light falling from the top to the bottom of a 22.5-meter tower. Their experiments indirectly confirmed the effect of gravity on time to within 1 percent of the prediction of the general theory of relativity.

There have been other gravitational redshift tests, confirming the effect of gravity on time since the 1950's. Gravitational redshift tests for the sun were performed by French astronomers in 1962. Another study was a redshift analysis of the spectra of light from more than one dozen white dwarfs, conducted jointly by American and British astronomers in the 1970's. The results of these two sets of tests indicated the slowing of time by, respectively, about one minute per year on the surface of the sun and more than one hour per year on white dwarfs.

Among massive objects in the universe, black holes have the most amazing effect on time. Because of the tremendous gravity of black holes, the slowing of clocks in their vicinity would be extreme. In fact, time within a black hole comes to a complete stop; a hypothetical observer outside the event horizon could not detect any passage of time inside the black hole.

Applications

The relativity and subjectivity of time become pronounced in situations involving vast distances, speeds near the speed of light, and the presence of strong gravitational fields.

The relation of time and space has some very interesting implications concerning high-speed and long-distance space travel. The slowing of time and the foreshortening of distances at high speeds make possible long journeys to distant reaches of the universe. According to the relativity theory, the size of the universe and the travel time are related to the speed of the traveler. The faster one travels, the more distances along the travel direction shrink, and the more slowly clocks on the spaceship run. At high speeds, this combination would result in tremendous reductions in travel time between the point of departure and the destination. To illustrate, consider that it takes light from the sun about eight minutes to reach the earth. Despite the fact that no speed can exceed the speed of light (as special relativity reveals), it is still possible to make the trip from the sun to the earth in less than eight minutes. At a speed one-tenth of the speed of light, the trip would take about eighty minutes. To make the trip in eight minutes, the speed need be only about 70 percent of the speed of light. At 99 percent of the speed of light, the trip would take about one minute. In principle, if it were possible to travel at the speed of light, one could make the trip in zero time. Indeed, traveling at the speed of light, one could cross the entire universe in no time, and at such a speed, the words "time" and "space" would become meaningless. In practice, however, such travels will likely remain in the realm of science fiction, because as the speed of the spaceship increases, so does its effective mass and the energy required to accelerate it to such high speeds. It should also be noted that it is only by the earth's clocks that life in the spaceship is different. Based on the time in the spaceship, the traveler lives a normal life and ages at the normal rate.

One practical application of the effect of gravity on time is in modern navigation. Navigation systems operate on the basis of data on time and distances, which are communicated to them by satellites in low-gravity space outside the earth's atmosphere. Ignoring the predictions of general relativity concerning the effect of gravitation on time could amount to an error of several kilometers in calculated distances.

The general theory of relativity also has implications regarding the beginning and the end of time. This characteristic of time is usually studied in cosmology, to which general relativity has made its greatest contributions so far. A major discovery in cosmology in the present century was Edwin Powell Hubble's observation that distant galaxies are moving away from the earth. This observation brought about the notion of the "expanding universe" and the "big bang" theory. According to the big bang theory, based on the present expanding rate of the universe, at some time, 15 to 20 billion years ago, all the matter in the universe was concentrated in one incredibly dense "point." The universe then began as a result of a big explosion—the big bang. There have been other theories of the origin and the behavior of the universe, but more and more observational evidence supports the big bang theory of the expanding universe. In 1965, Arno A. Penzias and Robert W. Wilson of Bell Laboratories made a discovery that strongly supports the big bang theory. The discovery was the detection of cosmic microwave radiation that fills the entire universe and is believed to be the remnant of an initial explosion.

The general theory of relativity supports the big bang expanding model of the universe. In fact, the theory was first used by Alexander Friedmann (a Russian meteorologist and mathematician), before Hubble's observation, to model the big bang. In Friedmann's models, space-time and matter came into existence at the time of the big bang. Therefore, in the context of the general theory of relativity, the big bang was the beginning of time for our universe, and the question "What happened before the big bang?" becomes meaningless. According to the general theory, time and space are possible only when matter is present, and therefore, within the physical world, it is matter and energy that ultimately determine the nature of time.

It is possible that the presently expanding universe will stop expanding and begin contracting at some time in the future. Should this occur, the whole universe will collapse into a state of infinite density. In cosmology, this is referred to as the "big crunch," which would correspond to the end of time in our universe. The general theory of relativity also models the big crunch as a later state of the universe that heralds the end of time.

General relativity and existing cosmological data imply the existence of black holes within the universe. Black holes can be thought of as small local "crunches." Because of the many similarities of black holes to the big crunch and the big bang, they are presently under extensive study by astrophysicists and cosmologists who wish to answer questions about the beginning and the end of time.

The beginning and the end of time, and the relationships of time with space and matter, have also concerned philosophers and theologians for centuries. Among the notable philosophers who speculated and reasoned about such issues are Immanuel Kant, Leibniz, and Saint Augustine. When Saint Augustine was asked, What did God do before he created the universe? he replied that time was a property of the universe that God created and that time did not exist before the beginning of the universe. To the question Why did God not create everything a year sooner? Leibniz responded that the question would be meaningful "if time was anything distinct from things existing in time." In fact, this view of the intimate association of time and the universe can be traced back to Plato, the Greek philosopher who in 400 B.C. believed that time was produced by the universe.

The theory of relativity and the concepts of time, space, and matter in relativity have allowed for the first time a re-evaluation of such philosophical questions from scientific viewpoints. In particular, the model of the expanding universe, with the big bang and the big crunch phenomena, has brought the concepts of "genesis" and "revelation" into the realm of science in general, and theoretical physics in particular.

Context

Inquiries concerning the nature of time tend to become less speculative and philosophical and more scientific and practical as scientists develop better physical theories to predict its nature and make more refined instruments to verify such predictions.

The role of timekeeping devices is detrimental in this endeavor. With the sand clocks of the fourteenth century, with error rates of roughly one hour per day, it would have been meaningless to speak of the slowing or speeding of time by less than an hour in a day. With the advent of seventeenth century pendulum clocks that were accurate within minutes per day, the time effects of minutes per day became measurable, and hence meaningful. Today, using atomic clocks, it is possible to measure the effects of matter and speed on time to within one-millionth of a second per year.

A major part of what is presently known about the characteristics of time as a physical entity is a result of the predictions of the theory of relativity. On the basis of various verifications of Einstein's theory of gravity, it now appears that this theory can explain the behavior of the universe at large. It is also known, however, that this theory cannot be used to describe the behavior of the physical world on subatomic scales. Fortunately, in the earlier part of the twentieth century, along with the theory of relativity, another branch of modern physics that dealt with the study of matter and universe on small scales was developed. This branch of physics was called quantum mechanics.

According to the quantum theory, for distances on the atomic scale, the behavior of physical objects cannot be modeled by large-scale theories such as the general theory of relativity. In particular, close to the big bang, the quantum effects become considerable, and general relativity is no longer applicable. In fact, close to the big bang, the quantum theory has its own implications concerning the state of the universe.

Obviously, a theory that could explain the universe on both a large scale and a small scale would provide better explanations concerning the nature of time than those now used. Einstein spent the last thirty years of his life looking for such a unified theory. At that time, however, there was little or no knowledge of the state of the universe on the large scale (expanding universe, black holes, and the big bang) or on the small scale (the nature of subatomic particles and nuclear forces). Despite decades of efforts by many physicists, there is still no link between the (macro) theory of general relativity and the (micro) quantum theory. Thanks to the perseverance of physicists, however, especially since the 1970's, many breakthroughs concerning the nature of the universe have been made. As a result, physicists seem to be considerably closer to achieving the goal of a unified theory.

Presently, efforts are under way to develop more powerful telescopes that will make looking deeper into the universe (and hence farther back in time) possible, and also to build the Superconducting Supercollider, which will be used to study the nature of subatomic particles. With the success of such efforts, much will be learned about the behavior of the physical world on both scales. It is hoped that, with new information, a unified physical theory of the universe will be formulated in the coming decades. Should such a theory become available, long-standing questions concerning the origin of the universe, the beginning of time, the relationships of time with matter and energy, and the end of time will be better answered.

Bibliography

Calder, Nigel. *Einstein's Universe.* New York: Penguin Books, 1979. A splendid popular science book and a national bestseller. Written in plain language, it discusses concepts of relativity and their implications to the universe at large. Easily understood by the layperson.

Davies, P. C. W. *Space and Time in the Modern Universe.* Cambridge, England: Cambridge University Press, 1977. Discusses the scientific discoveries about space, time, and the universe in relatively simple language. It should be especially appealing to people who have some knowledge of physics.

Einstein, Albert. *Relativity: The Special and the General Theory.* 15th ed. New York: Crown, 1952. An excellent book for the beginner. A popular exposition of the subject, it presents the sequence of ideas as they unfolded in the development of the theory.

Good, R. H. *Basic Concepts of Relativity.* New York: Reinhold, 1968. An introductory textbook for individuals with some knowledge of physics and mathematics who are interested in a more quantitative treatment of the subject.

Hawking, Stephen W. *A Brief History of Time.* New York: Bantam Books, 1988. A very interesting book for the general reader. It presents in simple language the findings of modern science on the origin and possible end of the universe.

Russell, Bertrand. *The ABC of Relativity.* 5th ed. New York: New American Library, 1958. First published in 1925, this book is still a very good source of information for the layperson on the basic principles of relativity.

Smart, J. J. C. *Problems of Space and Time.* New York: Macmillan, 1979. This excellent book is a collection of a number of classic papers on the nature of space and time. Treats the concepts of space and time mostly from a philosophical viewpoint.

Weinberg, Steven. *The First Three Minutes.* New York: Basic Books, 1977. An interesting and detailed description of the big bang by a leading physicist who is active in cosmology and quantum physics.

Whitrow, G. J. *The Nature of Time.* New York: Holt, Rinehart and Winston, 1971. A comprehensive discussion of the nature of time from scientific and philosophical points of views.

Badie Rowshandel

Cross-References

The Big Bang, 210; Black Holes, 238; General Relativity: An Overview, 950; Tests of General Relativity, 957; The Mössbauer Effect, 1467; Space-Time: An Overview, 2318; Space-Time Distortion by Gravity, 2325; Special Relativity, 2333; The Twin Paradox, 2608; The Expansion of the Universe, 2650.

TITRATION

Type of physical science: Chemistry
Field of study: Chemical methods

Titration enables chemists and many other scientists to identify how much of a chemical of interest is present in a test solution. It is very important to the chemical and pharmaceutical industries, as well as to most other industries.

Principal terms

ACID-BASE REACTION: a chemical reaction in which an acid and a base react to neutralize each other by producing water and a salt; also called a neutralization reaction

ENDPOINT: the point at which a titration being carried out is complete, as signaled by the method used for its analysis

INDICATOR: a chemical that changes color at the endpoint of a titration, signaling completion of the procedure

NEUTRALIZATION REACTION: another name for an acid-base reaction

OXIDATION-REDUCTION REACTION: a chemical reaction in which oxidation (the loss of electrons) and reduction (the gain of electrons) both occur

QUALITATIVE ANALYSIS: the analytic chemical method by which the chemicals present in a test sample are identified

QUANTITATIVE ANALYSIS: the analytical chemical method that identifies the amount of a chemical of interest present in a test sample being examined

STANDARD SOLUTION: a solution of known composition, containing the titrant, used in the titration of the test solution being analyzed

TITRANT: the chemical in the standard solution used to titrate the test solution being examined

UNKNOWN (TEST) SOLUTION: the solution being titrated with a standard solution of titrant, to identify its chemical content

Overview

Titration, which is often called volumetric analysis, is an analytical method by which chemists of many types—and many other scientists who study chemical reactions—identify how much of a chemical of interest to them is present in a solution that they wish to examine. Such analysis, to measure how much of a substance is present in a given test sample, is called quantitative analysis, in contrast to qualitative analysis, which simply identifies the chemicals that are present in a given test sample without determining how much of each chemical is present. Titration is applicable to the study of many different kinds of useful chemical processes, especially acid-base reactions and oxidation-reduction (redox) reactions.

The methodology of titration is essentially quite simple, and it is one of the most frequent operations that chemists and many others in chemical sciences and industry perform. Operationally, it is carried out in several steps. First, a measured volume of the unknown (test) solution to be titrated is placed in a container that will hold about 2.5 times its volume, and a stirring device is added to the container. The stirring device is used to keep the solution continuously mixing and homogeneous during the course of the titration procedure.

Next, a standard solution of a chemical (titrant) that will react with the test solution is added and allowed to react with it. This standard solution is prepared so that it contains titrant at a known concentration. Because of this, the measurement of the volume of standard solution used for titration allows the calculation of the concentration of the chemical that is undergoing quantitative analysis. It is for this reason that titration is said to be a volumetric procedure. While it is very desirable for chemical reactions studied by titrations to produce visible indications that they have occurred, titration is often monitored by use of indicators, which change color when the chemical reaction being carried out is over.

A simple example of a frequently carried out titration is that of a sample of an acid like hydrochloric acid (HCl). Such a titration is often carried out, because HCl is a common industrial waste that must be neutralized to prevent it from destroying equipment in any chemical plant that produces it as a waste. The data obtained from a titration show plant operators how much acid is being produced per day; then they can design the appropriate neutralization reaction to render it harmless.

The titration of HCl is usually carried out by reacting it with a standard solution of a base, like sodium hydroxide (NaOH). At the endpoint of such a titration with NaOH, all the acid is used up in a neutralization reaction that produces water (H_2O) and harmless NaCl, table salt. Use of other bases will produce water and other salts (for example, the use of potassium hydroxide, KOH, will produce potassium chloride, KCl). For every molecule of hydrochloric acid consumed in a titration, one molecule of the base has to be added. The endpoint of the reaction is also called its equivalence point, to underline the fact that equal amounts of acid and base are used up in the process.

The real participants in the reaction are hydrogen ions from the acid (protons, or H^+'s) and hydroxide ions from the base (OH^-'s). The titration is actually carried out with just enough NaOH (or another base, such as KOH) to produce a small excess of hydroxide ions. To identify when this point is reached, an acid-base indicator is often added to the reaction solution before the titration is started.

Acid-base indicators are strongly colored dye chemicals that are themselves weak acids and bases but that exist as different-colored forms when there is an excess of hydrogen ions present (at the start of the titration) and when there is an excess of hydroxide ions present (at the end of the titration), because the indicator exists as an acid form, HIn, in the presence of hydrogen ion excess, and as a basic form, In^-, in the presence of excess hydroxide ions.

An acid-base indicator that is often used for titration of hydrochloric acid is phenolphthalein, which is colorless in acid solutions and red in basic solutions. It may be said to exist as a colorless acid form (Hph), and as a red basic—or alkaline—form (ph^-). At the start of a titration of a sample of hydrochloric acid with base, all the phenolphthalein present exists as the colorless acid form. At the endpoint of the titration, a little of the alkaline red form is produced, signaling the completion of the process. If one wished to carry out the titration of an unknown solution of sodium hydroxide with hydrochloric acid as the titrant, phenolphthalein would again be used; however, the endpoint would be signaled by a color change from a red solution to a colorless one.

It may also be of interest that the use of phenolphthalein indicator and acidic and basic solutions is the basis for the wine-to-water trick carried out in some parlor magic acts. In addition, the color of some vegetables—for example, red and blue cabbages—results from excess acid or base in the soil reacting with acid-base indicator in their leaves. Many different acid-base indicators exist that can be utilized for titrations of different acids and bases.

Sometimes, titrations are carried out by the use of machines called pH meters, which use electrochemistry to identify the endpoint of the reaction carried out. Such potentiometric titrations use electrodes standardized against a solution of known acidity. For titration, the electrodes are then immersed in the test solution, and as acid is added, the pH (related to the hydrogen ion concentration by the equation pH = logarithm H^+ concentration) is used to identify the equivalence point of the titration. The pH meter has several advantages over indicators. These include the fact that titration of strongly colored acid or base unknowns can be carried out. Furthermore, measurements taken with pH meters are not affected by the fact that indicator solutions fade when they are allowed to stand.

Other types of titrations, including those of redox reactions, are carried out by very similar methods, modified by the fact that changes in parameters other than H^+ and OH^- concentrations are measured.

Applications

There are a tremendous number of applications of the process of titration, including study of industrial plant effluents for waste management; monitoring of the water leaching out of toxic dump sites to assure that hazardous wastes have not escaped; testing of the content and quality of foods and chemicals, from perfumes to plastics to pharmaceuticals; examining environmental contaminants in the atmosphere (for example, ozone); identifying the suitability of reaction conditions used in industry to prepare chemicals; and the design of appropriate conditions to carry out recombinant DNA research. Consequently, virtually every industry and aspect of science relies heavily on some aspect of titration. This is why, as noted earlier, virtually every chemist and most scientists carry out titration more often than almost any other analytical technique.

Among the most common titrations that are carried out are the acid-base titrations already described. Another often-used type of titration is the redox (or oxidation-reduction) titration. Here, the titrant of choice is usually an oxidizing agent, utilized to identify the content of a substance that can be oxidized (lose electrons), rather than a substance that can lose or gain protons in an acid-base reaction. Examples of chemicals studied via redox titrations include medicines and chemicals used in the storage and preservation of foods.

A third type of common titration is that used in the study of combination reactions: reactions that occur when two chemicals combine to produce a single chemical reaction product. For example, metal ions that can lead to food spoilage can be titrated by use of chemicals that combine with them to produce combination products called chelates. Combination reactions are also often studied by titrations accompanied by precipitation of reaction products (gravometric titrations).

In many cases, titration is carried out by procedures that use an indicator that changes its color (as already described) at the end of an acid-base titration, as a consequence of a redox reaction, or when a combination reaction is carried out. Other procedures utilized are coulometric titrations, conductometric titrations, and spectrophotometric titrations. Note that the equipment used and the basic procedures are similar in all such methods, and that additional discussion here will be limited to coulometric and spectrophotometric titration.

The electrochemical laws that Michael Faraday defined in the nineteenth century laid the groundwork for coulometric titration by recognizing that the extent of any redox reaction is proportional to the amount of electrical current passed through a reaction system studied. Consequently, a constant amount of an electric current applied for a specific time period can be used to produce a known amount of a titrant usable for a redox titration.

In such a process, the current passed through the test system is analogous to the titrant solution used in an acid-base reaction. When such a process is carried out for the purpose of titration, it is called a coulometric titration. Such titrations are very useful for study of redox reactions because chemically unstable titrants can be prepared electrochemically, just when needed, and because of the simplicity of operation (the opening or closing of an electrical switch). Note, however, that such titrations represent a case in which the titration is not a volumetric process, because the volume of the solution involved does not change.

Spectrophotometric titrations are carried out in optical machines called spectrophotometers. These machines can be caused to select a band of light of very narrow wavelength in the visible, ultraviolet, or infrared region of the electromagnetic spectrum. Such machines are capable of measuring the amount of color in a colored sample much more precisely than the human eye. They are useful because many chemicals absorb light of a specific wavelength, and these absorptions are often proportional to the amount of a particular chemical present in a test sample. Therefore, spectrophotometers can be used for the exact quantitative analysis of many chemicals, via a spectrophotometric titration.

Usually, a spectrophotometric titration is carried out in a fashion similar to an acid-base titration in that the reaction vessel contains a sample that is continuously mixed. The reaction vessel, however, is a special container (a cuvette) in the light path of the spectrophotometer, and at the start of the reaction, the sample titrated does not absorb in the color band chosen. In contrast, the reaction product absorbs this light well. Consequently, as the titration proceeds, the absorption of the solution rises until it reaches a maximum, and the amount of the chemical analyzed can then be calculated from the data. Such procedures are particularly important in analysis of the products of combination reactions; they also can be adapted for other uses.

Context

Modern titration techniques arose largely from the work of many researchers, beginning in the nineteenth century, in a wide variety of areas of chemistry. For example, the electrochemical endeavors of Michael Faraday produced much of the background for coulometric titration and other aspects of redox titrations. In addition, the elucidation of the theory of acids and bases by Svante August Arrhenius, and by J. N. Brønsted and Thomas M. Lowry, followed by Sören Sörensen's development of the concept of pH, produced the milieu for the design of many techniques used in acid-base titrations.

By the middle of the twentieth century, acid-base, redox, and many other titrations were widely used by chemists in research and in industrial applications ranging from examination of the composition of metal alloys to ensuring that perfumes, therapeutic drugs, and other manufactured products of the chemical industry were made correctly. In addition, many companies by that time were using titration techniques to assure themselves that the waste products produced by their boilers and furnaces were kept under control, so that acids, bases, and other chemicals arising as by-products of combustion did not corrode feed and exhaust pipes, or other plant components that would have to be replaced and add to company costs.

In the 1960's and 1970's, with the development of great societal interest in, and fear of, environmental contamination (resulting in part from Rachel Carson's *Silent Spring*, 1962), the extent of use of titration methodology burgeoned. The industrial wastes pumped into the environment began to be scrutinized by titration of air, water, and earth samples. The need for such scrutiny has led to many new techniques for examination of test samples that contain tiny amounts of chemicals believed to be hazardous. In addition, titration techniques were developed that allowed practitioners of analytical chemistry to carry out automated examination of very large numbers of test samples at once, rather than using labor-intensive, older methodologies.

Finally, it should be noted that the development of modern biochemistry and related molecular biology techniques (for example, recombinant DNA research) required wider use of titration techniques, to assure optimum experimental results, to facilitate analysis of results, and to provide exact methodology for maintenance of H^+ concentrations and concentrations of other reactant species at desired levels. Even maintaining a home aquarium appropriately to keep its occupants alive and healthy requires use of several different titration-related techniques.

Many aspects of science, technology, and everyday life, as well as both evaluation and prevention of environmental pollution, are just a few areas in which titration in its many forms makes relevant, important contributions to our lives and to society.

Bibliography

Carson, Rachel. *Silent Spring.* Boston: Houghton Mifflin, 1962. This well-crafted, controversial book describes many environmental pollution problems that result from industrial waste emission, inappropriate waste dumping, and related aspects of societal poisoning of the environment. Possible outcomes are also predicted. The reader will probably be able to envision areas in which titration techniques are useful in assessing the pollutant content of the environment.

Harris, Daniel C. *Quantitative Chemical Analysis.* 2d ed. New York: W. H. Freeman, 1987. This analytical chemistry textbook deals with titration at the level of the sophomore or junior science major. Many types of titrations are covered. Many practical examples and sample data are also included. Aspects of indicator choice and other choices in titration are detailed. The book is most useful for those desiring coverage by an analytical chemist.

Hill, John W. *Chemistry for Changing Times.* 5th ed. New York: Macmillan, 1988. This chatty, liberal arts chemistry textbook is particularly useful for basic concepts related to titrations, including acids and bases, chemical names, and redox reactions. These concepts are dealt with in a pleasant, disarming fashion that will educate the beginning reader in the basic concepts attempted. It is a good beginning for development of a factual base.

Lehninger, Albert L. *Principles of Biochemistry.* New York: Worth Publishers, 1982. This excellent undergraduate biochemistry textbook includes sophisticated coverage of aspects of pH maintenance, biochemical redox reactions, and other phenomena that underscore the importance of titration techniques to biochemical processes. The text is written at a level for the reader who wishes detailed information.

Manahan, Stanley E. *General Applied Chemistry.* 2d ed. Boston: PWS Publishers, 1982. This clear, simple text, aimed at liberal arts chemistry courses, does a nice job of explaining acids and bases, redox reactions, and titration at a basic level. Other fundamental concepts relevant to this article are developed throughout the book, and it is a good beginning for less-technical readers.

Miller, G. Tyler, and David G. Lygre. *Chemistry: A Contemporary Approach.* 3d ed. Belmont, Calif.: Wadsworth, 1987. This liberal arts chemistry text is particularly useful for its basic coverage of acid-base reactions and redox reactions (chapter 6). It handles the concepts well and provides other references in a section on further readings. Other concepts related to titration are addressed well here, too.

Radel, Stanley R., and Marjorie H. Navidi. *Chemistry.* St. Paul, Minn.: West, 1990. This excellent text, written for college science majors, covers acids, bases, redox reactions, pH, pH meters, and titration in depth. The coverage is valuable, but most suited to those wishing more detailed basic coverage than that found in texts aimed at college liberal arts majors. Other useful concepts related to titrations are included.

Willard, Hobart R., N. Howell Furman, and Clark E. Bricker. *Elements of Quantitative Analysis: Theory and Practice.* 4th ed. Princeton, N.J.: D. Van Nostrand, 1956. This older college text on quantitative analysis provides excellent coverage of the theory and practice of acid-base, redox, and combination reaction titrations. Many specific examples are included, and explanations of the processes involved are lucid. The book is best suited for readers who wish to obtain detailed information.

Sanford S. Singer

Cross-References

Acids and Bases, 16; Biological Compounds, 224; Chelation, 359; Electrochemistry, 721; Ox-Redox Reactions, 1725; Oxidizing Agents, 1732; Salts, 2126.

TOPOLOGY

Type of physical science: Mathematical methods
Field of study: General topics in mathematics

Topology is the study of the deformation of objects in spaces and the properties of those objects that are preserved under a very broad set of transformations.

Principal terms

BOUNDARY: the boundary of a set S is the set of all points that border the set S and its complement S'. Neighborhoods of points on the boundary will have points in S and S'

CONTINUOUS: a curve or surface is intuitively continuous at a point P if successively smaller neighborhoods of a point can be chosen to guarantee closeness of numerical values to that point P's values; basically, no breaks, gaps, or sudden jumps occur

HOMEOMORPHISM: a function that maps set to set if both its function and its inverse are continuous

MAPPING: a function that associates each point x in one set with a unique point $f(x)$ in another set

TOPOLOGICAL SPACE: a set with a collection of subsets

Overview

Topology is often popularized as rubber-sheet geometry, the study of deformable objects in space. This is an interpretation of topology as the study of the properties of spaces that are invariant when those spaces are stretched or contracted. In reality, topology permits other transformations, such as rotations, translation, and reflections.

Analytic geometry uses angles and distance as yardsticks of equivalence and therefore is quite rigid. Affine geometry is more lax. Topology permits any transformation that preserves the essential substructures of objects and the relationship between objects. Characteristics that are retained after those operations are studied and used to classify objects. A different value for any of those characteristics gives a topologically different object. First, it is necessary to define the topological equivalence of two objects.

Formally, a homeomorphism is a function $y=f(x)$ that maps set A to set B if every point x in A has a corresponding unique point $f(x)$ in set B (called the one-to-one property), every point y in B has a unique point x in A such that $y = f(x)$ (called the onto property), and continuity is preserved when mapping (both ways) between A and B. Two sets are homeomorphic if a homeomorphism exists between them. This is best illustrated and understood by means of examples.

A straight line and a parabola are homeomorphic. Imagine a straight line such as the x-axis and a parabola, $y = x^2 + 1$, which is above it. Clearly, each point on the *x*-axis can be mapped to a unique point above it on the parabola, and vice versa. Showing that continuity holds both ways requires proof. The continuity condition may be difficult to verify in many cases. In a similar case, a straight line and sine curve can be shown to be homeomorphic.

In fact, many simple curved (but non-self-intersecting) line segments on a plane can be shown to be homeomorphic. The appropriate mapping is to map endpoints to endpoints. Any intermediate points can be mapped to corresponding intermediate points by using proportions of the arc length. In other words, if *PR* and *SU* are the two segments (with endpoints *P*, *R* and *S*, *U*, respectively), then, for example, the point *Q*, one-third of the way from endpoint *P* to endpoint *R*, will map to the point *T*, which is one-third of the way from *S* to *U*. This exercise is more general than the *x*-axis trick. Again, continuity requires formal proof.

All circles are homeomorphic. The mapping required can be imagined by putting one circle inside another (so they are concentric) and then drawing a radius from the center to the outer circle; the point that is intersected in the inner circle can be mapped to the outer circle's point of intersection.

A circle and a square are homeomorphic. The required mapping can be constructed by inscribing one in the other. As in the previous example, it is necessary to draw a radius from the center. That radius will intersect the circle and the square at one point on each. That one point on the circle is mapped to the one on the square. This method will work for triangles, ellipses, and many polygons. Nonconvex polygons, however, are trickier.

This technique can also be applied in three dimensions to show that a cube and a sphere are homeomorphic. In more informal terms, two objects are homeomorphic if one can be deformed into the other without breaking neighborhoods. Therefore, spheres and their simple deformations into cylindrical shapes, tetrahedrons, and busts of Napoleon are all homeomorphic.

Not all objects are homeomorphic. Circles and line segments are quite different; deleting one point in a circle will still leave it a connected set, but deleting a non-endpoint point from a line segment leaves it as two unconnected subsets.

Similarly, a sphere and a torus ("doughnut") are not homeomorphic. A sphere can be cut by "drawing" any circle on it and deleting those points. With a torus, this is not the case. In this sphere-torus comparison, since the nonhomeomorphic diagnosis exists, there is a property that distinguishes spheres from tori. The genus, as it is called, can be used to classify surfaces. The genus will remain invariant under homeomorphism; therefore, it is a topological characteristic and can be used to classify objects. A more popular classifying attribute is dimension. A line segment and a cube are absolutely not homeomorphic.

Finding homeomorphisms is fun when the system can be intuitively or geometrically visualized. In dimensions higher than three or in complicated systems, purely formal proof methods and remarkable ingenuity are required. Rubber sheets, which are lacking in power and generality, are not adequate. The abstract approach is essential. Intuition can also be erroneous. Hard, often extensive proofs are not so faulty and can be exactly verified. Scrupulous logical consistency is more important than intuitive conceptualization.

Topology has often been called the study of homeomorphisms. When invariants are not obvious, showing that two objects are not homeomorphic is quite difficult. While showing that a homeomorphism exists often requires only a relatively easy construction, showing nonhomeomorphism necessitates showing that none of the many possible mappings will be a homeomorphism, a difficult prospect. Sometimes it is possible to determine whether a given case is homeomorphic; at other times, the situation may prove hopeless. Even worse, there is no general way to determine whether a little more effort may solve a particular problem. This does not diminish efforts along these lines, but demonstrates the complexity of the discipline and the progress that has been achieved.

These objects exist in spaces—not merely the standard metric spaces of Euclidean geometry, but more general abstract spaces. A topological space is a set S composed of elements (called points) such that: Each point P in S has a collection of subsets of S assigned to it ("around it") that are called a neighborhood of S. P is always a member of its own neighborhood; any set that contains a neighborhood of P is also one; if A and B are two neighborhoods of P, then their intersection is a neighborhood of P; if A and B are neighborhoods of P and A contains B, then A can serve as a neighborhood to each point in B. Several alternate definitions can be formulated, but the basic idea is that these subsets define the structure of the space at a "microlevel."

The most common topological spaces are conventional Euclidean spaces. As a broader example, n-dimensional manifolds form topological spaces. The "objects" referred to in this article are topological spaces (spheres, cubes, and so forth). Products of topological spaces can form topological spaces. For example, the Cartesian product of two line segments could form a two-dimensional rectangular surface. Defining a concept of distance can be used in creating neighborhoods that fabricate a topological space (called a metric space) in disciplines that do not involve the classical concept of point sets.

Topology has grown as a discipline and is subdivided into areas according to the objects considered, the methodologies used, the theorems proved, and tools for construction and decomposition. These branches are point-set topology, algebraic topology, combinatorial topology, and differential topology. Although they are distinct, the theorems, lemmas, and techniques of any one domain can be applied to any other.

Point-set topology is primarily concerned with topology from the perspective of sets of points, their connectivity, the nature of metric spaces, dimension theory, and "microproperties" (such as the openness of sets, space filling curves, and limit points).

One major subdivision of algebraic topology is homotopy theory. Assume that X and Y are spaces and f and g are two mappings from X to Y. Basically, f and g are homotopic if there exists a parameterized "super"-function $h(x, p)$ such that $h(x, 0)$

$= f(x)$ and $h(x, 1) = g(x)$, where x has every value in X and p varies from 0 to 1 continuously. The parameterized mapping h, which is called a homotopy, describes a gradual sequence of mutations that convert f into g. This basic tool can then be used as an equivalence relation (that is, a definition of equivalence) that produces classes of functions that are homotopic within each class. These are called homotopy classes.

Other directions of development are the construction of homotopically equivalent spaces and the group theoretic topological invariant, the fundamental group, the first in a sequence of homotopy groups. The fundamental group view is conducive to computation and extremely fruitful. The approach, which is oriented toward the investigation of connectivity characteristics of spaces, has a wonderfully combinatorial nature. Topological spaces are represented by means of generators and defining relators. To oversimplify, problems in topology can be reduced to problems in group theory for solution by algorithmic means. Another class of groups used to study the nature of topological spaces is that of homology groups. Unfortunately, concepts in and examples of differential topology are beyond the scope of this level of presentation.

Applications

A common adage states that any area of mathematics, no matter how pure and abstract, will eventually have physical applications. Even previously distrusted illusory, bizarre, or wild concepts such as non-Euclidean geometry, infinite dimensional spaces, divergent series, and complex numbers have become standard tools of science. Around the time of Sir Isaac Newton, the rates of progress in mathematics and physics were approximately equal, and the two fields were quite intertwined as models. In the 1800's, the situation shifted, and mathematics became more abstract, formal, rigorous, and axiomatic. Tools such as Lie groups, differential geometry, tensors, and abstract groups were being developed, but they had no extensive immediate application. With the passage of time, these tools formed the building blocks of new scientific interpretations and the framework of modern theories.

Topology can be considered in terms of the framework that has just been described. In the spectrum of mathematics, topology and number theory seem to have the least application in other sciences. Abstraction apparently removed application. A deeper analysis is required. First, topology is the fundamental tool for describing the structure of space. Perhaps humans have not crossed the threshold of extensive applications to space in general. Perhaps new higher-dimensional models and symmetries will require the incorporation of topologies that are not so obvious. Perhaps modern theories will be greatly simplified by reinterpretation along topological guidelines.

The speculation is already confirmed. The fractal structures that occur in nature are but one area of application. Several monographs (notably, Benoit Mandelbrot's *The Fractal Geometry of Nature*, 1982, and Michael Barnsley's *Fractals Everywhere*, 1988) show applications of fractals in the modeling of dynamical systems, turbulence, physiology, geomorphology, imaging, snowflakes, galaxies, spray phenomena, and countless other areas. These are only examples from observation; future applications may be based on unobservable structures whose topological nature affects observable factors.

Topology has been applied to the study of molecular structure in chemistry. In a *Scientific American* article entitled "Predicting Chemistry from Topology," Dennis Rouvray cites several topological numerical characteristics of molecules that are used to predict chemical characteristics. Several complex formulas that are used for calculations are cited, notably, the Wiener index, the Randic molecular-connectivity index, the Balaban centric index, and the hydrogen-deficiency number. Properties predicted involve carcinogenicity (the "cancer-inducing" measure), fuel combustion, corrosion, drug effects, pollution, and many others. The wealth of applications, the power of these methods, and the availability of computers to conduct the calculations make this a potential area for a scientific revolution.

Topology's influence is quite subtle. Assumptions about spaces often go unmentioned or are buried in implicit usage of conditions required for the solution of equations on surfaces or spaces.

Topological methods have been applied to the study of knots and braids. While applications exist in physics, they have not been widespread in the literature.

Combinatorial topology has multitudes of applications in graph theory. The concept of connectivity in graphs, Eulerian paths, Hamiltonian paths, and many other definitions are topological—hence the area of topological graph theory. Map-coloring theorems, traversals, matchings, planarity of graphs (and more generalized embeddings), and network designs are some aspects of graph theory that are influenced by topological considerations. In the map-coloring problem, the goal is to determine the minimal number of colors needed to color countries on an arbitrary map so that no two adjacent countries have the same color. The solution of the problem depends on the nature of the connectivity of the underlying space. On the Euclidean plane, the celebrated result was that a minimum of four colors is required, but on a torus the minimum has been shown to be seven colors. With networks, the connectivity criterion directly affects features such as fault tolerance and reliability. Planarity also influences the design of chips on silicon surfaces. Topology has also been directly applied to switching-circuit theory.

Topology has had a profound history of interchange with algebra. Groups that meet the conditions of topological spaces form topological groups, provided that the group operation and its inverse are continuous. Rings and other algebraic structures with appropriate conditions likewise have topological theories.

Invertible matrices form a group, with the operation being matrix multiplication. Simultaneously, this group also constitutes a topological space in a Euclidean space of dimension equal to the number of entries in the matrix. Continuity and neighborhood conditions can be formally verified for this operation.

Topological vector spaces arise in theoretical physics as the foundation for further work. Among the most important types of topological vector spaces, Hilbert spaces and Banach spaces are used as the canvas of quantum mechanics and the groundwork for operator theory and functional analysis. Topological fixed-point theorem conditions can actually determine the existence of differential equation solutions that occur in several areas of physics. In dynamical systems, phase spaces are examples of topological spaces.

Topology can be used to explain more whimsical phenomena. For example, a buttoned jacket can be removed without unbuttoning if one uses a familiar vertical motion. If one clasps both hands together, however, removing the jacket becomes impossible (without tearing). The body becomes a torus instead of the "sphere" it commonly "is." The clasping of hands changed the topological structure. Other concepts such as one-sided surfaces (Möbius strips) and Klein bottles, although they are scientifically valid examples and pedagogically useful, are not the substance of topology's power and generality.

Context

The earliest works in topology were threads of independent research that were interwoven by later mathematicians who reinterpreted, refined, and unified the area. In 1679, Gottfried Wilhelm Leibniz called it analysis situs, which he initially described as a geometry without numbers. In 1751, Leonhard Euler formulated his famous convex polyhedron equation, which states that $V - E + F = 2$, where V, E, and F are the vertices, edges, and faces of a convex polyhedron (like a tetrahedron in which $4 - 6 + 4 = 2$). In 1811, Augustin-Louis Cauchy extended Euler's equation to other spaces.

Euler also solved the Königsberg bridge-crossing problem by introducing the Eulerian path, the launch point of graph theory and network topology. The four-color conjecture for planar maps came from Francis Gutherie in 1852.

The term "topology" first appeared in an 1848 monograph by Johann Listing. The work of Carl Friedrich Gauss on surfaces and Georg Friedrich Bernhard Riemann on manifolds, connectivity, and genus provided the basic groundwork for later inquiry. With progress, older proofs and definitions were superseded by more encompassing and rigorous ones.

Henri Poincaré (1895 and later papers) conceived and developed the theory of fundamental groups and progenitors of the homology groups, the indispensable invariants of algebraic and combinatorial topology. Refinements by Heinrich Tietze and Max Dehn in the early 1900's established a rigorous group theoretic foundation for these concepts, the key problems of the area, and algorithmic methods for attacking them. Eventually, this would pave the way for inquiry and solutions by high-speed computer.

In the nineteenth century, Georg Cantor formulated the theory of point sets. Work on continuity and analysis long predated his research, but a reinterpretation was brewing. As an extension of the concept of a metric space (in which distance is the exclusive focus for defining neighborhoods), Felix Hausdorff conceived of topological space as a set with an associated family of subsets with several neighborhood properties. Again, abstract models have generalized and redefined the older methods and constructs.

The second half of the twentieth century witnessed some remarkable progress in classification theorems, major conjectures, and open problems (some reportedly on the verge of solution). Advances in applications such as fractals and molecular topology are significant, motivating many mathematicians to conduct further research.

Bibliography

Aleksandrov, A. D., A. N. Kolmogorov, and M. A. Lavrent'ev. *Mathematics: Its Content, Methods, and Meaning.* Cambridge, Mass.: MIT Press, 1969. Although not as well done as most articles in this set, the topology chapter (18) is a pleasant introduction to the barest basics. The style is interesting, but the coverage is a bit weak and lopsided. The vector section is the most beneficial.

Arnold, B. H. *Intuitive Concepts in Elementary Topology.* Englewood Cliffs, N.J.: Prentice-Hall, 1962. Among the easiest of textbooks on the basics of topology, this book covers all the foundations of the area in a leisurely way but still retains some degree of theorem-proof-exercise organization. Great for building rubber-sheet intuition and a fine stepping-stone to the more standard texts.

Blackett, Donald W. *Elementary Topology: A Combinatorial and Algebraic Approach.* New York: Academic Press, 1967. A readable introductory textbook with minimal material on point-set fundamentals. The topics selected are stimulating, with extensive explanations and useful tools for further work.

Francis, George K. *A Topological Picturebook.* New York: Springer-Verlag, 1987. With enough illustrations, even complex concepts become intuitive. This text has among the highest ratio of pictures to concepts of any book on topology and higher mathematics. These are not trivial illustrations used to popularize; each is expertly done and elucidates deep concepts. This visual key opens the portals of higher dimensions and complexities that are often buried in notation and formalism. Paradoxically, it is not elementary but is still accessible.

Griffiths, H. B., and P. J. Hilton. *A Comprehensive Textbook of Classical Mathematics.* New York: Springer-Verlag, 1970. A contemporary synopsis of topology is presented in chapter 25. Key theorems and concepts are presented with a refined brevity. The level is not totally elementary, the goal being an overview rather than an introduction; however, the immaculate and professional style easily compensates for it. The occasional asides and remarks are thought-provoking treats, with philosophical, unifying, or applicational ramifications.

Hilbert, David, and S. Cohn-Vossen. *Geometry and the Imagination.* New York: Chelsea, 1952. A reprint of a classic by two of the forefathers of modern mathematics, this book devotes chapter 6 to the exploration of topology in an original and intuitive manner. Unlike some popularizers, the authors are true masters of eloquence, clarity, and imaginative brilliance. Aesthetes will appreciate the wire-frame photographs and illustrations.

James, Robert C. "Combinatorial Topology of Surfaces." *Mathematics Magazine* 29

(September/October, 1955): 1-39. Pleasant, readable, concrete introduction to combinatorial topology basics.

Lang, Serge. *The Beauty of Doing Mathematics*. New York: Springer-Verlag, 1985. Chapter 3 is a dialogue-style exposition of topology and geometry by a most respected and prolific mathematician. A rare treat from an author noted for axiomatic formality and meticulous rigor. The diagrams are highly instructive.

Lord, E. A., and C. B. Wilson. *The Mathematical Description of Shape and Form*. Chichester, England: Ellis Horwood, 1984. This is a wonderful collection of perspectives on the mathematical study of surfaces and shapes. Devoid of exercises, this is not a textbook. The examples and numerous illustrations are effective. The diverse and interesting topics include catastrophe theory, chaos, and fractals. Topology is dispersed throughout. In the same series by the publisher is *Surface Topology*, by P. A. Firby and C. F. Gardiner, which presents a more cohesive and consistent view of topology (with exercises).

John Panos Najarian

Cross-References

Algebra: An Overview, 82; Differential Geometry, 663; Euclidean Geometry, 852; Group Theory and Elementary Particles, 1022; Matrices, 1346; Non-Euclidean Geometry, 1547; Nonlinear Maps and Chaos, 1556; Sets and Groups, 2191.

TOPS AND GYROSCOPES

Type of physical science: Classical physics
Field of study: Mechanics

The unusual motions of tops and gyroscopes are explained through the concepts of rigid body motion. These simple devices have diverse applications in modern technology.

Principal terms

ANGULAR MOMENTUM: the product of an object's rate of rotation and its moment of inertia around the axis of rotation

MOMENT OF INERTIA: a property of an object's mass distribution around a given axis of rotation

PRECESSION: the motion of a rotating rigid object in which the spin axis changes directions because of the presence of an applied torque

TORQUE: the effect of a force applied so as to cause a rotation

Overview

Gyroscopes and spinning tops are specific instances of objects whose behavior falls under the category of mechanics known as "rigid body motion," in which it is assumed that each object can be treated as perfectly rigid, that is, having no flexure. The underlying theory applies to all rigid objects during periods of rotation.

A critical quantity in the characterization of rotational motion is an object's moment of inertia, which does not depend on the object's mass directly, but rather on how the mass is distributed. When a mass rotates, it moves around an axis of rotation, which may be a physical object (as with a wheel rolling about an axle) or an imaginary line (as with a spinning ball tossed into the air). The object's moment of inertia depends on its axis of rotation; the more distant the mass is from the axis of rotation, the greater the moment of inertia relative to that axis. For example, in the case of a solid sphere, the moment of inertia about any axis that passes through the sphere's center is $2mr^2$ (where m is the sphere's mass and r is the radius), while for a hollow sphere with the same mass (the mass is concentrated in a spherical shell), the moment of inertia is $2mr^2/3$.

A second quantity that affects rotational motion is torque, a force acting either to cause rotation or to change some property of the rotation (that is, a turning effect). Torque has both magnitude and direction; it is a vector quantity. The magnitude is defined as the product of an applied force times the distance from the object's axis of rotation to the imaginary line along which the force acts. Torque also has a direction, along the line that a screw would progress if this same torque were applied, through a screwdriver, to drive the screw into a surface. For example, the torque required to start a large wheel turning is given by the radius of the wheel times the force acting tangent to the wheel's edge.

An equation analogous to Sir Isaac Newton's second law of motion ($F = ma$)

relates the torque T to the object's moment of inertia and the rotational acceleration that results: $T = I\alpha$, where $I\alpha$ is the product of the moment of inertia and the angular (rotational) acceleration around a given axis of rotation. It is crucial to note that this equation applies *only* when the axis of rotation has a fixed orientation relative to both the object and to the environment surrounding the object. Returning to the example of the two spheres, the solid sphere will have 3/5 times the moment of inertia of the hollow sphere, so that if the same torque is applied to both, the solid sphere will undergo 5/3 times the acceleration of the hollow one.

A further analogy exists between rotational motion and particle motion (as described by Newton's laws of motion): In the absence of torques, an object undergoing rotational motion will continue indefinitely without decelerating. Therefore, the object's energy of rotation will remain constant unless a torque acts to remove some or all of this energy.

As with particles in motion, a rotating body also has momentum, but it is referred to as "angular momentum," and its magnitude is given by the product of the body's moment of inertia around an axis of rotation and the body's rate of rotation around that same axis. The angular momentum is assigned a direction as well, along the axis of rotation in a direction away from an observer who sees the object rotating clockwise. For example, each wheel of an automobile (moving forward) has angular momentum directed outward to the left of the car.

In the late eighteenth century, the Swiss mathematician Leonhard Euler systematized the analysis of rotational motion for rigid bodies. A full understanding of the theory requires knowledge of differential equations, but its essence can be stated as the following: The time rate of change of a body's angular momentum (in magnitude and/or direction) is equal to the applied torque. The corresponding mathematical formulation is known as Euler's equations of rigid body motion. The equations not only explain increases or decreases in rotational speed when a torque is applied but also imply that with no torque present, the object maintains constant angular momentum (both in magnitude and in direction). A further implication is that when a torque is applied to a rotating object, after a small interval of time, the object's angular momentum will have changed direction so as to be more nearly aligned with that of the torque.

Euler's equations explain the behavior of gyroscopes. A gyroscope, in its simplest form, consists of a solid disk that is allowed to spin about an axle perpendicular to the flat face of the disk. This axle is supported at each end by a rigid frame and sets of bearings that allow the axle and disk to turn freely within the frame.

When the disk is spinning, it is evident to one holding the frame and attempting to turn it about a direction that is not aligned with the axle that the turning motion is strongly resisted. This phenomenon, which is simply the reaction of the gyroscope to the applied torque needed to change the angular momentum's direction, is called "gyroscopic stiffness." Higher angular momentum (created by spinning the disk faster) will give greater stiffness, but any applied torque will cause some rotation of the frame. One of the great fascinations with gyroscopes is that this resulting rotation of the frame is not in the direction that one would expect. For example, if the spin axis is horizontal and the frame is held so that one views the disk rotating clockwise, then trying to turn the frame so that the far end would move downward and the near end upward (applied torque directed to the left) results instead in the entire frame rotating about the vertical axis. (Viewed from above, the direction of this rotation will be counterclockwise.) This phenomenon is explained by Euler's equations: The angular momentum changes in such a way as to align itself with the applied torque. As long as one continues to apply the torque, the rotation of the frame continues; removing the torque stops the frame's rotation.

A similar phenomenon occurs when one end of the frame is suspended (for example, by a string) and the other is released. With the disk spinning, gravity creates a torque in the horizontal direction (because it exerts a force that tries to rotate the gyro about a horizontal axis), but the gyro rotates around a vertical axis for the same reason. In both situations, the rotation about the vertical axis is called "precession." The precession rate (rate at which the frame rotates) is given approximately by the applied torque divided by the angular momentum of the disk.

A spinning top is supported by its point of contact with a horizontal surface. If perfectly balanced and set on a frictionless surface, the top would spin upright indefinitely; however, friction and imperfections in the surface cause it to tilt over. Gravity then exerts a torque in the horizontal direction and the top reorients itself as its angular momentum changes to align with the torque. Interesting behaviors of a top are caused by the additional torque created by the actual friction between the top's point and the surface. As the top loses angular momentum and begins to tilt further, the motion of its point creates frictional torque that changes the top's orientation in somewhat unexpected ways.

Applications

Even before Euler's elegant treatment of rigid body motion, Newton had analyzed the motion of simple gyroscopes and applied the same basic principles to the gyroscopic motion of Earth. Because Earth bulges slightly at the equator and because its spin axis is tilted relative to its plane of motion in the solar system, the sun's gravity pulls slightly harder on the part of the bulge closer to it, exerting a small torque on Earth (the torque direction is approximately in the earth's plane of motion). As a result, Newton predicted that Earth's spin axis must precess at a rate of about one circle every twenty-six thousand years.

In modern times, one of the principal uses of gyroscopes has been for navigation. To generate a reference direction for navigational purposes, the frame of the simple gyroscope is attached to two or more outer frames via a series of pivoting joints called "gimbals," which allow the inner frames to rotate inside the outer frames. This system provides a means of suspending the rotating disk, while allowing it to orient itself in any direction. The less friction there is in the gimbals, the more closely this approaches a torque-free system. By driving the rotor (the disk) with a small electric motor attached to the inner frame, the rotor speed can be held constant. Placing this apparatus in a spacecraft provides a reference direction for the vehicle, since the spacecraft's motion will not affect the orientation of the rotor's spin axis. Electrical sensors at the gimbal joints provide input to the spacecraft's onboard computer, indicating how much the spacecraft has rotated relative to the reference direction. For the computer to determine completely the vehicle's orientation, there must be three such reference directions, mutually perpendicular orientations. Therefore, a minimum of three gyroscopes is required; in many spacecraft, a fourth gyroscope is included for redundancy in case of a failure. Such a system of gyroscopes constitutes part of a spacecraft's inertial reference assembly, from which ideally the onboard computer can determine the vehicle's position and orientation without having to use optical readings of star positions or other information. Realistically, though, even the small amount of friction in the gimbal bearings exerts torques on the rotor as the vehicle changes orientation, and the reference direction drifts away from its desired position. Periodically, the onboard computer must employ optical sightings of stars or other external information to determine the new reference directions of the rotors in the assembly. Nevertheless, this approach provides useful intervals of completely independent navigational capability between star sightings.

While the constant direction of a rotor axis makes the simple gimballed gyroscope ideal for navigation in space, this property is undesirable for Earth-bound navigation. As an aircraft, surface ship, or submarine crosses lines of longitude, the position of the geographic North Pole changes relative to the vehicle. For a gyroscope to indicate the direction to the North Pole, it is then necessary to make the rotor axis precess at exactly the rate that accounts for the apparent motion of the North Pole. This is accomplished by means of a spring mechanism attached to the gimbal frames and adjusted to provide the necessary torque for the current speed of the vehicle. The rotor axis of this device, called a gyrocompass, then always points toward geographic North.

The property of gyroscopic stiffness is also applied to controlling the orientation of various objects. Historically, one of the first uses of this principle was in stabilizing the motion of a bullet. The cylindrical shape of a bullet makes it susceptible to tumbling as it experiences torques because of uneven atmospheric forces in flight. This tumbling not only causes a greater atmospheric resistance (and consequent reduction in range) but also can skew the bullet's path. By cutting spiral grooves in a gun barrel ("rifling"), the bullet is made to spin about its long axis, giving it a high value of angular momentum and stiffening it against the torques created by the air.

It is for precisely the same reason that one deliberately imparts a spin to a football as it is thrown. Also, satellites are frequently spin-stabilized to resist the torques exerted by Earth's gravitational and magnetic fields or even by the extremely rarefied atmospheric gases at altitudes of 500 kilometers.

Context

The principles of rigid body motion have been applied widely since Newton's time. From toy tops to modern high-speed rotating machinery, these concepts are crucial to a full understanding of how such systems behave, and in the case of engineering design, how to make the systems work in some desired fashion.

One of the most vexing problems for navigational gyroscopes has always been the elimination of friction in the gimbal joints. Improved lubricants in mechanical joints gave way to air bearings (in which the bearing elements are actually separated by a cushion of compressed air that prevents mechanical contact and thus reduces friction). That approach has, in some cases, been superseded by the use of magnetic bearings, which use electromagnets in one frame to keep it from coming into contact with the gimbal axle of the adjacent frame.

A technological innovation that will eventually replace all mechanical gyroscopes for use in inertial reference assemblies as well as gyrocompasses is the ring laser gyroscope. Although misnamed (it has no rotating masses), this device operates by sending the two halves of a split laser beam in opposite directions around a ring (using mirrors). If the vehicle that houses this apparatus is undergoing a rotation about an axis perpendicular to the ring-plane, then the two beams will create an optical interference pattern on the screen where they are recombined. This pattern shifts on the screen at a rate proportional to the rotation rate of the vehicle, and an onboard computer (using an optical counter) can then determine in a given time interval how much rotation has occurred. As with a mechanical reference assembly, at least three ring laser gyroscopes are needed for navigation, but their great advantages are an absence of moving parts and consequent lack of friction.

Bibliography

Asimov, Isaac. *Understanding Physics.* Vol. 1. New York: Walker, 1966. This volume treats both classical and modern physics using almost no mathematics. The explanations are lucid and witty, a hallmark of this author. His examples are drawn from common experience and serve to illustrate the concepts quite effectively.

Feynman, Richard P., Robert B. Leighton, and Matthew Sands. *The Feynman Lectures on Physics.* Vol. 1. Reading, Mass.: Addison-Wesley, 1963. These lectures were intended to challenge some of the brightest physics students at the California Institute of Technology, yet they have exceptional clarity and appeal for general audiences. One of the few books that effectively treats the approximate nature of physical laws.

Frautschi, S. C., R. P. Olenick, T. M. Apostol, and D. L. Goodstein. *The Mechanical Universe: Mechanics and Heat.* New York: Cambridge University Press, 1986. This is the companion text to *The Mechanical Universe*, a series of videotaped lectures produced by the Public Broadcasting System and based on Goodstein's lectures at the California Institute of Technology. The historical development of these subjects, carefully integrated with the superb graphic analysis, is one of the best treatments available.

Gamow, George. *Matter, Earth, and Sky.* Englewood Cliffs, N.J.: Prentice-Hall, 1965. Although written as a textbook for freshman physics courses, this work is based on the author's long experience in lecturing and writing for general audiences interested in science. Successfully combines physics, chemistry, geology, and astronomy, including numerous helpful diagrams and some of Gamow's own cartoons.

Perry, John. *Spinning Tops and Gyroscopic Motions.* New York: Dover, 1957. This fine treatment of the dynamics of rotation presents the concepts of rigid body dynamics without any equations. There are numerous examples and detailed illustrations, many of them taken from everyday experience, to assist the nontechnical reader.

Rothman, Milton A. *The Laws of Physics.* New York: Basic Books, 1963. Using a minimum of mathematics, this book introduces the basic principles of classical and modern physics, with many excellent examples. It gives considerable attention to the philosophical aspects of human knowledge of physical reality.

Walker, Jearl. *Roundabout: The Physics of Rotation in the Everyday World.* New York: W. H. Freeman, 1985. This is a collection of ten articles from "The Amateur Scientist" column in *Scientific American*, covering not only a variety of tops but also boomerangs, ballet, and amusement park rides. Each article includes many examples of rigid body motion taken from common experience and gives some suggestions for simple experiments.

Robert G. Melton

Cross-References

Conservation Laws, 546; Differential Equations, 658; Mechanics, 1367; Newton's Laws, 1509; Vectors, 2679.

TRAJECTORIES

Type of physical science: Classical physics
Field of study: Mechanics

The trajectory of a particle can be predicted using Newton's second law of motion if the mass of the particle and the total force acting on it are known.

Principal terms

ACCELERATION: the rate at which an object's velocity changes in speed and/or direction

APOGEE: the point of an orbit at which the object is the farthest from the earth

FORCE: the amount of push or pull exerted against an object

MASS: the amount of matter in an object

PERIGEE: the point of an orbit at which the object is the nearest to the earth

VELOCITY: the speed and direction of a moving object

Overview

The trajectory (path) of a moving object is dictated by Sir Isaac Newton's second law of motion, first published in *Philosophiae Naturalis Principia Mathematica* (1687; *Mathematical Principles of Natural Philosophy*, 1729), which states that a body of mass m will undergo an acceleration a when an external force F acts on it. (An external force is one that does not act between two components of the mass itself.) Newton's second law is concisely stated as $F = ma$ (where F is measured in newtons, m in kilograms, and a in meters/second^2); however, it does not directly indicate either the shape of the trajectory or where the object will be at some future time. These are the subjects of trajectory analysis, which generally requires the use of the calculus and a detailed knowledge of the net force acting on the mass.

Newton's second law applies only to a given particle of mass (a geometric point without measurable dimensions). It is always possible to model a real object as a particle of equivalent mass placed at the object's center of mass (the average location of all the body's mass). In addition, the quantities F and a are vectors: Each has both a magnitude and a direction. When several forces are present, they are added together as vectors to give the single net force that acts on the center of mass of the object. Each force vector generally has components along each of three mutually perpendicular directions, and all the components along a particular direction are added to give the component of the net force along that direction. If a particle experiences a force of constant magnitude and direction, then it accelerates in the same direction as the force and with a constant acceleration. If the force is changing (either in direction or magnitude), then at each instant in time, the particle's acceleration will be in the direction of the force at that instant and will have a magnitude determined by Newton's second law.

The quantity acceleration is the rate at which the object's velocity (speed and direction) is changing. In order to compute a particle's trajectory, it is necessary to know its initial velocity. A positive acceleration along a given direction means that the object is moving with ever-increasing speed (in that same direction). A negative acceleration indicates that the object is slowing down, or even starting to move in the opposite direction. A further consequence of Newton's second law is that an object which is already moving at some constant speed cannot change its velocity (that is, experience any type of acceleration) unless an external force acts on it. Thus, the simplest of all types of trajectories is a straight line. For example, a mass which has started to move in some direction in deep space, far from the significant gravitational attraction of stars or planets, will not experience any forces. Therefore, the mass will continue to move in a straight line along the original direction until it does encounter some force, such as the result of a collision with another mass. On the earth, the object would typically experience not only the force of gravity but also forces that are attributable to friction, such as atmospheric drag forces or contact friction while sliding along a surface. These forces will tend to decrease the object's speed by removing energy from it, and they may change the shape of its trajectory as well.

Because the relation $F = ma$ indicates the rate at which a body's velocity is changing over time, it is called a differential equation. In general, such an equation requires special methods of calculus for its solution and indicates where an object will be at any future time. Indeed, another useful way to view a trajectory is as the set of all positions that an object has during its motion.

An interesting property of trajectories is that their shape depends on the frame of reference in which they are viewed. For example, if a mass is dropped inside a train car that is moving with constant speed, then the passengers will see the mass fall straight downward. This motion results from the fact that both the passengers and the mass all have the same horizontal speed; gravity is the only external force present, and it causes the purely vertical motion. To an observer standing on the ground and looking through a window into the train, however, the object moves forward and downward on a curved path. This motion is seen because, relative to the ground, the object has some initial horizontal speed. The difference in trajectory shape in this example is relatively simple. A more complex difference occurs when one observer is in a rotating frame of reference. For example, a cannonball fired horizontally at high speed from the North Pole actually moves along a path which would otherwise be straight except for the force of gravity causing it to follow the curvature of the earth. As viewed from a stationary point in space, the trajectory is due south. As the cannonball flies, however, the earth rotates underneath it, giving the appearance to an observer on the ground that the cannonball is curving to the west. Because of the relatively slow rotation rate, the earthbound observer does not recognize intuitively what is causing the deflection of the trajectory to the west and concludes that there must exist a special force not associated with gravity which is causing this effect (known as Coriolis acceleration).

Applications

Trajectories are best illustrated by several examples that show different behaviors of motion. An object released from a high platform will fall straight downward (neglecting any horizontal wind) with increasing speed. The simplest model of this motion assumes that there is no atmospheric friction (drag), in which case the speed of the body increases constantly according to the equation $v = gt$, where v is the velocity in meters/second, g is the acceleration caused by gravity (9.81 meters/second^2), and t is the elapsed time in seconds. The distance d through which the body falls is found from the equation $d = (1/2)gt^2$. A more sophisticated model would include the effects of atmospheric drag, which would not change the purely vertical shape of the trajectory but could significantly alter the speed and distance formulas, depending on the particular object's shape.

A more interesting trajectory results from casting the mass from a high platform, with its velocity initially in the horizontal direction. Gravity immediately begins to pull the object downward but, according to Newton's second law, the velocity change is only in the direction of the external force (downward). Therefore, the horizontal speed of the body does not change, again neglecting air friction. As a result, the trajectory is the familiar downward curve that is characterized mathematically as a parabola. This same type of trajectory is also produced by a projectile fired from the earth's surface with both horizontal and vertical components to its initial speed. The body arcs upward, with gravity acting only to slow its vertical motion, while its horizontal speed remains constant. Eventually, the projectile reaches the apex of its trajectory and then starts to fall back. If air friction is significant, then the shape of the trajectory is not precisely parabolic, but still qualitatively similar.

It is also possible to apply a force to a particle in such a way that only the direction of motion changes while the speed remains constant. This situation is easily demonstrated by swinging a small object at the end of a string in a horizontal circle about one's head. By keeping a constant tension in the string (that is, by applying a constant magnitude of force), the object can be kept moving in a circle at constant speed. Its direction of motion is constantly changing, however, as predicted by Newton's second law. The direction of the force is always from the object back toward the center of the circle. An important example of such a trajectory is that of subatomic particles being forced to move in a circular path by means of a magnetic field, which is a method used by nuclear physicists in particle-collision experiments. Because of their motion in the magnetic field of the accelerator, these electrically charged particles experience what is known as the Lorentz force, which is always directed toward the center of a circular path. The magnetic field holds the particles on the path, their energy is increased, and they are allowed to collide with other particles as part of the experiment being performed.

Satellite trajectories (orbits) are even more complex. The simplest is a circular path, such as about the earth, on which the satellite moves at constant speed, with gravity providing exactly the right force to change only the direction. If the orbit is high enough, generally greater than 1,000 kilometers, then no atmospheric drag is present and the satellite will continue on the circular trajectory forever. To initiate such a trajectory, the booster rocket for the satellite must stop its engines just as it reaches the condition $v^2 = GM/r$, where v is the speed (in meters/second), r is the distance (in meters) from the center of the earth, and GM is the product of the universal gravitational constant (6.67×10^{-11} meters^3/kilograms − seconds^2) and the earth's mass (5.98×10^{24} kilograms). If the satellite is moving somewhat faster or slower than the particular speed needed for the given circular radius, then the resulting trajectory is an ellipse with the earth's center located at one focus. The gravitational force from the earth still regulates the motion, but in this case, the satellite will move alternately closer to and farther from the earth, with its speed becoming respectively greater and then lesser. Special terminology used for this most common type of satellite orbit includes "apogee," which is the point on the orbit farthest from the earth, and "perigee," which is the point nearest to the earth. If the perigee is located at least 1,000 kilometers above the earth's surface, then no significant atmospheric drag occurs and the satellite will never alter from its orbit.

Two other types of satellite trajectories are possible, both of which result in the satellite leaving the earth and never returning. If the initial speed of the satellite at injection is such that $v^2 = 2GM/r$, then the resulting trajectory will be a parabola with a perigee close to the earth but with its two open ends an infinite distance away. This situation is the simplest of all escape trajectories and must be used to start a space probe on a mission to another planet. If the initial speed is greater than that needed to produce the parabola, then the resulting path is a hyperbola. The only qualitative difference between parabolas and hyperbolas is that the two legs of the parabola eventually become parallel at great distances from the earth while those of a hyperbola do not. As for the spacecraft's motion, it will be greater (on the average) on a hyperbola than on a parabola; that is, the spacecraft has more than enough energy to get away from the earth without gravity slowing and eventually pulling it back.

For satellites in relatively low orbits (circular or elliptical), the effects of atmospheric drag on the trajectories can be quite pronounced. Qualitatively, the drag reduces the satellite's energy, causing it to fall steadily closer to the earth in a form of spiral motion. Unless corrective action is taken, such as using the satellite's own engines to counteract the drag and raise it back to a higher altitude, the spacecraft will eventually burn up from the intense heat of friction and crash on the earth.

Context

The motion of objects has been a subject of fascination since antiquity, beginning with the first awareness of the apparent motion of the stars. Much of the early effort in studying motion attempted to characterize *how* objects moved, without considering the causes. For example, Plato believed that each celestial body must follow a circular path, since that was the perfect trajectory. Aristotle modeled the universe as a set of natural places, with the earth being in the lowest natural place. He suggested that every object moved toward its natural place. Unfortunately, this view was not subject to verification in the sense that it did not predict in detail how a body moves, that is, what its position would be at some future time. It was not until the Middle Ages that serious attempts to quantify motion were made. Galileo Galilei (1564-1642) was the first to understand that the trajectory of a body is determined by both its speed and its direction of motion. His experiments and writings about motion brought considerable precision to the human understanding of how objects move under the influence of gravity.

Newton's formulation of the laws of motion for particles remains the greatest single advance in knowledge of the mechanical world. These laws provide a means to predict or understand the trajectory of a particle in the presence of any type or combination of forces. Further, Newton's invention of differential and integral calculus (which was also done simultaneously and independently by Gottfried Wilhelm Leibniz) was and still remains indispensable for predicting and analyzing trajectories.

In everyday applications, where the masses and forces are within the realm of human experience, Newton's laws are adequate for such analysis. Near the end of the nineteenth century, however, it was found that, if exceedingly small (atomic-sized) or large (stellar-sized) masses are involved or if the speeds between frames of reference are significant fractions of the speed of light, then more sophisticated theories are required to describe the trajectories with any accuracy. These situations led to the development of quantum mechanics and the special and general theories of relativity, which came to be used routinely by scientists for the study of those environments.

Bibliography

Asimov, Isaac. *Understanding Physics*. Vol. 1. New York: Walker, 1966. This two-volume set treats both classical and modern physics using almost no mathematics. The explanations are lucid and witty, which is one of Asimov's hallmarks. His examples are drawn from common experience and serve to illustrate the concepts quite effectively.

Barbour, Julian B. *Absolute or Relative Motion?* Vol. 1 in *The Discovery of Dynamics*. Cambridge, England: Cambridge University Press, 1989. Primarily a philosophical study of the basic problem of frames of reference, but also contains many interesting historical insights.

Brancazio, Peter J. "Playing It by Ear." *Scientific American* 248 (April, 1983): 76. Although brief, this article contains a number of useful findings about the effects of atmospheric drag on the trajectory of a baseball, as well as how an outfielder uses different sensory information to predict intuitively where the ball will land.

Feynman, Richard P., Robert B. Leighton, and Matthew Sands. *The Feynman Lectures on Physics*. Vol. 1. Reading, Mass.: Addison-Wesley, 1963. These lectures were intended to challenge some of the brightest physics students at the California Institute of Technology, but they have exceptional clarity and appeal for general audiences. One of the few books that effectively treat the approximate nature of physical laws.

Frautschi, S. C., R. P. Olenick, T. M. Apostol, and D. L. Goodstein. *The Mechanical Universe: Mechanics and Heat.* New York: Cambridge University Press, 1986. The companion text to *The Mechanical Universe* film series (Public Broadcasting System), which is based on Goodstein's lectures at the California Institute of Technology. The historical development of these subjects, carefully integrated with the analysis, is one of the best treatments available.

Gamow, George. *Matter, Earth, and Sky.* Englewood Cliffs, N.J.: Prentice-Hall, 1958. Written as a textbook for introductory physics courses, this work is based on Gamow's long experience in lecturing and writing for general audiences interested in science. Successfully combines physics, chemistry, geology, and astronomy, including numerous helpful diagrams and some of the author's own cartoons.

Newton, Isaac. *Mathematical Principles of Natural Philosophy.* Translated by Andrew Motte and edited by Florian Cajori. Berkeley: University of California Press, 1934. In this book, Newton puts forth his laws of motion and the law of universal gravitation. While the language is flowery, the mathematics are fairly easy to follow: Newton presents his theories using geometry, making the developments sometimes tedious but always comprehensible.

Rothman, Milton A. *The Laws of Physics.* New York: Basic Books, 1963. Using a minimum of mathematics, this book introduces the basic principles of classical and modern physics, with many excellent examples. Gives considerable attention to the philosophical aspects of the human knowledge of physical reality.

Robert G. Melton

Cross-References

Centrifugal/Centripetal and Coriolis Accelerations, 332; Differential Equations, 658; Mechanics, 1367; Newton's Laws, 1509; Numerical Solutions of Differential Equations, 1643; Vectors, 2679.

TRANSFORMERS

Type of physical science: Classical physics
Field of study: Electromagnetism

A transformer is an electrical device that uses electromagnetic induction to convert the voltage in a circuit either to a higher or lower value. Their wide variety of applications includes power distribution systems, appliances, and toys.

Principal terms

ALTERNATING CURRENT: an electric current that smoothly and periodically changes direction in a circuit

CORE: internal material in a transformer that helps to contain the magnetic field

EDDY CURRENTS: electrical currents induced in a transformer's core by the changing magnetic field generated by the primary coil

ELECTROMAGNETIC INDUCTION: a phenomenon in which a changing magnetic field induces a changing voltage in a circuit; a changing voltage generates a changing magnetic field

Overview

Transformers serve the useful purpose of conveying electrical power between two circuits without the circuits having to be directly connected. Simultaneously, the transformers can convert the voltage level of this power to a higher or lower value (or simply at a constant value). Transformers operate on the principles of electromagnetic induction, discovered by Michael Faraday in 1831 at the Royal Institution in London. An electric current (flowing in a wire) always produces a magnetic field that surrounds the current; a current whose magnitude changes with time produces a magnetic field whose magnitude also changes with time. Conversely, a changing magnetic field in the vicinity of a wire induces a voltage that causes a changing current to flow in the wire. (A constant magnetic field does not induce a constant voltage.)

In the case of transformers, at least two wires are involved. Each is usually formed into a coil of some form, and the two coils are placed close together, but not necessarily touching. There is one primary coil, and one or more secondary coils; the coils are often called windings, and their wire loops are called turns. Placing a constant voltage (for example, using a battery) across the primary winding causes a current to begin to flow. Before this current reaches a steady value, it creates a magnetic field that is also growing in magnitude, which consequently induces a voltage in the secondary winding. Once the primary current reaches its steady value, so does the magnetic field, and the secondary voltage disappears. Therefore, the primary voltage is usually made to oscillate sinusoidally (that is, vary smoothly and periodically between two values). The resulting oscillations in primary current produce an oscillating magnetic field, which in turn induces an oscillating secondary voltage. Both the primary and secondary voltages (and the corresponding currents), as well as the magnetic field generated by the primary winding, oscillate at the same frequency. A simple relationship governs the operation of a transformer: The ratio of primary to secondary voltage equals the ratio of secondary to primary turns of wire. For example, if a voltage that varies sinusoidally between 100 volts and −100 volts is applied to the primary winding of a transformer with 400 primary turns and 100 secondary turns, the secondary voltage will be 25 volts.

The physical form of a transformer depends upon several factors, including the amount of inductive coupling required and the potential range of oscillating frequencies involved. Generally, the coils are wound on an object called the core, which not only serves as physical support for the winding but also can help to contain the magnetic field. Certain magnetic materials serve this latter purpose quite well, notably the ferromagnetic compounds (for example, iron). The better the magnetic field is contained, the more complete the transfer of electrical power between the coils. In an ideal transformer, the magnetic field is completely contained in the core, allowing all the power to be transferred to the secondary circuit.

Realistically, however, four different phenomena prevent this from occurring. Both coils have some electrical resistance and therefore a small amount of power is lost as heat in the windings (referred to as copper losses). Second, some of the magnetic field generated by the primary current will be outside the core, where it cannot contribute to the induction in the secondary coil (referred to as inductive losses). Third, the core itself (usually an electrical conductor) will experience an induced voltage and, consequently, electric currents (called "eddy currents") will flow within the core. The electrical resistance of the core material then dissipates the energy of these internal currents as heat (referred to as eddy losses). Finally, the oscillating magnetic field in the core causes a reversing magnetization of the core material, which removes energy from the magnetic field (referred to as hysteresis losses). The manifestation of this loss is an additional warming of the core. Each of these four losses reduces the amount of power available in the secondary winding, typically by 3 to 5 percent.

Improvements in performance can be obtained by employing low-resistivity wire and by using the largest possible wire diameter to lower the electrical resistance further. The choice of core material depends largely on the range of oscillation frequencies of the current involved. For example, iron compounds work well for high-power applications and low frequencies such as the 60 hertz (cycles per second) used by power companies for distributing electrical power; however, at radio frequencies (for example, 1,000,000 hertz), the hysteresis and eddy losses may be unacceptable. Generally, hollow ceramic or plastic forms (air cores) work well at these high frequencies. Improving technology has continued to make ferromagnetic cores perform better at higher frequencies, making it possible to employ their superior field-containment properties.

In low-frequency applications, eddy losses are successfully reduced through the use of laminated iron cores. The core is divided into a number of thin layers (laminae) and separated by thin layers of electrical insulation, usually a baked enamel. This insulation prevents the flow of eddy currents between laminae and thus reduces the overall current in the core and the corresponding eddy losses. Dividing the core into even thinner laminae reduces the eddy losses significantly, but the enamel layers must maintain a minimum thickness to prevent electrical arcing, and the total core thickness can thus become too great for practical use.

At high frequencies, the capacitance between laminae can adversely affect the circuit in which the transformer is placed; therefore, laminated cores are avoided. Powdered ferromagnetic compounds, mixed with electrically insulating cement, have low eddy losses and good magnetic field containment at high frequencies but are more expensive to manufacture and somewhat less durable than iron cores (that is, the powdered cores are more brittle and subject to cracking).

In general, transformers are designed so that the core forms a closed loop, either as a simple torus or sometimes with more complex geometry. This reduces leakage of the magnetic field but makes construction of the transformer somewhat more complicated. It is important to keep the core path as short as possible to reduce eddy losses and magnetic field leakage.

Two common forms of cores are the shell, which consists of an E-shaped piece (often laminated), with both the primary and secondary coils wound on the short center section, and an I-shaped piece (also usually laminated), which is cemented to close the E-section. The result effectively gives two cores that share a common side (on which the coils are wound); however, because the I-section is separated by a thin layer of cement, the magnetic path is not completely closed and some leakage is inevitable. The ease of winding on the open E-section provides some compromise between performance and cost. A second form avoids the leakage problem by using two C-shaped laminated stacks with a separate winding on the middle section of each. The two stacks are joined by interleaving the laminae as the open sides of the C-sections are joined.

In cases where electrical isolation between two circuits is not required but a voltage change is needed, an autotransformer may be useful. This device uses the same principles of electromagnetic induction but employs only one winding, to the middle of which is connected a third wire. The primary winding is the entire coil, but the secondary is only that part of the winding between one end of the original coil and the third wire. This arrangement can result in substantial reduction in the total amount of wire needed (and also reduce the copper losses).

Applications

The most common and well-known use of transformers is in the power distribution system that provides electricity to homes, buildings, industry, and the like. Power is initially generated at a central plant in an oscillating form (alternating current), with a frequency of 60 hertz. The voltage is stepped up using transformers that have large secondary-to-primary turn ratios, resulting in electrical power with high voltage (100,000 volts or more) but relatively low current. Because the resistive losses in the distribution network are proportional to the square of the current, this approach reduces the loss of power during distribution to the users. An added advantage is that the distribution wires can be considerably smaller in diameter than if no transformers were used, since the reduced current requires smaller cross-sectional area in the conducting material of the wire. Power can be transmitted over many kilometers in this way with only minor loss because of resistance; however, the extremely high voltages necessitate the use of elevated transmission lines for safety.

Even with the small percentage of power lost because of hysteresis, eddy currents, and the like, these step-up transformers dissipate considerable heat and must be cooled by keeping them immersed in oil or by circulating cooling liquids through the cores to transport the heat to a radiator, where it is emitted into the air. Near the user's end of the network, a step-down transformer converts the power back to a lower-voltage (and higher-current) form that is suitable for use in homes and offices. Some industries require much higher voltages, and the corresponding step-down transformers at those locations can be designed with smaller turn ratios. In many networks, the wiring takes the form of a tree, with each major branch carrying power to a group of users in an area. Since the wires must now be brought lower to the ground for users to access them, the voltage levels must also be lowered for safety. A series of transformers is employed, each at one of the forks in the wiring network to step down the voltage for that branch. The network for a metropolitan community may contain hundreds of thousands of these intermediate step-down transformers, with a final voltage reduction occurring at a local transformer in each neighborhood, giving a typical level of approximately 110 volts.

While many appliances appear to use this form of electrical power directly, frequently they contain internal transformers that either step up or step down the voltage. For example, most solid-state electronic equipment operates at less than 10 volts. A step-down transformer can convert the 110 volts to 10 volts (and then a power conditioning circuit must convert the power to direct current). Circuits that require several different voltages can use a transformer with one primary and several secondary windings, each producing the desired voltage. Indeed, a single transformer often has some secondary windings that step up the voltage and some that step it down. This is the case with televisions and computers that use cathode-ray tubes as displays (requiring up to 25,000 volts), while employing low-voltage integrated circuits for the signal processing or computing actions.

A second routine use of transformers is in audio amplifiers, where the output signal must go to an audio speaker and produce sufficient sound volume level. The speaker also operates via electromagnetic induction: The input signal to the speaker is an electric current that creates a magnetic field in the speaker coils. The current changes in proportion to the intensity of the audio signal that is to be reproduced and causes the speaker's membrane (which has attached a small permanent magnet) to vibrate at a frequency and amplitude proportional to that of the audio signal. Good performance of the speaker requires a significant current in the speaker coils in order to generate a strong magnetic field. By inserting a step-down transformer between the audio circuit and the speaker, the output current to the speaker is made greater than it would be without the transformer.

In many circuits, the power being output to a second circuit can be maximized if the second circuit has a particular electrical resistance or can be made to appear effectively as though it has that resistance. This is called "impedance matching," and a transformer can be employed for this purpose. Impedance is analogous in alternating current circuits to resistance in circuits that use direct current. The relationship between current, voltage, and resistance is given by Ohm's law, which states that voltage equals current (measured in amperes) times resistance (measured in ohms). For example, if the first circuit has an output impedance of 100 ohms, then the second circuit (which receives the output power from the first) should be made to have, or appear to have, the same impedance. If the second circuit has an input impedance of 10 ohms and the first circuit outputs 100 volts at 200 ohms output impedance, then a transformer with 45 primary windings and 10 secondary windings will give 100 volts/4.5 = 22.2 volts at the second circuit. By Ohm's law, the second circuit will draw 22.2 volts/10 = 2.22 amperes of current. Therefore, the secondary coil of the transformer delivers 49.3 volt-amperes; however, the input volt-amperes to the transformer must be nearly equal to the output, so the transformer must be drawing 49.3 volt-amperes. Since the first circuit outputs at 100 volts, it now sees an impedance of 100 volts/0.493 amperes = 202 ohms, which is approximately what was desired.

Context

Following Faraday's discovery of electromagnetic induction, a number of successful experiments demonstrated the feasibility of stepping-up and stepping-down voltages. In 1884, the first commercial transformers were operated by the Hungarian team of Max Deri, Otto Blathy (who introduced the term "transformer"), and Karl Zipernowsky. Their transformers, which had iron cores, were used in a simple power distribution system for lighting at an exposition in Budapest in 1885. In the United States, George Westinghouse purchased patent rights in 1886 on the invention of an induction coil system developed previously by Lucien Gaulard and Josiah Willard Gibbs. Under Westinghouse's encouragement, one of his engineers, William Stanley further developed the ideas of Gaulard and Gibbs, introducing the use of laminated and continuous cores. Westinghouse was the first to suggest the use of a shell core to ease the construction. His plant in Buffalo, New York, which began operation in November, 1886, was the first for distributing power using step-up and step-down transformers.

The capability to transmit electrical power over great distances made it possible to locate power-generation plants remotely from populated areas. This was particularly important for hydroelectric generation plants located on dams, sometimes far removed from metropolitan centers. Nuclear power plants must be located near bodies of water (for cooling purposes), which also are frequently rather distant from the users of the power.

Transformers will continue to serve a crucial role for electrical isolation. In many appliances, it would be possible to achieve the necessary low-voltage by simply using resistors to drop the voltage from the customary 110 volts to the desired value. This poses a hazard, though, because overheating of a resistor could result in a failure, placing the full 110 volts across parts of the low-voltage circuitry. Such a voltage level is life-threatening and so the use of "transformerless" power supplies in appliances is too dangerous. A step-down transformer also serves to isolate the high-voltage wires from the low-voltage circuits. Failure of a transformer in such a way as to cause high voltage to appear on the secondary side is far less likely than the dangerous failure of the resistor system.

Bibliography

Asimov, Isaac. *Understanding Physics.* Vol. 2. New York: Walker, 1966. This two-volume set treats both classical and modern physics with minimal mathematics. The explanations are lucid and witty. His examples are drawn from common experience and serve to illustrate the concepts quite effectively.

Borowitz, Sidney, and L. A. Bornstein. *A Contemporary View of Elementary Physics.* New York: McGraw-Hill, 1968. An introductory physics text, this contains an excellent treatment of electromagnetic induction and some of the history of its discovery. A most appealing aspect of this book is that it allows the reader to see the mathematical descriptions of the physical phenomena without relying on calculus.

Feynman, Richard P., Robert B. Leighton, and Matthew Sands. *The Feynman Lectures on Physics.* Vol. 2. Reading, Mass.: Addison-Wesley, 1963. These lectures challenged some of the brightest physics students at the California Institute of Technology, yet they have exceptional clarity and appeal for general audiences. Effectively treats the approximate nature of physical laws.

Gamow, George. *Matter, Earth, and Sky.* Englewood Cliffs, N.J.: Prentice-Hall, 1965. Geared for freshman physics courses. Based on Gamow's experience in lecturing and writing for general audiences interested in science. Combines physics, chemistry, geology, and astronomy, including numerous helpful diagrams and some of the author's own cartoons.

Wilson, Mark J., ed. *The ARRL Handbook for the Radio Amateur.* 65th ed. Newington, Conn.: The American Radio Relay League, 1988. Intended as a reference source for amateur ("ham") radio operators, this remarkable book also quite successfully teaches about a wide variety of electrical components and circuits, requiring only a modest knowledge of electricity and mangnetism. Contains an extensive treatment of the theory and uses of transformers, along with numerous tables of design data.

Robert G. Melton

Cross-References

Charges and Currents, 351; Conductors and Resistors, 540; Electric and Magnetic Fields, 699; Electrical Circuits, 706; Electromagnetism, 750; Inductors, 1102; Magnetic Materials, 1296.

TRANSISTORS

Type of physical science: Classical physics
Field of study: Electromagnetism

Transistors, semiconducting circuit elements, opened the tremendously powerful possibilities offered by solid-state electronics. The transistor provides amplification of voltage or current without many of the disadvantages of vacuum tubes.

Principal terms

ACCEPTOR: an impurity atom added to a crystalline lattice to accept electrons

BASE: the thinnest of three parts of a transistor that is doped opposite to the other two parts; it acts as the input portion of the transistor

COLLECTOR: one of three parts of a transistor, doped opposite to the base part and usually connected to the chassis for thermal energy dissipation

DONOR: an impurity atom specifically introduced into a crystal lattice to donate electrons

DOPING: the process of intentionally introducing specific impurities into a semiconducting material to alter its conducting characteristics

EMITTER: one of three parts of a transistor, doped similar to the collector

HOLE: the moving absence of an electron that would make a given lattice atom neutral with respect to the crystal lattice

MAJORITY CARRIER: the carrier of electric charge, positive or negative, that is in the majority in a portion of a semiconductor

MINORITY CARRIER: the carrier of electric charge, positive or negative, that is in the minority in a portion of a semiconductor

Overview

Negatively charged particles, electrons, are the current carriers in vacuum tubes. In semiconducting devices, such as transistors, the concept of conduction (via negative charge carriers) must be modified to include the contribution of positive charge carriers, called holes, to current flow. Holes are basically the absence of electrons necessary to maintain electrical neutrality in the crystalline lattice. Holes can be considered to have mass, they are mobile, and they can propagate through the lattice. Their movement, thus, constitutes a current. (In physics, a current is defined in terms of the movement of positive charge, a throw-back to the time when it was believed that positive charge was responsible for conduction in ordinary metals. Thus, a negative charge moving to the left is equivalent to a positive charge moving to the right in the presence of an electric field. Both charge motions contribute to a total current moving to the right in this simple example.)

Semiconducting materials that are relatively pure are not particularly useful as circuit elements. Silicon and germanium are the most commonly used semiconducting materials in solid-state devices. To make a semiconducting component, impurities are purposely introduced into the crystalline lattice in a process known as doping. Commonly used impurities are indium, gallium, arsenic, and antimony. These materials alter the conduction of silicon and germanium by changing the number of negatively charged free electrons. Silicon and germanium are group IV elements. Antimony and arsenic are group V elements, and indium and gallium are group III elements. Thus, introduction of antimony or arsenic in large concentrations makes a semiconducting material enriched in free electrons. Such a material is called *n* type. Introduction of indium or gallium would increase a deficit of free electrons (hence an excess of holes). Such a material is called *p* type. In an *n*-type material, free electrons are the majority carriers and mobile holes minority carriers. In a *p*-type material, mobile holes are the majority carriers and free electrons the minority carriers. Holes and free electrons have an attraction for one another and can recombine to restore charge neutrality to the lattice locally. In electron-hole recombination, both the hole and free electron are lost as contributions to the current flow. While this occurs at one place in a semiconductor, however, new current carriers are formed at other places in the semiconductor to replenish them. Random movements of mobile charge carriers can be controlled by the application of an external voltage source. This sets up an electric field within the material and dictates a direction of flow for hole and electron movement.

If a semiconductor crystal is doped with group III impurities in part and then doping is abruptly changed to group V elements, a *p-n* junction diode is created. A diode will allow conduction current flow in one direction and only minority carrier flow in the other. Thus, a diode can restrict current flow direction and act as a rectification element.

A transistor consists of three layers of impurity-doped semiconducting material. The middle layer is doped differently from the two ends. A single transistor, thus, is either a *p-n-p* or *n-p-n* type. (All discussions that follow will refer to *p-n-p* transistors. The *n-p-n* transistors operate similarly, but holes and electrons reverse roles in the amplification process.) The middle section of a transistor is called the base. Its boundaries with the two outside layers, called the emitter and collector, are very abrupt. If the change in doping were gradual, a transistor would not function properly. The collector portion is somewhat larger than the emitter and is thermally attached to the transistor housing for power dissipation. Excess heat inside the semiconducting crystal could degrade its amplification capability and provide enough energy for impurity atoms to become mobile and destroy the abrupt boundaries between the emitter and base and the collector and base.

At a *p-n* junction, current will flow in the forward-biased direction but will be suppressed in the reverse direction. A *p-n* junction, however, is not an ohmic conductor, meaning the voltage applied is not proportional to the resultant current flow. In a *p-n* junction that is forward-biased, the current increases exponentially as the voltage is increased. A transistor has two *p-n* junctions. For a *p-n-p* transistor to function properly, one junction is forward-biased (the base-emitter junction) and the other junction is reverse-biased (the base-collector junction). With this arrangement, the *p* type material is positive in voltage with respect to the *n* type. A large current will flow into the base from the emitter. Little flow of current goes from the collector into the base.

The reverse-biased junction between the base and the collector forms a depletion zone with an electric field pointing from the base to the collector. The reverse-biasing indicates that the depletion field sweeps majority carriers from the base and collector away from the base-collector junction rather than toward the junction. Outside this depletion zone, there is no significant electric field within the semiconductor crystal. Thus, except at the depletion zone, the motion of charge carriers (holes and electrons) is determined exclusively by diffusion.

Charge is a conserved quantity; therefore, by one of Kirchhoff's laws of circuit analysis, the current flowing into the emitter must equal the current leaving the base plus the current going into the collector. Holes, remember, are the majority carrier in the *p*-type emitter. Three fates are possible for holes injected into the base through the forward-biased emitter-base junction: They can recombine with mobile majority carrier electrons in the *n*-type base; they can diffuse through the base and recombine with free electrons injected into the base from the external circuitry; or they can diffuse across the base into the depletion zone, where they will be swept across the reverse-biased base-collector junction into the *p*-type collector. Once there, they diffuse until recombining with free electrons injected into the collector from external circuitry.

It is the thinness of the base that makes a transistor a useful device for amplification. If the base is sufficiently thin, the most probable fate for holes injected into the base is that they will diffuse through the base into the depletion zone. Most of the holes injected into the base are transported into the collector, usually 98 to 99 percent in a typical low-cost transistor. The ratio of collector current to emitter current—the alpha rating—is thus 0.98 to 0.99 typically. Often, transistors are specified by their beta rating, defined as the alpha rating divided by one minus the alpha rating. Typical beta ratings are 20 to 200 or more.

One other typical feature of transistor construction is that doping of the emitter is more concentrated than that of the base. Therefore, the concentration of holes injected into the base from the emitter is higher than the concentrations of mobile electron majority carriers in the base. That ensures high values for both the alpha and beta ratings.

Applications

One of the most important uses of a transistor is amplification. Amplification occurs when the voltage output is greater than the voltage input. The gain is defined as the ratio of output voltage to input voltage.

In a typical *p-n-p* transistor, the base current represents the input and the collector current represents the output. (The discussion will focus on what is called a common emitter amplifier configuration.) Since a transistor has three elements, one must be common to both input and output. Thus, there is a common base, common collector, and common emitter configuration. Each has its own particular utilities, but a common emitter is typically used for voltage amplification. For example, the common collector, often referred to as the emitter-follower configuration, has a voltage gain of approximately one, but can have a sizable current gain. The common base configuration has a current gain that is approximately one, but can have a sizable voltage gain. (It must be emphasized that it is not possible to have both a high voltage and high current gain from the same transistor. That would violate conservation of energy. Power is the product of current and voltage and is equal to the rate at which electrical energy is delivered. Thus, if voltage is high, current must be low, and vice versa.)

One must properly bias and set an operating point for the transistor. The base-emitter junction must remain forward-biased by at least 0.5 volt for silicon and 0.2 volt for germanium. Since the base-collector junction is reverse-biased, the voltage across this junction is essentially equal to the voltage from collector to emitter. A base current—typically in the range of 10 to 100 microamperes—is selected to set operating conditions for the transistor. With the base current constant, the collector current is essentially only a function of the voltage across the collector to emitter. A power supply or voltage divider often is used to set that negative voltage. The operating point needs to be set at a point somewhere in the middle of a line of constant base current in a graph of collector current versus voltage across collector to emitter. This is necessary for two reasons. First, it permits alternating current (AC) input signals to swing negatively and positively equally about the transistor's operating point. Second, it keeps the transistor's operation away from the forbidden zone, where the product of collector current and voltage across the collector to emitter would exceed the maximum power rating of the transistor. Entry into the forbidden zone for even a short time can cause irreparable damage to the transistor crystal and shorten useful lifetime.

A transistor amplifier must be made stable against thermal runaway. Any rise in temperature will cause an increase in collector current as a result of increased hole mobility. That, in turn, will raise the temperature of the collector by joule heating of dissipated energy, thereby increasing collector current in a repetitious cycle that will eventually destroy the transistor. To prevent this from happening, a resistor in parallel with a capacitor is placed between the emitter and electrical ground. With the emitter resistor present, if the emitter voltage goes negative, the base voltage will also go negative by the same amount, thereby maintaining the forward voltage of the base-emitter junction constant, regardless of the nature of the induced changes in the base and emitter voltages. The presence of the emitter resistor will decrease the amplifier gain unless a capacitor is attached in parallel to it. This emitter capacitor provides an AC path to ground if it has a smaller capacitative reactance than the resistance of the emitter resistor. Since capacitative reactance depends inversely on the frequency of the signal, the presence of the emitter capacitor sets a lower limit on the frequency of signals that can be effectively amplified by the transistor when operating properly biased and stably against temperature and input voltage changes.

Transistors can be used in multiple stages, one transistor's output serving as the input for the next. In this way, an extremely large effective gain can be accomplished. The net gain of the multistage amplifiers would be the product of the gains of each individual stage.

Context

Many of the functions of transistors in circuits were possible prior to the development of transistor units. Although quite novel in their approach, the early transistor circuit did not represent the development of new electronic capability. Early transistors had advantages over vacuum tubes that made them more attractive as rectifiers, amplifiers, and detectors of electrical signals and as oscillators and switching elements. All those functions were possible previously, using vacuum tube technology. Because of the advantages transistors offered, transistor development proceeded rapidly from the early crude working models to the miniaturized versions available in the 1990's.

The first transistor was developed at Bell Telephone Laboratories, invented by John Bardeen and Walter H. Brattain. Although years of development were involved in the effort, the invention of the transistor is often historically placed in 1948. Compared to modern versions, this first point-contact transistor was quite bulky.

The advantages of transistors over vacuum tubes are numerous and their disadvantages few. Vacuum tubes can handle higher currents flowing through them and larger voltages applied across them than transistors. Vacuum tubes could, in principle, be repaired, whereas transistors must be replaced if they go bad. Nevertheless, that quality is not much of an advantage considering the low cost of most transistors.

The advantages of transistors over vacuum tubes include the following: Semiconducting elements are small and lightweight, permitting miniaturization of electronic equipment. Second, microminiaturization on integrated circuit chips can contain circuits composed of transistors and other elements and their leads that do the work of vacuum tubes and other components that are 100 to 100,000 times larger in size. Third, semiconducting elements, such as transistors, are solids and therefore relatively vibration-free; element vibration in vacuum tubes was the origin of troublesome microphonics in amplifiers. Fourth, transistors dissipate far less thermal energy than vacuum tubes and do not need to warm up; they operate as soon as power is applied to them. Fifth, semiconductor elements, such as transistors, are rugged and can be easily shielded from external environmental conditions. Finally, semiconductor elements, such as transistors, do not experience chemical deterioration of elements; vacuum-tube cathode deterioration will degrade tube performance and will eventually lead to early replacement of expensive components. As a result, the development of the transistor quickly led to an explosion of technological innovation in the realm of electronics.

Bibliography

Brophy, James J. *Basic Electronics for Scientists*. 5th ed. New York: McGraw-Hill, 1990. Despite the title, this is an excellent text on basic electronics for anyone with a decent command of high school mathematics and physics. Higher-order mathematics sections are well explained to be understandable for the layperson. Filled with practical circuits.

Fry, Jim. *Electronic Circuits*. Benton Harbor, Mich.: Heath, 1987. Excellent for the electronics hobbyist and person beginning to investigate electronic circuits involving solid-state semiconducting components such as diodes, transistors, and integrated circuits. Includes a workbook.

Halliday, David, and Robert Resnick. *Fundamentals of Physics*. 3d ed. New York: John Wiley & Sons, 1988. A classic for introductory physics instruction. Contains descriptions of breaking areas of modern physics. The section on electricity and basic electrical components is thorough and easy to understand. Good text for self-study.

Murr, Lawrence E. *Solid-State Electronics*. New York: Marcel Dekker, 1978. Highly descriptive and readable, well-illustrated discussion of the theoretical physics of semiconductors. Explains energy bands, crystal structure, conduction processes, and defects in solids.

Simpson, Robert E. *Introductory Electronics for Scientists and Engineers*. Boston: Allyn & Bacon, 1975. Presents basic electronics and transistor circuits in a manner understandable for the introductory college student. Much of the book is readable by the average reader. Practical circuits are useful to the electronics hobbyist.

Sprott, Julian C. *Introduction to Modern Electronics*. New York: John Wiley & Sons, 1981. Complete treatise on basic electronic circuits. Particularly strong on discussion of basic transistor design, characteristics, and amplification properties. Most sections, including problems, are accessible with high school mathematics.

Suprynowicz, V. A. *Electrical and Electronics Fundamentals: An Applied Survey of Electrical Engineering*. St. Paul, Minn.: West, 1987. Most sections are understandable with high school mathematics. Lucid explanation of electrical conduction, transistor circuits, amplification, and feedback. Provides practical circuits and thorough description of electrical instrumentation, such as oscilloscopes.

David G. Fisher

Cross-References

Conductors and Resistors, 540; Electrical Circuits, 706; Forces on Charges and Currents, 907; Insulators and Dielectrics, 1139; Integrated Circuits, 1153; Rectifiers, 2086; Semiconductors: Atomic-Level Behavior, 2177; Defects in Solids, 2235; Electrical Properties of Solids, 2253; Vacuum Tubes, 2663.

TRANSITION ELEMENTS

Type of physical science: Chemistry
Field of study: Chemistry of the elements

Transition elements are a group of twenty-seven chemical elements with similarities in chemical reactivity and electronic structure. They are widely used in chemical processes and are components of several important biomolecules.

Principal terms

ATOMIC ORBITAL: the region of space occupied by a particular pair of electrons having a characteristic shape, energy, and orientation

COORDINATION COMPLEX: a chemical compound with a central transition metal ion surrounded by several chemical species via covalent bonds

COVALENT BOND: a chemical bond in which a single pair of electrons is shared between two atoms

ION: an atom possessing a positive or negative electrical charge resulting from loss or gain of electrons, respectively

LIGAND: a neutral molecule or charged ion that donates a pair of electrons to form a metal-ligand bond in a coordination complex

METAL: an element possessing high luster, malleability, and excellent thermal and electrical conductivity

OXIDATION STATE: an indication of the electrical charge possessed by a metal ion in a compound

Overview

Transition elements are the largest subgroup of chemical elements. They are characterized by similarities in electronic structure and chemical reactivity that distinguish them from non-transition elements. Without exception, the transition elements are metals. Compared to metals that are not transition elements, such as lead or tin, transition elements are hard and strong, possess a high luster, typically have very high boiling and melting points, and are excellent conductors of both heat and electricity. In addition, transition elements readily form alloys with other metals, both transition and non-transition. Most transition elements dissolve readily in mineral acids, with the exception of noble metals such as silver, gold, or platinum.

Additional properties of the transition elements clearly distinguish them from other types of elements. All metallic elements form positive ions, or cations, in compounds via loss of electrons from the metal atom, resulting in a positive electrical charge on the metal atom. Nontransition metals typically display only one stable electrical charge, or oxidation state. For the transition elements, without exception, there is more than one stable oxidation state possible. Compounds containing transition metal ions display interesting magnetic properties and are often colored, with color typically dependent on the metal and its oxidation state.

Iron is the most abundant transition element. Its properties are typical of transition metals, as iron displays two stable oxidation states in its compounds. Iron is also found in brightly colored compounds such as hemoglobin. The red and blue of oxygenated and deoxygenated blood are caused by the presence of iron in hemoglobin.

To understand the chemical properties of the transition elements, an understanding of their electronic structure is necessary. In a simple picture of the atom, the nucleus is surrounded by a series of electron-containing shells at increasing distances from the nucleus. The farther a shell is from the nucleus, the higher the energy of electrons placed in that shell. Within a shell, there exist one or more subshells, which are sets of atomic orbitals. Electrons occupy these orbitals, which describe the region of space in which a particular electron is localized. Atomic orbitals within a subshell have identical energies and shapes, differing from one another only by their orientation in space. Subshells within a shell differ slightly in energy, so that in actuality a shell includes a range of orbital energies.

To understand how electrons in an atom occupy the available orbitals, a concept called the aufbau principle must be employed. *Aufbau* means, in German, "to build up." In the aufbau process, the electronic arrangement, or configuration, of an atom is described by hypothetically placing electrons one at a time in available orbitals so that each electron is placed in the lowest-energy orbital available on the atom. Ideally, in so doing, one shell would be completely filled with electrons before any were placed in the next, higher energy shell. It is found, however, that shells become less separated farther from the nucleus, and overlapping of different shells can occur. For instance, the highest energy subshell in the third shell, called a *d* subshell, is at a slightly higher energy than the lowest energy subshell in the fourth shell.

Overlapping of shells is critical to understanding the electronic arrangement of the transition elements. In the aufbau process for transition elements, this arrangement is characterized by the placement of two electrons in the outermost, or valence, shell, followed by placement of the remaining electrons in an inner *d* subshell containing five orbitals. A maximum of ten electrons can be placed in this subshell, as each atomic orbital can contain a maximum of two.

There are three sets of elements in the periodic table that display this type of configuration. The transition elements are those in which this inner *d* subshell is partially filled. In the periodic table, the transition, or *d*-block, elements are in the center and include the three rows from scandium to copper, yttrium to silver, and lanthanum to gold.

The unique properties of the transition metals are explained by their typical electronic configuration. It is easy to remove the two electrons chemically from the outermost shell. As a result, virtually all transition elements have a stable oxidation state corresponding to the loss of these electrons. Additional stable oxidation states correspond to loss of electrons from the inner *d* subshell. The partially filled nature of this subshell also explains the magnetic properties of the transition elements.

An electron, which is an electrically charged particle, behaves as though it were spinning about an axis, which generates a magnetic field. There are two possible directions of electron spin, which generate opposing magnetic fields. An orbital can contain a maximum of two electrons with opposite, or paired, spins. When all the electrons in an atom are spin paired, the magnetic fields generated by the electrons are all negated. If one or more electrons are not paired with another electron of opposite spin, however, the atom will produce a net magnetic field and thus be attracted to a magnet. Transition elements form a class of compounds called coordination compounds, in which a diversity of magnetic behavior may be displayed by the same element in different compounds.

Coordination complexes are formed when several chemical species, called ligands, form coordinate covalent bonds with the metal. Covalent bonds are bonds in which an electron pair is shared between two atoms. In a coordinate bond, one atom donates both of the shared electrons to the bond. In a complex, the ligands are arranged about a central metal ion in such a way that the ligands maximize their separation. As a result, the metal is surrounded by several ligands that occupy the corners of an imaginary polyhedron about the metal. These geometries are characteristic of certain numbers of ligands, or coordination numbers. The most common coordination numbers are six and four, and corresponding geometries are octahedral and tetrahedral, respectively.

Complexes often absorb visible light. When exposed to white light, they selectively remove one color so that the reaming light appears to have the complementary color. For a given metal, the frequency of light absorbed depends on the type of ligand present. The chloride ion and water, as ligands, form complexes that absorb light of lower frequency than cyanide or carbon monoxide complexes of the same metal. Coupled with this trend is the magnetic behavior of complexes. High-spin complexes are complexes that have the same number of unpaired electrons as the free transition metal ion. Low-spin complexes contain fewer unpaired electrons. Chloride and water complexes tend to form high-spin complexes, while cyanide and carbon monoxide complexes are low-spin.

Two theories of bonding in coordination complexes are used to explain these observations. Crystal field theory is an electrostatic model; its main postulate is that the ligands cause the orbitals in the partially filled *d* subshell to split into different energy levels. The resulting pattern of *d* orbitals depends on the geometry of the complex, and different ligands cause different degrees of splitting. Strong field ligands cause large splitting, while weak field ligands cause slight splitting. Complexes of strong field ligands tend to be low-spin as a result of the great energy gap caused by large splitting of the *d* orbitals. When electrons in the lower subset of split orbitals absorb light of the correct frequency, they absorb energy and jump to the higher orbital subset. The greater the splitting, the higher the frequency of light required to cause this transition. Thus, the difference in color of complexes caused by ligands can be explained in terms of the degree of splitting caused by the ligand. Ligands can be ranked according to increasing crystal field splitting. Such a listing is called a spectrochemical series and may be used to make predictions about the properties of different complexes.

Molecular orbital theory is another theory of bonding in complexes, in which the atomic orbitals of the metal and ligands combine to form a new set of molecular orbitals in the complex. This is the more accurate picture of bonding, but the results obtained in this theory are essentially the same as those of crystal field theory.

Applications

Gold, silver, and platinum are three transition metals almost universally recognized. Throughout history, these precious metals have been valued for their rarity and beauty. This is reflected in the use of these elements as jewelry, currency, and backing for paper currency. Gold and silver are also excellent conductors of electricity, more so than copper at high temperatures, and are often used in electrical components and wiring that will be subjected to such an extreme.

Historically, copper and iron have been the most useful of the transition elements. Bronze, an alloy of copper and tin, and steel, an alloy of iron and carbon, are vitally important in the history of civilization. Steel is indispensable to modern society because of its strength and availability.

Many of the transition elements that are not highly abundant in the earth's crust are still vital to modern metallurgy. Titanium is produced in large quantities and utilized in the aircraft industry, primarily because it is an extremely light but also extremely strong metal, especially when alloyed with small amounts of aluminum or tin. Manganese is useful in the steel industry, both as an agent for sulfur removal and as a hardening agent in the alloy.

Use of the transition elements is not restricted to metallurgy, however. Cobalt historically was used to add blue tint to glass, and presently is still utilized to remove the tint in clear glass caused by the presence of trace iron impurities. Trace amounts of certain transition elements are also responsible for the characteristic colors of certain gemstones. For example, the red color of a ruby is caused by trace amounts of chromium in the basic crystalline structure of the stone.

Nickel is extremely versatile, even for a transition element. It is used to make hardened steel and also utilized in the formation of nonferrous alloys. Pure nickel is chemically inert and thus used to coat materials in certain manufacturing processes that require the use of highly corrosive materials. An important example of this application was the use of nickel to line the miles of pipe utilized in a gaseous diffusion plant in Oak Ridge, Tennessee, designed to separate uranium, in the form of highly corrosive uranium hexafluoride, into its isotopes for the isolation of material to use in the atomic bomb.

The importance of transition elements in their metallic states should be obvious from the preceding paragraphs. Equally important, however, are the uses of these elements when they exist in chemical compounds rather than as free elements. Transition metal compounds are important in both chemical and biological processes.

In industry, the most important use of transition element compounds is as catalysts. A catalyst is a material that causes a reaction to proceed to completion more rapidly than it would in the absence of the catalyst. The catalyst undergoes no net change in the chemical process. Often, chemical reactions form more than one product. Certain selective catalysts are useful because they cause a reaction to form only one of the possible products.

By far the most familiar transition metal catalyst is the platinum complex in catalytic converters, used to remove pollutants from automobile exhausts. This is only one of many catalytic applications of platinum. One particular advantage of platinum is that it can easily be isolated in a very fine, powdery form called platinum black. This finely divided platinum has a tremendous surface area, which is of primary importance in catalytic processes. As the surface area of the catalyst in contact with the reaction mixture increases, the length of time required for complete reaction decreases. Platinum in this form is also utilized as an electrode material in certain electrochemical cells, for the same reason. While platinum is probably the most versatile bulk catalyst, other transition elements are utilized, as well. For example, an iron oxide catalyst is used to produce ammonia from nitrogen and hydrogen in the Haber process.

Many coordination complexes are useful in catalysis. They are often more desirable than bulk-metal surfaces for two reasons. First, complexes can dissolve in solution and so are in intimate contact with the reaction mixture, which increases the reaction rate. In addition, complexes tend to be more selective as catalysts leading to higher product yield and purity. Many of the transition elements form complexes of use in industry as catalysts. For example, a ruthenium complex is utilized to form synthesis gas, a term used to describe mixtures of hydrogen and carbon monoxide, which have a variety of applications. A rhodium complex is used in the production of acetic acid, the main ingredient of vinegar, from methanol.

Increasing utilization of coordination complexes in industry has followed much fundamental research focused on understanding the structure and bonding characteristics of complexes. Coordination chemistry has begun to bridge the gap between organic and inorganic chemistry with the isolation of what are known as organometallic complexes. These are materials in which an organic, or carbon-based, ligand is attached to a metal atom or ion. These complexes display reactivities that reflect the unusual nature of the metal-carbon bond.

Biologically important molecules incorporating transition metal ions are examples of naturally occurring organometallic compounds. Copper, cobalt, iron, and molybdenum all are examples of biologically vital transition elements. The function of vitamin B_{12} depends on the presence of a cobalt-carbon bond. Hemoglobin, which transports oxygen in the human body, contains four heme units, each of which includes an iron atom. These iron atoms can each bind one molecule of oxygen. Iron complexes are also vital to the electron transfer processes in respiration, while copper is found in a variety of proteins in both plants and animals.

Context

Copper, silver, and gold are metals that can be found in their pure, metallic state in nature. This being the case, these are undoubtedly the first known transition elements. Meteorites provided a limited amount of a fourth pure metal, iron. The history of metallurgy is in many ways a history of the transition metals. However, the discovery of new metals and alloys, as well as the refinement of techniques to obtain and purify them, led to many advances without which modern civilization could not exist. Two great epochs in human history, the Bronze and Iron Ages, resulted from utilization of transition elements. Improvements in iron recovery developed at the end of the eighteenth century opened the way to the industrial revolution by providing abundant, inexpensive iron and steel. This resulted in an outburst of new machines and tremendous improvement in construction capabilities, so that complicated bridges and buildings could be erected.

Discovery of additional transition elements continued throughout the nineteenth century. Rhenium, the last naturally occurring transition element to be discovered, was first reported in 1925. One transition element was not isolated until the nuclear age, as it is radioactive with all isotopes having such a short half-life that no hope of finding a natural deposit exists. Technetium was finally isolated in 1937 by bombarding a molybdenum sample with deuterons, which are heavy hydrogen nuclei.

Coordination complexes of transition elements were studied intensively by Alfred Werner, who would later receive a Nobel Prize for his work. Werner synthesized many complexes and elucidated much information about their structure and bonding. Much of modern inorganic chemistry focuses on coordination complexes and their structure and reactivity. Understanding bonding in complexes and other properties facilitated development of crystal field and molecular orbital theories to explain the bonding in these compounds.

With the development of complexes containing predominantly organic ligands, a totally new area of chemistry opened up. Organometallic complexes are interesting chemically in terms of the structure and bonding present, and in industry are quite useful because of the special properties of the metal-carbon bond.

With the increased emphasis in the chemical industry on specialty chemicals, undoubtedly transition metal complexes and organometallic compounds will continue to be utilized in novel manufacturing processes and chemical syntheses.

While much has been learned about biologically important molecules containing transition metals, much remains to be learned about their specific functioning in biological systems. This is especially so for those elements that are necessary in the body only in trace amounts.

Bibliography

Asimov, Isaac. *Asimov's Chronology of Science and Discovery.* New York: Harper & Row, 1989. Geared to a general audience, this work places significant advances in science and technology in historical perspective. Includes information about the discovery of most transition elements. Discusses significant technological advances associated with some of the transition elements.

Greenwood, Norman N., and Alan Earnshaw. *Chemistry of the Elements.* Oxford:

Pergamon Press, 1984. This comprehensive book describes every element, including discussions of terrestrial abundance, methods of isolation, properties, uses, and chemical reactivities. The sections on isolation and uses are extremely well written and easy to understand. Also included are special sections detailing elements with particular importance.

Grubbs, Robert H., and William Tumas. "Polymer Synthesis and Organotransition Metal Chemistry." *Science* 243 (February 17, 1989): 907. Outlines recent work with transition metal complexes and their application in the production of polymers. A good summary of a rapidly expanding field.

Kostić, Nenad M. "Organometallics: Workhorses of Chemistry." In *Yearbook of Science and the Future: 1990*, edited by David Calhoun. Chicago: Encyclopaedia Britannica, 1989. Well-illustrated summary of organometallic chemistry, with much information about transition metal complexes and their uses.

Nance, Lewis E. "Electronic Structure Prediction for Transition Metal Ions." *The Journal of Chemical Education* 61 (April, 1984): 339. Prediction of the electron structure of transition metal ions is difficult because of the overlapping electron shells. This article deals with the prediction of these configurations.

Craig B. Lagrone

Cross-References

Actinides, 30; Biological Compounds, 224; Catalysis, 317; Chelation, 359; Electronegativity, 771; Inorganic Compounds, 1125; Lanthanides, 1220; Metals, 1380; Organometallics, 1697; Photon Interactions with Molecules, 1812.

TRANSMISSION ELECTRON MICROSCOPY

Type of physical science: Condensed matter physics
Field of study: Surfaces

Transmission electron microscopes are magnifying instruments that use a beam of electrons as an illumination source. Electrons passing through the specimen are focused into a magnified image by magnetic lenses. The high resolution obtained by these instruments allows visualization of specimen details at molecular and near-atomic levels.

Principal terms

ELECTRON GUN: a device consisting of a filament and anode that emits a beam of electrons

FOCAL LENGTH: the distance between the center of a converging lens and the point at which parallel rays entering the lens are brought to a point of focus

MAGNETIC LENS: a magnetic field, capable of focusing an electron beam, that is generated by current passed through a massive wire coil and shaped by an iron pole piece inserted into the field

RESOLUTION: a measure of the ability of a microscope to render small details in the image as separately distinguishable points

SHADOWING: a method for preparing specimen surfaces for examination in the electron microscope in which the surfaces are coated with a layer of metal evaporated from a source placed to one side

WAVELENGTH: the distance between corresponding points, such as peaks or troughs, in successive waves of radiation

Overview

A transmission electron microscope is a magnifying instrument that uses a beam of electrons as a source of illumination for examining specimens. The word "transmission" means that these microscopes form an image from electrons that pass through the specimen rather than being reflected from the surface or emitted. The primary reason for the development and application of transmission electron microscopes is that the electron beams used for illumination are of relatively short wavelengths, allowing objects approaching the dimensions of individual atoms to be visualized. This is in contrast to light microscopes, which use the comparatively long wavelengths of visible or ultraviolet light as an illumination source; such microscopes are capable of visualizing objects only if their dimensions are hundreds of times larger than those that can be seen with a transmission electron microscope.

Transmission electron microscopes resemble light microscopes in overall design. The primary differences between the two instruments are in the illumination source and the types of lenses used in the instruments. A light microscope uses an incandescent light or the sun as an illumination source; transmission electron microscopes use an electron "gun" that emits a beam of electrons at a very short wavelength. In a light microscope, beams of light are focused by glass lenses; in transmission electron microscopes, the focusing of the electron beam is accomplished by magnetic lenses. The magnetic fields of the lenses are generated by massive coils of wire through which a precisely controlled electric current is passed. The magnetic fields are shaped into the correct three-dimensional form for focusing electrons by iron pole pieces that are inserted into the wire coils. Although the pathways followed by electrons passing through the magnetic fields are complex, the net effect is the same as that of a glass lens on light rays: magnetic lenses focus electrons to a point, or focus electrons leaving a point into an image of the point.

A typical transmission electron microscope is inverted as compared to a light microscope. The gun that emits electrons is at the top of the instrument, a series of magnetic lenses is in the middle, and the image is formed at the bottom. The electron gun at the top of the microscope consists of two major parts: a filament and an anode. The filament is a thin tungsten wire that is heated to a high temperature by an electric current. The heat drives electrons from the surface of the filament. The filament and its holder, which are electrically insulated from the rest of the column, are maintained at a high negative voltage (−50,000 to −100,000 volts in most transmission electron microscopes). The anode, which consists of a metal plate placed a few centimeters below the filament, is grounded and is thus positive with respect to the filament. As a consequence, electrons from the filament are strongly attracted to the anode. As they traverse the distance from the filament to the anode, the electrons are accelerated to a velocity that depends upon the voltage difference between the two locations—the greater the voltage difference, the greater the velocity. The wavelength of the electron beam, in turn, is inversely proportional to the velocity obtained—the higher the velocity, the shorter the wavelength. In transmission electron microscopes with a difference of 50,000 to 100,000 volts between the filament and anode, the velocity attained produces electron wavelengths in the range between 0.054 and 0.037 angstrom.

The electron beam passes through a small hole in the anode and is focused by a series of magnetic lenses located below the gun. Just below the gun, a condenser lens focuses the electron beam into a very small, intense spot. The specimen is placed at the level of the focused spot. Electrons passing through the specimen are focused into an image by a series of three lenses: the objective, intermediate, and projector lenses. Each of these lenses produces a magnified image, with a total magnification that varies from a few thousand to 300,000 times or more.

Rather than focusing by means of moving lenses, as is done in the light microscope, the magnetic lenses of the transmission electron microscope are focused by altering the current that is applied to the wire coils of the lenses. Focusing in electron microscopes is therefore accomplished by altering the magnifying power rather than the positions of the lenses. In more technical terms, changes in the current applied to the lens coil alter the focal length of a magnetic lens. Electrons entering the lens in a parallel beam converge at a point just beyond the lens, which is called the focal point. Increasing the current applied to a magnetic lens moves the focal point closer to the lens and thus decreases the focal length. As the focal length becomes shorter, the magnifying power of the lens increases. A reduction in the current applied to magnetic lenses has the opposite effect and decreases the magnification. All the magnetic lenses in a transmission electron microscope can be varied over wide ranges of focal length and magnification.

The final lens of the three below the specimen, the projector, focuses the magnified image onto a fluorescent screen at the bottom of the lens column. This screen, which is similar to the screen of a television tube, is coated with a layer of crystals that respond to electron bombardment by emitting visible light. By this means, the electron image is converted to a visual image. Because photographic emulsions can be directly exposed by electrons as well as light, a photograph of the electron image can be made by placing a photographic plate at the level of the screen. The entire microscope, including the electron gun, the vertical column that houses the lenses, and the viewing screen is roughly a meter in length.

The effect of the specimen on the electron beam and the basis for image formation differ significantly from the equivalent processes in a light microscope. In the light microscope, subparts of the specimen absorb or delay the passage of light rays differently. These differences produce differences in color and contrast in the image. In transmission electron microscopes, atoms in the specimen deflect or scatter individual electrons from their paths in the beam. Atomic nuclei in the specimen, which are positively charged, scatter electrons by attracting them from their paths; negatively charged electron clouds around atomic nuclei scatter electrons by repelling them. Atoms of relatively high atomic number, such as those of heavy metals, including lead, gold, and uranium, scatter electrons to the greatest degree.

Electrons scattered widely enough by subparts of the specimen collide with the sides of the lens-pole pieces and are eliminated from the beam. This elimination of scattered electrons creates "holes" in the beam that contain reduced numbers of electrons. The positions of the holes correspond to points in the specimen that contain heavier atoms. After magnification by the objective, intermediate, and projector lenses, and focusing on the specimen screen, the specimen points that contain heavier atoms appear darker because fewer electrons strike the screen in these regions. The entire collection of darker areas produced by electron scattering traces out the image of the specimen on the screen as differences in contrast. Because passage through the specimen does not significantly alter the wavelengths of electrons, the image seen in the transmission electron microscope is monochrome; that is, no differences are produced that are equivalent to the color variations in light microscope images. Although any single color could be used, electron images are usually printed in black and white for practical reasons.

Applications

The ability of a microscope of any kind to visualize small objects is called the resolution or resolving power of the instrument. The resolving power of electron microscopes is equivalent to 0.61 (a factor that evaluates the conditions necessary for visualization by the human eye) times the wavelength of the electron beam, divided by one-half of the angle of the cone of electrons entering the objective lens. For the very best resolution, this relationship requires that the wavelength of the electron beam be as short as possible and that the angle of the cone of electrons entering the objective lens (called the angle of illumination) be as large as possible. Practical difficulties with presently uncorrectable faults in magnetic lenses limit the angle of illumination to a very small value, 0.004 radian, or about 0.23 degree. Using this value for the angle of illumination and electron wavelengths of 0.054 to 0.037 angstrom obtained in microscopes operated at 50,000 to 100,000 volts gives a practical resolving power of about 5 to 10 angstroms. This means, for example, that two points in the specimen separated by this distance can be seen as separate, individual points. These dimensions lie near atomic levels; in fact, large atoms such as those of uranium have actually been resolved and made visible in transmission electron microscopes.

The shortest wavelengths of visible light that can be used as an illumination source for light microscopes are near 400 nanometers. Using this value as the basis for calculating the best possible resolution of light microscopes, which follows a relationship similar to that of electron microscopes, gives an ultimate resolving power of about 0.2 micrometer, or 2,000 angstroms. Although about a thousand times larger than the resolution of the unaided human eye, this value is about 400 times smaller than the resolving power of a typical transmission electron microscope.

The resolution of transmission electron microscopes can be improved beyond the 5- to 10-angstrom level by increasing the operating voltage to 1 million volts or more. An operating voltage of 1 million volts, for example, shortens the wavelength of the electron beam to 0.009 angstrom and gives a resolving power of about 1.3 angstroms. This advantage, however, is offset by the fact that the degree of scattering by specimen atoms is inversely proportional to the velocity of electrons. As the velocity increases as a result of increases in the accelerating voltage, the contrast of specimen details decreases. In instruments operated at 1 million volts, contrast is often so poor that specimen points, although theoretically resolvable, cannot be distinguished. In addition, the specimen becomes subject to destruction by heating as a result of the absorbance of electrons. As a result of these difficulties, the practical limitations of transmission electron microscopes for most applications remain at operating voltages of 50,000 to 100,000 and resolutions of about 5 to 10 angstroms.

Electrons, even those traveling at relatively high velocities, are easily deflected by gas molecules or specimen atoms. For this reason, the entire pathway traveled by electrons inside an electron microscope must be kept at a near-perfect vacuum, and the specimen must be very thin. The image on the fluorescent screen is viewed through ports that consist of thick leaded-glass plates. Interlocks allow specimens and photographic plates to be exchanged without disturbing the vacuum in the instrument. In order to protect the operator from X rays generated when stray electrons strike metal surfaces in the microscope, heavy metal shielding is placed around the column housing the gun, magnetic lenses, and viewing area. The shielding and wire coils for the magnetic lenses make a transmission electron microscope a bulky and heavy instrument that must be assembled and installed in a permanent location.

Transmission electron microscopes are provided with controls for adjusting current to the magnetic lenses. These controls focus the condenser lens precisely on the specimen and allow the total magnification produced by the objective, intermediate, and projector lenses to be varied at will. Other controls are connected to ultrafine screws that allow the specimen to be held stationary or to be moved over distances as small as a few angstroms. By means of these controls, the specimen can be scanned for areas of interest and held without movement for photography. Photographs are taken by a control that raises the viewing screen, allowing the electron beam to fall on a photographic plate held just below the screen.

Specimens to be viewed in transmission electron microscopes must be dry and nonvolatile in order to avoid disturbance of the operating vacuum. In order to make specimens thin enough to transmit electrons, they must be dried to a thin film or cut into ultrathin slices. In fact, specimens to be examined in the transmission electron microscope cannot be thicker than about 1,000 angstroms; ideal thicknesses are about 400 to 600 angstroms. At these levels, specimen thickness lies well below the 2,000-angstrom resolving power of the light microscope.

These operating limitations place severe requirements on specimens, particularly those examined in applications of the transmission electron microscope to biology and medicine. Living specimens are far too thick and volatile to be examined directly in the microscope. Instead, specimens are prepared by one of two major techniques. In one, specimens are fixed, embedded in plastic, and cut into very thin sections. Fixation involves reaction with reagents such as formaldehyde or glutaraldehyde that introduce chemical cross-links between specimen molecules. The cross-links stabilize the molecules and hold them in position against the rigors of embedding, sectioning, and examination in the microscope. As part of the embedding procedure, water and other volatile materials in the specimen are replaced by the molecules of the plastic in liquid form. Hardening or polymerization of the plastic converts it and the specimen into a stable, nonvolatile form that can be cut into ultrathin slices. Epoxy plastics are routinely used for this purpose. Sections are cut by a device called an ultramicrotome; this precision instrument is capable of advancing a plastic-embedded block of tissue against a knife edge in increments of a few hundred angstroms. Freshly broken glass edges or polished diamond knives are used to cut the sections.

In the other preparative technique, a solution of molecules or structures is spread in a thin layer over a plastic film. After it has dried, the material is ready to be placed in the microscope.

Biological specimens prepared by either method produce little contrast in the electron microscope because the primary atoms of living tissues—carbon, oxygen, hydrogen, and nitrogen—do not scatter electrons to any appreciable degree. In order to increase the contrast, biological specimens are usually "stained" by being exposed to a solution of a heavy metal compound such as lead citrate, osmium tetroxide, or uranyl acetate. Atoms of the heavy metal deposit in or around specimen details, greatly increasing electron scattering by these regions. The scattering outlines specimen details with a dark contrast that makes them clearly visible in the electron image.

The contrast and surface details of isolated specimens are also increased by shadowing, in which a thin layer of a heavy metal is deposited on specimen surfaces. For shadowing, molecules or structures dried on a plastic film are coated under a vacuum by a heavy metal such as gold or platinum. The coating is made by heating a small quantity of the metal to the boiling point on an electrical filament located to one side of the specimen. Atoms of evaporated metal deposit on raised surfaces of the specimen facing the filament. The raised surfaces, because of their coating of heavy metal atoms, produce high contrast in the electron microscope image. Depressions in the specimen, located in the shadows of higher points, are not coated and appear transparent. In the electron microscope, the specimen surface appears to be thrown into sharp relief by a bright light placed to one side. These methods, now applied routinely, produce specimens that reveal details at the molecular and even the atomic level.

Context

Light microscopes capable of maximum resolution at 0.2 micrometer were constructed during the latter half of the nineteenth century. As it became obvious that light microscopes are fundamentally limited to resolution at this level, physicists identified forms of radiation with shorter wavelengths as possible sources of illumination. The greatest possibilities were offered by electron beams, which could readily be produced at wavelengths much shorter than those of visible light. Intensive research into the problems of designing lenses capable of focusing electrons began in Germany during the late 1920's, leading to the development of the first practical transmission electron microscope by Ernst Ruska and B. von Borries in the 1930's. Commercial production of their microscope began in Germany shortly before the onset of World War II. These developments were paralleled in the United States, where J. Hillier and W. W. Vance designed a commercially successful transmission electron microscope by 1941. By the late 1950's, specimen preparation methods were sufficiently advanced to allow effective use of transmission electron microscopes in physical, chemical, and biological research.

Applications of transmission electron microscopes have been particularly effective in metallurgy, the biological sciences, and medicine. During the period from about 1958 until 1970, the resolving power of the transmission electron microscope revealed details of biological structure that were unimagined by earlier investigators who were limited by light microscopy. The transmission electron microscope pushed structural investigations to the molecular level and was instrumental in starting the molecular revolution in the biological sciences.

Bibliography

Bradbury, Savile. *The Evolution of the Microscope*. New York: Pergamon Press, 1967. This highly readable book presents a complete description of the theory and development of the light microscope and continues with a historical survey of events leading to the transmission electron microscope. Many illustrations of light and electron lenses, their action, and microscopes of both types from earliest models to contemporary instruments are included.

Sjostrand, F. S. *Electron Microscopy of Cells and Tissues*. Vol. 1, *Instrumentation and Techniques*. New York: Academic Press, 1967. Written by an individual who was involved in the development of instrumentation and preparative techniques in electron microscopy, this book describes the theory of the instrument and its applications in the biological sciences in detail. Although written at a technical level, it can easily be understood by readers who do not have a background in physics or mathematics.

Watt, I. M. *The Principles and Practice of Electron Microscopy*. New York: Cambridge University Press, 1985. A readable work that presents the theoretical background of electron microscopy in clear and understandable terms and outlines its applications, including methods used to prepare specimens for examination.

Wilson, Michael B. *The Science and Art of Basic Microscopy*. Bellaire, Tex.: American Society for Medical Technology, 1976. Although concerned with light microscopes, this short book begins with basic optical principles and then deals with lens theory and the construction, operation, and applications of various types of microscopes. Clearly and simply written and is easily understood by a nontechnical reader.

Wischnitzer, Saul. *Introduction to Electron Microscopy*. 3d ed. New York: Pergamon Press, 1981. This book explains the theory and operation of electron microscopes and gives step-by-step details of the major specimen preparation techniques. This is a practical, understandable, and readable book that has served as a valuable introduction to electron microscopy for laypeople as well as scientists.

Stephen L. Wolfe

Cross-References

Charges and Currents, 351; Diffraction, 671, Electromagnetic Waves: An Overview, 735; Electromagnetism, 750; Lenses, 1235; Optics, 1673; Scanning Electron Microscopy, 2143; X-Ray and Electron Diffraction, 2749.

TRANSURANICS

Type of physical science: Chemistry
Field of study: Chemistry of the elements

The transuranic elements are those elements that are more massive than uranium. They are synthesized in the laboratory and are rarely found in nature unless involved in a supernova. These elements on Earth have been used in nuclear energy and weapons.

Principal terms

ALPHA RADIOACTIVITY: helium nuclei of two protons and two neutrons with no electrons, carrying a +2 charge, emitted as a radioactive decay; may be used to bombard elements to create heavier elements

ATOMIC MASS: the sum of the atomic number and the number of neutrons in the nucleus of an atom

ATOMIC NUMBER: the number of protons found in the nucleus of an atom; each element has its own atomic number

CYCLOTRON: a device that accelerates ions in a spiral path by the influence of electrical and magnetic fields until they emerge at great speed, at which the ions can then bombard other elements for the creation of other atoms

DEUTERON: the nucleus of a deuterium, which is an isotope of the hydrogen atom with one proton and one neutron; it carries a positive charge

ELECTRONVOLT: the unit of energy equal to the energy acquired by an electron falling through a potential difference of 1 volt, which is a very small amount of energy

HALF-LIFE: the time required for the concentration of a reactant to decrease to half the initial concentration

ISOTOPES: atoms having the same atomic number but different atomic masses; different isotopes of the same element have different physical and chemical properties; the atomic mass of an isotope is denoted as a number following the element's symbol, name, or atomic number

NUCLEON: those particles in an atom's nucleus, namely, the protons and neutrons

OXIDATION: the process in which an atom loses one or more electrons in a reaction, with the electron transferred to an adjoining atom in the chemical reaction; the atom that gains an electron is said to be reduced

Overview

The advent of particle accelerators has made it possible to synthesize elements with atomic numbers greater than 92, called the transuranium elements. Since neptunium, with atomic number 93, was first isolated in 1940, more than thirteen additional transuranium elements have been discovered. These transuranium elements are all radioactive and made in the laboratory. They very rarely, if ever, are found in nature. The effect of transuranium elements on civilization has been as profound as the atomic bomb. Plutonium, the first synthetic element found, was later used in the atomic bombs dropped on Nagasaki.

In 1934, Enrico Fermi and others had found that neutron irradiation of uranium leads to a considerable number of radioactive substances. The chemical investigation of these radioactivities led, however, to the discovery of the fission process rather than transuranium elements. In their original work, Fermi and his coworkers were led, on the basis of chemical experimentation, to assign these radioactivities to the transuranium elements. Early in 1939, however, experiments were described that made it certain that what was observed as the result of the bombardment of uranium with neutrons was radioactive isotopes of barium and other "light" elements. Subsequent work showed that practically all the radioactivities previously ascribed to transuranium elements were actually the result of fission products.

Several experimenters measured the energies of two main fission fragments by observing the distances they traveled from each other as a result of their mutual recoil as the nucleus exploded. Edwin McMillan of the University of California noted another radioactive product of the reaction, with a half-life of 2.3 days, that did not escape from the thin layer of fissioning uranium.

In the spring of 1940, it was deduced by chemical means that this product was an isotope of an element with atomic number 93, arising from the emission of an electron from uranium 239. This new element was named neptunium, as Neptune is the next planet beyond Uranus, for which uranium had been named. Neptunium was found to be very much like uranium in chemical properties.

Neptunium is unstable and will eject another electron to become plutonium with atomic number 94, as was discovered in 1941. Plutonium would later be used for the second atomic bomb. Neptunium was the second synthetic element isolated, after plutonium. The early history of neptunium, like that of plutonium, was limited to studies by using traces of the short-lived isotope neptunium 239. Neptunium 237 was discovered in 1942 and found to be sufficiently long-lived to make work with weighable amounts possible. This isotope is the decay product of uranium 237, which has a half-life of seven days and emits electrons. This in turn is formed as the result of a reaction of uranium 238 and is an alpha emitter of relatively long half-life, 2 million years.

The first weighable amounts of neptunium 237 were produced by bombardment of large amounts of uranium with high-speed protons from a cyclotron. In 1944, neptunium was isolated, and the element in the form of pure compounds was used to study a number of its chemical properties. Neptunium resembles chemically promethium, which is directly above it on the periodic table.

Plutonium was first synthesized by bombardment of uranium oxide with 16-million-electronvolt deuterons from a cyclotron on December 14, 1940. Alpha radioactivity was found to grow into the chemically separated element 93 (neptunium) fraction during the following weeks, and this alpha radioactivity was chemically separated from neighboring elements, especially the elements 90 to 93 inclusive. These experiments, which gave solid proof and identification of element 94, showed that this element had at least two oxidation states. These oxidation states were distinguishable by their precipitation chemistry and by the fact that it required stronger oxidizing agents to oxidize element 94 to the upper state than was the case for element 93. The first successful oxidation of element 94 took place on February 23 and 24, 1941.

The public announcement of the discovery of plutonium was withheld until after World War II. The announcement was delayed because of the use of plutonium in the nuclear weapons project. Much of the work was done in secret. Plutonium was named for the planet Pluto.

Plutonium in its highest oxidation state is similar to uranium in its (VI) oxidation state, and in a lower oxidation state is similar to thorium (IV) and uranium (IV) oxidation states. Normal, stable plutonium in its (IV) oxidation state as an ion would form insoluble compounds or stable complex ions analogous to those of similar ions.

Oxidation-reduction cycles (in which the element is said to be oxidized if it loses one or more electrons, and reduced if it gains one or more electrons) have been applied to the separation process for plutonium and other transuranium elements. In the case of plutonium, a substance was used that carried plutonium in one of its oxidation states but not in another. By use of this principle, a carrier could be used to carry plutonium in one oxidation state and thus separate it from uranium and its fission products. Then the carrier and the plutonium could be dissolved, the oxidation state of plutonium changed, and the carrier reprecipitated, leaving plutonium in solution. The oxidation states of plutonium could again be changed, and the cycles repeated. With this type of procedure, only a contaminating element having chemistry nearly identical with plutonium would fail to separate if a large number of oxidation-reduction reaction cycles were employed.

The successful operation of the reactor and plutonium extraction plant at Oak Ridge, Tennessee, led to the availability of the first milligram, then the first gram amounts of plutonium in 1944. The availability of milligram amounts of plutonium led to the discovery of the plutonium (III) oxidation state. Earlier tracer work at the University of California in 1941 had established the existence of a lower oxidation state (IV and/or III state) and a higher state (VI). Ultramicrochemical work in late 1942 and early 1943 defined the existence of the IV and VI oxidation states. The existence of the V oxidation state was established in the late summer of 1944.

Plutonium, having four oxidation states, has a chemistry as complex as any other element. It is unique among elements in that these four oxidation states all can exist simultaneously in aqueous solution at appreciable concentration. As a metal, its properties are also unique. Its six forms have melting points that range from room temperature to 640 degrees Celsius. Their appearance resembles that of lead. Plutonium has isotopes with atomic masses ranging from 232 to 246 inclusive.

Applications

The most important transuranium element in human affairs, plutonium has enormous nuclear properties. Nuclear reactors use plutonium 239 as breeders, which are used in conjunction with the abundant uranium 238. Such a system "burns" uranium 238 indirectly through the medium of fissionable plutonium 239, formed by the absorption of neutrons in uranium 238. A fissionable isotope such as plutonium 239 gives rise to an amount of heat energy equivalent to approximately 10 million kilowatt-hours per pound when it completely undergoes the fission reaction.

Plutonium is one of the most dangerous poisons known as a result of its alpha radioactivity. Plutonium 239 undergoes fission with thermal neutrons, showing that all neutrons emitted in the process are eligible to cause further fissions, establishing the great value of this isotope. This led to the wartime Plutonium Project for the production of plutonium on a large scale for nuclear weapons. The Plutonium Project sought to cause the chain reaction from uranium to plutonium on a large scale, and to separate plutonium 239 from uranium and from tremendous fission product radioactivities that resulted from the many fission product elements present within it. The first successful operation of the chain reaction, opening the atomic age, took place under the western stands at the Stagg football field at the University of Chicago.

Thus, plutonium was discovered in December, 1940, the first compound isolated in August, 1942, and the unusual properties of bismuth phosphate as a carrier for plutonium described in December, 1942. The Bismuth Phosphate Separation Process was placed in successful operation in the pilot plant in Clinton Laboratories in Oak Ridge, Tennessee, in December, 1943. The first uranium from this pile in Oak Ridge entered the separations plant in December, 1943. By the following month, metal from the pile was being processed in the plant at a rate of 0.3 ton per day. Several grams of plutonium were produced over the next months.

In remote Washington State, there were two principal types of plants at the Hanford Engineering Works, those that form plutonium within uranium by a nuclear chain reaction, and those that separate the plutonium from the uranium and the fission products. The responsibility for making a weapon out of the plutonium produced at Hanford and the uranium 235 produced at Oak Ridge fell to the Los Alamos National Laboratory, in New Mexico. Here, the plutonium 239 and uranium 235 were purified for a nuclear weapon. Many fields of work were involved with this process. Only four and a half years after the discovery of plutonium, it was used in a nuclear bomb.

After the completion of the wartime Metallurgical Laboratory, where the most essential part of the investigations was concerned with the chemical processes involved in the production of plutonium, attention turned to the problem of synthesizing and identifying the next transuranium elements.

Curium 242 was produced in the summer of 1944 as a result of the bombardment of plutonium 239 with 32 million electronvolts of helium ions. Curium was first isolated in the form of a pure compound, the hydroxide, of curium 242 in the fall of

1947. The isotope curium 242 is highly radioactive and has a short half-life. Thus, chemical investigations with it in macroscopic concentrations are difficult. The identification of element 95 followed, in late 1944, as a result of the bombardment of plutonium 239 with neutrons. These two elements had stable (III) oxidation states that greatly resembled the rare earth elements in chemical properties; thus, they were most difficult to study. Americium was first isolated in the form of a pure compound as a hydroxide in the fall of 1945, at the Metallurgical Laboratory. Americium 243 has a half-life of 7,950 years, and curium 244 has a half-life of 19 years, sufficient for study of these isotopes.

Transuranic elements up to number 103 all are members of a chemical family similar to the series of rare earths running from lanthanum (atomic number 57) to lutetium (atomic number 71) known as the lanthanides, in which the 4*f* electron shell is filled. The second rare earth series, the actinides, in which the 5*f* electron shell is filled, begins with actinium (atomic number 89) and includes the natural elements thorium, protactinium, uranium, and the transuranic elements up to lawrencium, atomic number 103.

Since the transuranic elements are all of synthetic origin, their atomic masses, generally, depend on the source material because this determines what isotopes are involved. Suitable isotopes are available for study of neptunium through einsteinium (atomic number 99). Study of the chemical properties of the elements fermium (atomic number 100) and above is limited to the use of trace amounts.

Metallic protactinium, uranium, neptunium, and plutonium have complicated structures. Americium was the first actinide to display a similarity to lanthanide metals in its crystal structure. As mentioned above, the metallic properties of plutonium are unique. The interatomic distances deduced from crystal structure data indicate the presence of 6*d* electrons in these metals; however, from the standpoint of chemical reactivity, they behave more like the lanthanide metals than like metals that fill their outer 5*d* electron shell, such as tantalum, tungsten, rhenium, osmium, and iridium.

The most important prerequisite to the process for making transcurium elements was that sufficiently large amounts of americium and curium had to be available for neutron bombardment.

Berkelium, with atomic number 97, was discovered in December, 1949, as the result of the bombardment of milligram amounts of americium 241 with 35 million electronvolts of helium ions accelerated in the 152-centimeter cyclotron. Berkelium 243 has a half-life of 4.5 hours. Californium, atomic number 98, was discovered in February, 1950, by the bombardment of milligram amounts of curium 242 with a 35-million-electronvolt stream of helium ions. The isotope with an atomic mass 243 has a half-life of forty-four minutes.

Elements 99 and 100 were discovered in debris from the "Mike" thermonuclear explosion, which took place in the Pacific in November, 1952. Debris from the explosion was collected first on filter paper attached to drone airplanes that flew through the clouds, and later, in surface material from nearby atolls. This led to positive identification of isotopes of elements 99 and 100. Element 99, einsteinium 253, has a half-life of twenty days. Element 100, fermium 255, has a half-life of sixteen hours. Mendelevium, atomic number 101, was formed in 1955 by bombardment of einsteinium 253 with a helium ion beam. Mendelevium 256 decays by electron capture, with a half-life of 1.5 hours, to isotope fermium 256.

Context

The entire set of elements containing more than ninety-two protons are the transuranic elements. Because their lifetimes are all short compared with the age of the earth, they are not found on earth naturally and need to be synthesized for their discovery. Both plutonium 239 and neptunium 237 have been found in nature in very small concentrations in uranium-containing ores. These isotopes are formed continuously as the result of the interaction of neutrons with uranium 238. Plutonium 239 has been found in pitchblende, a uranium ore, at concentrations of one part per trillion of uranium. This concentration is approximately that to be expected by the absorption of the available neutrons by the uranium ore. Similar, although somewhat smaller, concentrations of neptunium 237 are also found by the action of neutrons on the uranium.

It has been speculated that all nuclei with more than about 60 nucleons were produced by absorption of up to 180 neutrons by iron nuclei in the very short time of the explosion of a supernova. A supernova is a massive star that explodes at its death. In the process, elements more massive than iron, with atomic number 26, are created.

Supernovas may create some transuranic elements. This theory comes from observing how long supernova explosions last from the study of Chinese records of ancient and contemporary observation of extragalactic supernovas. Several supernovas have lost half their luminosity in fifty-five days; thus, this is their half-life, the same as the half-life of californium 254. This may be produced in the supernova explosion as the result of a large flux of neutrons. Supernova 1987A had a half-life of seventy-seven days, the same as cobalt 56.

If a heavy nucleus is given enough energy to overcome electric repulsion of another nucleus, the collision could produce a compound nucleus containing as many nucleons as the sum of the colliding nuclei. This compound nucleus may contain many more nucleons than any naturally occurring nucleus. For example, some of the heavier transuranium elements have been produced by collisions between the lighter transuranium elements and boron nuclei stripped of their electron cloud. There are also proposals to produce some of the elements by collisions between heavy, stripped nuclei such as copper and uranium nuclei. Compound nuclei with more than three hundred nucleons could then decay to the metastable ground states, still more massive than the first transuranic elements.

The periodic table will be used to plot properties and names of future transuranium elements. Elements 93 through 103 chemically resemble actinium, protactinium, and uranium; thus, they are grouped together as the rare earth elements, the actinides. Element 104, discovered by the Soviets in 1965, will lie directly below hafnium on the periodic table. Based on recommendation by the International Union of Pure and Applied Chemistry, three-letter Latin symbols are to be used to name elements having atomic numbers greater than 103, and the names are to end in "-ium" for metals and nonmetals.

Bibliography

Chang, Raymond. *Chemistry.* 2d ed. New York: Random House, 1984. This undergraduate-level college text covers all areas of chemistry. Gives the reader a solid understanding of some of the simpler chemical processes, and an in-depth study of chemical reactions, including oxidation-reduction. Most useful in the nomenclature for the transuranium elements.

Feinberg, Gerald. *What Is the World Made Of? Atoms, Leptons, Quarks, and Other Tantalizing Particles.* Garden City, N.Y.: Doubleday, 1977. Addresses the building blocks of nature, including the atom. Briefly covers nuclear synthesis of elements. Feinberg brings up the issue of whether superheavy elements may exist as a result of collisions between heavy elements.

Goldsmith, Donald. *Supernova! The Exploding Star of 1987.* New York: St. Martin's Press, 1989. Goldsmith discusses creation of elements. All elements heavier than iron (atomic number 26) are believed to have been created in supernova explosions. Other scientists suggest that transuranium elements are created in these explosions as well. Goldsmith argues that cobalt 56 caused supernova 1987A to glow, as this isotope had the same half-life as the supernova's luminosity.

Ley, Willy. *The Discovery of the Elements.* New York: Delacorte Press, 1968. Using the periodic table, Ley shows how elements, including the transuranics, were discovered. Shows how each fits into the periodic table and uses this to predict properties of yet-to-be-discovered transuranium elements.

Seaborg, Glenn T. *The Transuranium Elements.* New Haven, Conn.: Yale University Press, 1958. The main reference used for this article and a valuable source. Seaborg was one of the scientists who worked with the Plutonium Project, and thus presents much of this book from a firsthand perspective. Very clear in much of his discussion. Some of the text is too technical for the general reader. Though dated, it is the text from the period of discovery of the transuranium elements.

Taylor, John G. *The New Physics.* New York: Basic Books, 1972. Taylor discusses some new concepts in physics—for example, the idea that californium may have been created in supernova explosions, as the half-life of historical supernovas is the same as californium 254.

Weart, Spencer R. *Scientist in Power.* Cambridge, Mass.: Harvard University Press, 1979. Weart discusses the early research into plutonium, and how nations, especially France, worked toward the atomic bomb.

David R. Teske

Cross-References

Actinides, 30; Breeder Reactors, 259; Cyclotrons, 635; Detectors on High-Energy Accelerators, 643; Fission and Thermonuclear Weapons, 892; Lanthanides, 1220; Metals, 1380; Ox-Redox Reactions, 1725; The Periodic Table and the Atomic Shell Model, 1763; Radioactive Nuclear Decay and Nuclear Excited States, 2057; Supernovas, 2434; Transition Elements, 2578.

THE TURING MACHINE

Type of physical science: Computation
Field of study: Computers

A Turing machine is a theoretical, highly generalized, abstract computer that encapsulates the notion of an automatic computer and all the operations that any computer could ever perform, without introducing any physical or technological limitations. A true Turing machine could never be built, but its complete generality makes it a valuable reasoning tool, since any facts deduced about general Turing machines will apply equally to all computers, even those that have not yet been invented.

Principal terms

ABSTRACT MACHINE: a hypothetical computing machine that is not intended to be built but has properties that make it useful for theoretical study

ALGORITHM: a fixed and describable method for achieving some result, usually a list of steps to be carried out in sequence

COMPUTABLE: a problem that might be solved by some possible form of computer, even though such a computer may not yet exist

FINITE STATE AUTOMATON: a restricted abstract machine that is limited to a fixed number of internal conditions and has no memory component

STATE: a recognizable configuration of the logical contents of the internal components of a computer, which may be used to characterize that computer's condition at some time (for example, "idle," "awaiting input," "finished")

TRANSITION: a change of state that constitutes a single atomic step in a computation

Overview

The Turing machine is the universally accepted archetype for the most powerful computer possible. Any problem that can ever be solved by any computer can also be solved by a Turing machine. Not surprisingly, Turing machines are abstract machines—that is, they are not intended ever to be physically constructed—but a study of Turing machines may reveal fundamental truths about the nature of computability. The Turing machine is such a successful encapsulation of the notion of a computer that "Turing computable" has become a standard phrase that means "potentially soluble by some kind of computer at some time," and the power of a computing machine is often expressed in comparison with the power of a Turing machine.

The Turing machine is a member of the group of abstract computing machines based on finite state automata with various added features, usually external memory components, that are required to give the machines some useful degree of computing power. Any understanding of Turing machines must be based on an understanding of finite state automata.

The finite state automaton is a basic kind of abstract machine. It is a simple, memory-free device that reacts to inputs from its environment in an explicitly predetermined way. It has a fixed, finite number of stable states; its initial state is known; it only changes state or produces a response (output) as a direct result of inputs from the environment; and most important, any change of state or production of output is determined solely by the state it is already in and the input it receives—there are no other hidden mechanisms or motivations. Inputs arrive one at a time, and the state transitions are considered to be instantaneous.

As an example, a finite state automaton that recognizes very simple correctly formed sentences may be defined with:

Eight inputs, representing possible words in the sentence:
  The, A, Cat, Man, Saw, Ran, Watch, Big.
Two outputs, representing the machine's judgment of the inputs:
  OK—The sequence of inputs represents a valid sentence.
  NOK—The sequence of inputs does not represent a valid sentence.
Six steady states, including:
  INIT—The initial state, in which the machine starts.
  END—The terminal state, which the machine reaches after making a decision.
  A, B, C, D—Intermediate states.

The machine's actions may be described by means of a simple table, defining the state that it will move into and the output it will produce, for each possible combination of current state and received input:

Current State	**Received Input**	**New State**	**Resultant Output**
INIT	THE or A	=> A	NOK
INIT	WATCH	=> C	NOK
INIT	any other	=> END	NOK
A	CAT or MAN	=> B	NOK
A	BIG	=> A	NOK
A	any other	=> END	NOK
B	RAN	=> END	OK
B	SAW	=> C	NOK
B	any other	=> END	NOK
C	THE or A	=> D	NOK
C	any other	=> END	NOK
D	CAT or MAN	=> END	OK
D	BIG	=> D	NOK
D	any other	=> END	NOK

Alternatively, finite state automata are frequently depicted graphically:

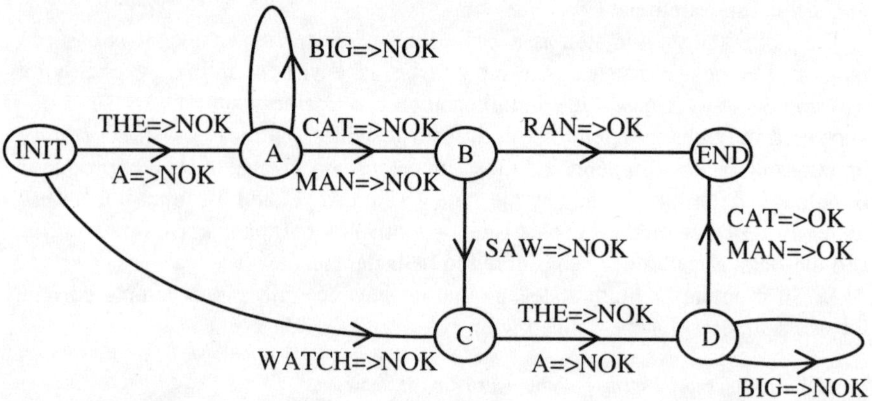

The operation of the machine may be illustrated by means of an example, such as the sentence "The man saw a big big cat." With the machine initially in state INIT, the first word, "The," is received, and the machine moves into state A while outputting "NOK," indicating that the input sequence so far does not form a correct sentence. The second word, "man," is received by the machine in state A; this combination results in the machine moving to state B and maintaining the output "NOK." Subsequent words take the machine through the sequence of states C, D, D again, and finally END, keeping the output as "NOK" until the final transition, when the word "cat" completes the correct sentence and results in an output of "OK."

The major and definitive limitation of a finite state automaton is that it has a fixed, finite number of states, and those states are its only form of memory. This means that although finite state automata may provide simple and convenient specifications of simple tasks, any operation that requires an unbounded (unlimited but not necessarily infinite) amount of memory will be quite impossible. In fact, the selection of tasks that may be performed by a finite state automaton is severely limited. Tasks that are impossible include the representation of numbers, the detection of palindromes (sentences that are the same backward and forward), and the archetypal nonfinite problem of checking for a correctly balanced sequence of parentheses ("(()())" is acceptable but ")()()(" and "((())" are not).

Turing machines are finite state automata augmented by an external memory component, which takes the form of an infinite tape (that is why a true Turing machine could never be built). The tape is divided into an infinite number of squares, in each of which may be written a single symbol chosen from a finite, predetermined set (the machine's alphabet). At each state-transition step, the Turing machine's controlling finite state automaton, instead of accepting input from the environment or the user of the machine, "reads" a symbol from the tape at a particular position (known as the tape-head position) and uses that as its input. Similarly, instead of outputting some value, it erases the symbol at the tape-head position, and replaces it with another. After rewriting the symbol on the tape, the machine is free to move the tape-head position by one square to the left or right.

Thus a Turing machine is a finite state automaton with its input and output operations redirected to an infinite storage tape, which it may use in much the same way as a worker uses a notepad, except that access is strictly sequential; it may not shift its attention unrestrictedly to any part of the notepad in a single operation.

The increased power of Turing machines may be illustrated by designing a Turing machine that solves a problem that could not be solved by finite state automata. An illustrative example is the problem of checking a string of parentheses for correct balance. This is impossible for a finite state automaton, because checking the balance requires the ability to store, in some form, a memory of how many unmatched parentheses have been read. There is no upper limit to the number of parentheses that could appear in a sequence; therefore there is no fixed finite number of states that could possibly record that number reliably in all cases.

A Turing machine may circumvent the problem of having a limited number of states by keeping the whole string on its tape and gradually eliminating matched pairs of parentheses from it as they are found. The computation continues either until the tape is empty, which indicates that the original string was correct, or until the elimination process fails to find a matched pair to eliminate, which indicates that the original string was incorrect.

This machine would have an alphabet of five symbols: (and), to form the string in question, #, with which it overwrites paired parentheses as they are eliminated, S, to mark the start, or left-hand end of the string, and E, to mark the end of the string. It would also have five states: INIT, the initial state in which the machine starts; OK, a terminal state that the machine will reach if the string is correct; NOK, a terminal state that the machine will reach for incorrect strings; and A,B,C, intermediate states that are occupied while the machine searches for pairs.

As with a finite state automaton, the machine's actions may be described by means of a simple table that defines the state that it will move into and the output it will produce for each possible combination of current state and received input. In this case, the input will be the symbol read from the tape and the output will be the symbol with which it is to be overwritten. There will also be an additional entry, specifying the direction in which the tape-head position is to move.

Current State	Received Input	New State	Output	Direction
INIT	S =>	A	A	R
A	(=>	A	(	R
A	# =>	A	#	R
A	E =>	C	E	—
A	) =>	B	#	L
B	# =>	B	#	L
B	(=>	A	#	R
B	S =>	NOK	S	—
C	# =>	C	#	L
C	(=>	NOK	(	—
C	S =>	OK	S	—
OK	any =>	OK	E	—
NOK	any =>	NOK	E	—

The design assumes that the machine will begin running in state INIT, with the tape-head position ready to read the "S" that marks the beginning of the sequence of parentheses. The machine then alternates between states A and B, which together search out and erase a matched pair of parentheses. State A performs half of the procedure by scanning the tape to the right until it finds a ")", and state B performs the other half by scanning to the left for a "(". When the symbol being searched for is encountered in either of these states, it is erased and thereby eliminated from future searches.

If state B (the left scanner) fails to find a "(", that means that a ")" previously found in state A was unmatched, and the string is unacceptable. If state A (the right scanner) finds the "E" that marks the end of the string before finding any ")" symbols, that means that every ")" has already been eliminated, and the string is correct if every "(" has also been eliminated. In this case, state C is entered and scans to the left, checking that there are no "(" symbols left. If there are none, it will eventually reach the "S" that marked the beginning and terminate.

Applications

Knowledge of what, if any, limits there are to the powers and abilities of computers is central to the study of computer science. It is easy to believe that the only limits are those imposed by the technology of the hardware and the creativity of the programmer. Even though a problem may now be completely impossible to solve by computer, at some time in the future, when faster and more powerful computers are available, and new algorithms have been discovered, it should be possible.

This reasoning has no true logical basis. It is pleasant but incorrect to believe that everything is within the grasp of human inventiveness, and similar reasoning has led to incorrect conclusions in the past. It seems logical that the only limit to the speed of a vehicle is determined by the power of its engine—with enough time and resources to develop new propulsion methods, it should be possible to cross the galaxy in a day. This is so obviously correct that nobody doubted it until Albert Einstein put aside the obvious and, through abstract reasoning, developed the theory of relativity. Relativity shows that there is an absolute limit to speed, which is not a product of inadequate technology; no matter what new discoveries are made, no object shall ever travel faster than light.

The Turing machine is not a real machine at all, but a theoretical device, another product of abstract reasoning, which, much like the theory of relativity, leads to results that have an enormous, fundamental, and rather surprising impact upon the understanding of what computers are capable of doing. The major result of the theory of Turing machines is that there are certain problems that can never be solved by any form of computing device, no matter how advanced technology and algorithm design may be. Just as there is an absolute limit to speed, there is also an absolute limit to computability.

The power of a Turing machine is not at all obvious, given its description, and it is quite clear that if Turing machines were built, they would be very inefficient, cumbersome, and unpleasant to program. If the nature of deterministic computation is considered, however, it becomes apparent that the Turing machine exactly captures the essence of deterministic processing.

Any physically constructable machine must have some finitely limited number of distinguishable internal states; an infinite machine is just not possible to build. Any computing device that uses an additional memory component (even a human "computer" writing on paper) is capable only of reading and writing a finite number of distinct symbols and transferring its attention from one portion of the memory to another from time to time. This is purely a description of the process of computation, without any mention of how the computation is performed or how the computer works. It is also the exact description of a Turing machine.

One of the most important results of the theory of Turing machines, which is usually referred to as "Turing's thesis," is the "halting problem." This is a proof that no computing machine is capable of inspecting other computing machines and determining whether they will eventually produce some complete result or "go into a loop" and never finish the computation. This has great significance for automatic programming (it leads to the result that programs in general cannot be checked reliably for run-time errors), artificial intelligence (there can be no all-powerful infallible computers), and even philosophy (without any assumptions about the existence of a "soul," the human brain is merely a complex finite state automaton, with eyes and hands to read and write symbols on some form of memory device, so humans are no more than restricted, living Turing machines).

Another result is that the number of different Turing machines that could possibly exist, although infinite, is smaller than the number of different real numbers (real numbers include the integers 0, 1, 2, 3, 4, and so forth, fractions, and irrational numbers such as pi and the square root of two). This means that there are numbers that can never be computed.

Other important results are in the area of complexity theory. Shannon's theorems (for Claude E. Shannon) show that for any Turing machine with A symbols in its alphabet and S states, an equivalent Turing machine (solving the same problem) may be constructed with 2 states and no more than 4AS + A symbols, or with 2 symbols in its alphabet and no more than 8AS states. This illustrates the trade-off between time and space in alternative solutions to the same problem and leads to a useful measure of the overall complexity of a computable problem.

Context

Alan Mathison Turing (1912-1954) was a mathematician, philosopher, and long-distance runner. While running, he occupied his mind by working on theoretical problems in various areas of scientific study. In 1935, he was preoccupied with the problem of finding an answer to Hilbert's third question, Is there any fixed method or algorithm that can be applied to any mathematical statement and determine reliably whether or not it is correct? This is one of the critical questions posed by the mathematician David Hilbert in 1900, on which a vast effort has been expended by many of the world's leading mathematicians and philosophers. While resting after a long run, Turing conceived the idea of a Turing machine as a universal algorithmic machine, which enabled him to become the first to prove that the answer to Hilbert's question is "No."

Since about 1900, mathematics has become a very theoretical and introspective discipline. Many practitioners are no longer satisfied with discovering new methods for solving mathematical problems and are more interested in deeper and more philosophical concerns, such as What does it all mean? How can we be certain that mathematics is correct and meaningful? and What are the ultimate limits to mathematical knowledge? The Turing machine, by providing an answer to Hilbert's third question and a tool for investigating other related problems, is therefore of great importance.

Computer science, as a recognizable field, only came into existence with the construction of the first electronic computer, the Colossus, in 1943. Turing's work, which is universally recognized as providing the fundamental theoretical core of computer science, is thus in the unusual position of predating the field to which it is most directly applicable.

The theory of Turing machines is so completely generalized that its position of central importance is perpetually guaranteed. Any advances in the theory, technology, and methodology of computer science must always be subject to its consequences.

Bibliography

Brady, J. M. *The Theory of Computer Science*. London: Chapman and Hall, 1977. This technical description of theoretical computer science covers Turing machines in depth. Written as a college textbook, with many illustrative examples and exercises, but not requiring significant prior knowledge, this volume is not difficult.

Hodges, Andrew. *Alan Turing: The Enigma*. New York: Simon & Schuster, 1983. This biography of Turing contains carefully crafted and easily understood descriptions of the central ideas of Turing machines and Turing's other important projects in the early development of computers. This well-written volume is suitable for both nontechnical readers and experts.

Shannon, Claude E., and John McCarthy, eds. *Automata Studies: A Universal Turing Machine with Two Internal States.* Princeton, N.J.: Princeton University Press, 1956. Shannon proposes, describes, and demonstrates a means of measuring the complexity of an algorithm, based on the possible trade-off between the number of states and the number of symbols in a Turing machine.

Turing, Alan M. "Can a Machine Think?" In *Computers and Thought*, edited by Edward A. Feigenbaum and Julian Feldman. New York: McGraw-Hill, 1963. This work on theoretical computer science and artificial intelligence explores the differences or lack of differences between humans and computers. Introduces the "Turing Test" for successful artificial intelligence.

____________. "On Computable Numbers with an Application to the Entscheidungsproblem." *Proceedings of the London Mathematical Society* 33 (1936). This article, in which Turing first describes Turing machines, argues that they are universal computing machines, and demonstrates the existence of uncomputable problems. A historic document.

Stephen Murrell

Cross-References

Artificial Intelligence: An Overview, 146; Cellular Automata, 324; Computer Architecture, 477; Neural Networks, 1481; Symbol Manipulation Programs, 2450; The Von Neumann Machine, 2690.

THE TWIN PARADOX

Type of physical science: Relativity
Field of study: Special relativity

The twin paradox (or clock paradox) of special relativity arises if one attempts to compare the ages of biological twin brothers when one of them is moving at constant velocity relative to the other. Each twin says that time is passing more slowly for the other brother than for himself; in fact, both are correct, a result of the relativistic effect known as time dilation.

Principal terms

ACCELERATION: the rate at which the velocity of a moving object changes with time; any change in speed or direction of motion

FRAME OF REFERENCE: an observer or coordinate system relative to which various physical or geometric properties can be measured

INERTIAL: refers to an unaccelerated state in which an object is moving with a constant velocity

PRINCIPLE OF RELATIVITY: a statement about how different observers agree on the laws of physics

PROPER TIME: time indicated on a clock which is stationary with respect to the observer

TIME DILATION: the slowing down of a moving clock relative to a clock which is stationary with respect to an observer

VELOCITY: the rate at which the position of a moving object changes with time; constant velocity means motion in a straight line at constant speed

Overview

The twin paradox—also called the clock paradox—is a famous conundrum, or word puzzle, which arises as a result of the special theory of relativity. In order to understand how the paradox comes about, and especially how it is resolved, it is necessary to reconsider very carefully the fundamental assumptions on which special relativity is based.

Albert Einstein (1876-1955) published his theory of special relativity in 1905. The theory examines how two observers, who are equipped with various sorts of measuring devices (such as rulers, clocks, and balance scales) and are moving with a constant velocity relative to each other, would agree or disagree on various measurements. Special relativity is based on the principle of special relativity, which states that the laws of physics are the same for both observers. More specifically, this principle, or axiom, says that two observers, each of whom appears to the other to be moving with a constant speed in a straight line, cannot actually discern which one of them is moving. In a practical sense, most people have, at one time or another, experienced something of this phenomenon when in a vehicle which is "stopped" at an intersection: They look out the window to see the vehicle in the next lane apparently drifting backward, only to discover that it is, in fact, their own vehicle which is drifting forward.

The laws of physics are based on experiment or observation, and among these laws is one which states that the speed of light in empty space is an absolute constant, independent of any motion of a light source or observer (detector). The speed of light in a vacuum (c) is about 300,000 kilometers per second, a result obtained by all observers. The fact that the measured speed of light is always the same would seem to defy common sense. Intuitively, one would expect that a measured light signal from a rapidly receding source would be moving slower than a light signal from a source which is rapidly approaching the observer, but such is not the case experimentally. This profound result, the constancy of the speed of light, has important consequences for the concepts of space and time. In his 1905 paper on special relativity, Einstein demonstrated that an intuitive notion of space and time as separate and distinct concepts must give way to a more unified concept of space-time as a single entity. It is this combination of the special relativity principle with the constancy of the speed of light that necessitates such a conceptual revision.

Suppose that two observers have a constant relative velocity between them. The situation may be treated as if one observer were at rest and the other observer were moving in a straight line at a constant speed. The two observers must agree on the speed of any light signal, but because speed equals distance divided by time, they must disagree on the measurements of both distance intervals and time intervals. The traditional example that is used to illustrate this concept is one observer riding in a vehicle, such as the car of a train moving uniformly along a straight track. A light source on the floor of the car sends a signal to a detector mounted directly above it on the ceiling. Therefore, to the observer in the car, the light seems to have traveled in a vertical line. To an observer who is stationary on the ground, however, the situation is somewhat different. While the light is moving from the floor to the ceiling, the car is moving forward at some constant speed v, and thus the light seems to have taken a longer, diagonal path. Since both observers must agree on the speed of light, the longer path as reckoned by the stationary observer on the ground must correspond to a longer time interval. Thus, for the same experimental situation, the time interval reckoned by the stationary observer is longer than that reckoned by the moving observer.

This disagreement concerning time intervals is called "time dilation" because a moving clock would seem to run slower than a stationary clock. How much slower? The answer is: Not very much, unless the speed of the moving observer relative to the rest observer, v, is an appreciable fraction of the speed of light, c. (The time dilation factor, or "relativistic factor," is $\sqrt{1 - v^2/c^2}$.) Since the speed of light is so large, this relativistic effect is essentially negligible at ordinary, everyday speeds. Even at one-tenth the speed of light ($v = 30{,}000$ kilometers per second), the time dilation factor is 0.995, which means that one hour on a stationary clock would correspond to about 59.7 minutes on the moving clock. At 50 percent the speed of light, the time dilation factor is 0.866; one hour on a fixed clock corresponds to about 52.0 minutes on the moving clock. At 99 percent the speed of light, however, the time dilation factor is 0.141; one hour on a fixed clock corresponds to about 8.5 minutes on the moving clock.

Time intervals measured on a clock at rest relative to the observer are called proper time intervals, as being at rest is the usual or "proper" situation. Because a moving clock runs slower than an identical stationary clock, a time interval on a moving clock will be longer than the corresponding proper time interval. It must be emphasized that the term "clock" can refer to any device which repeats itself at regular intervals; that is, the device must oscillate, vibrate, or beat. Also, all physical processes, including chemical and biological processes, slow down relative to a stationary clock when they occur in a moving frame of reference.

The paradox occurs because the relationship between the two observers is completely reciprocal. Both observers regard themselves as being at rest, and each observer believes that the other is the one in motion. In other words, each observer says that the other's clock is running too slowly. How can both be correct? It would seem that, if one clock is slower, then the other must then be faster, and that one can resolve this question by comparing the two clocks.

This situation is called the clock paradox, and it is most commonly illustrated as a hypothetical space travel problem involving a set of biological twins—hence the term "twin paradox." One twin, Tom, leaves the earth in a spaceship traveling at a large fraction of the speed of light, such as 99 percent ($v/c = 0.99$) and journeys to a nearby star system, such as one that is 3.5 light-years away. The other twin, Tim, stays home on the earth. Tom arrives at the nearby star system but immediately decides to return home.

Consider the situation from Tim's point of view. Tom's complete trip, over and back, covered a total distance of 7.0 light-years. He traveled at nearly the speed of light, and if one neglects the short time required to achieve this speed as well as the short time spent at his temporary destination, one sees that the total time for this journey must be approximately seven years. Since Tom is moving relative to Tim, however, Tom's clock (and his biological processes) must run slower than Tim's clock. At 99 percent of the speed of light, the relativistic factor is about 0.141, or about one-seventh. Therefore, while Tim has aged about seven years during Tom's trip, Tom himself has aged only about one year.

Now, look at the situation from Tom's point of view. Tom may consider himself sitting stationary in his spaceship while the earth (with Tim aboard) moves away at 99 percent of the speed of light for three and one-half years, after which time the earth reverses direction and makes a return trip to Tom at 99 percent the speed of light. In Tom's frame of reference, he is seven years older at the end of Tim's trip, while Tim (who is moving) has aged only about one year.

Which of these equivalent, but seemingly contradictory, points of view is correct? Who has aged more, and who has aged less? These questions constitute the twin paradox. The answer is that both points of view are correct and that it is the return trip that creates the problem. The paradox is resolved by recognizing the fact that the return trip (in order to make the comparison) requires an acceleration of the moving observer; that is, the moving observer must reverse direction. Therefore, the fundamental assumption of observers moving with constant velocity has been validated. It is important to remember that the predictions of special relativity are not consistent if the criteria on which that theory is based are not met. As long as the relationship between the two observers is strictly inertial (that is, unaccelerated), the rest observer will correctly deduce that the moving observer's clock runs too slow and either observer may regard the other as the one in motion.

In order to allow for accelerations, one must turn to the general theory of relativity. This theory predicts that it is the accelerating observer for whom time actually runs slower. Thus, in the space-traveling-twin problem, it is the accelerating twin, Tom, who actually ages less, but the twins cannot really be compared in the context of special relativity.

Applications

The twin paradox itself is merely a conceptual difficulty having to do with the assumptions on which special relativity is based, and it does not have any applications, per se. The time dilation effect on which it is based, however, is quite real and does have relevance for the physical world. Perhaps the most striking application has to do with subatomic particles called mu mesons, or muons. Muons are particles that have the same charge as an electron but a rest mass which is about 207 times greater, or about one-ninth the rest mass of a proton. They are produced in the earth's upper atmosphere as a result of the bombardment of air molecules by cosmic rays (high-speed particles, mostly protons, from space). They can also be produced in laboratory experiments, where it has been determined that muons have a very short lifetime, spontaneously disintegrating into other particles on a time scale of about 2.2 microseconds.

The muons that are produced in the upper atmosphere travel with relativistic velocities, some as high as 99.8 percent of the speed of light. Even at the speed of light, they could only travel a distance of somewhat less than a kilometer in their average lifetime of 2.2 microseconds. How is it, then, that they can be detected in large numbers near the surface of the earth, having traveled several kilometers? The answer is time dilation. In the muon's frame of reference (that is, to an observer riding along with the muon), they last about 2.2 microseconds. To an observer on the ground, however, relative to whom they are moving at large velocities, they survive much longer and thus can travel the distance to the earth's surface where they are detected in abundance. (At 99.8 percent of the speed of light, the 2.2 microsecond lifetime of a muon would dilate to about 35 microseconds, allowing the muon to travel about 10 kilometers relative to the ground.) Such observations provide strong confirmation of the validity of special relativity.

In 1971, a very important experiment was carried out by Joseph C. Hafele, of

Washington University, and R. E. Keating of the U.S. Naval Observatory, in order to obtain direct evidence for the phenomenon of time dilation. Hafele and Keating transported very precise cesium-beam atomic clocks on commercial jet flights, one eastward and one westward, in order to test Einstein's theory. The intervals measured on the clocks in flight were to be compared with reference clocks at the U.S. Naval Observatory. Although atomic clocks are accurate to about a billionth of a second (a nanosecond), the experiment was very difficult to carry out and analyze, since periods of acceleration and deceleration had to be considered, as well as the weaker gravitational field experienced by the flying clocks. Also, the reference clocks were not strictly unaccelerated because of the rotation of the earth.

On the eastward flight, the theory predicted that the flying clocks, compared with the reference clocks, should have lost 40 nanoseconds (± 23 nanoseconds) and should have gained 275 nanoseconds (± 21 nanoseconds) during the westward trip. The observed values were 59 nanoseconds (± 10 nanoseconds) lost during the eastward trip and 273 nanoseconds (± 7 nanoseconds) gained during the westward trip. The agreement is striking, and it provides unambiguous empirical evidence for time dilation.

One other example of time dilation can be illustrated by use of the Mössbauer effect. This effect refers to the emission or absorption of radiant energy in which the emitter or absorber does not recoil. As a result, the frequency of the energy involved is very precisely determined. In his 1905 paper, Einstein had suggested that a clock moving in a closed curve at a constant speed v would lose time relative to a stationary clock. The experiment, first conducted in Great Britain in the late 1950's, uses a turntable with an emitter (radioactive iron 57) on the spindle at the center and a resonant absorber of the same isotope mounted on the circumference. Each absorbed photon corresponds to one "tick of the clock." A stationary counter beyond the circumference of the turntable keeps track of those photons that pass through the absorber, that is, missed ticks of the clock. Because the resonant frequency is so sharply defined by the Mössbauer effect, it is possible to detect a significant difference in the number of ticks of the clock when the turntable is rotated at fifty revolutions per second and at five hundred revolutions per second. The experimental result agrees precisely with Einstein's prediction: Moving clocks slow down.

Context

In order for any physical theory, such as special relativity, to be successful and meaningful, it must yield results that are consistent with what one observes. This condition is the ultimate test of its validity, even if the assumptions on which the theory is based seem somehow strange. A successful theory of special relativity must, therefore, give the correct results for light and must also reduce to the familiar, well-verified results of Newtonian physics at low velocities. The prediction of time dilation, as well as other effects of special relativity, has been checked and rechecked in many experiments. Consequently, it must be regarded as a well-founded theory.

Therefore, there is something very "special" about the relationship between two observers who are moving in a straight line at constant speed relative to each other: They cannot determine who is moving and who is at rest. Such observers are said to be inertial observers or to be in inertial reference frames. Among other things, they disagree on time intervals. Observer A says that observer B's clock is running too slow; observer B says exactly the same thing about A's clock. It is precisely because they say exactly the same thing about each other that it is impossible to distinguish which one is moving.

This idea of the equivalence between inertial observers was not new with Einstein but, in fact, was the foundation of Galilean-Newtonian mechanics nearly three centuries earlier. Newton's "first law" states that a body will remain either at rest or in a state of uniform motion in a straight line unless it is acted on by some outside force. Again, for inertial observers, it is impossible to distinguish between who is moving uniformly and who is at rest.

The Newtonian-Einsteinian idea that the natural state of things is uniform motion in a straight line is a basic tenet of modern physics and contrasts strongly with the views of the Greek philosophers, who held that circular motion is natural. Where Einstein differed from Newton was his extension of the principle of relativity to include electromagnetic phenomena (light) as well. Requiring different observers to agree on the speed of light means that they must disagree on measurements of length and time. Yet, the fact that they disagree on time intervals, for example, does not mean that one is right and one is wrong. Both observers are correct in their own frames of reference.

Bibliography

Blatt, Frank J. *Principles of Physics*. 3d ed. Boston: Allyn & Bacon, 1989. An excellent and successful introductory college physics text which uses only elementary algebra and geometry. Chapter 27 contains a straightforward discussion of special relativity and specifically addresses the problem of the twin paradox with numerical examples.

Bronowski, Jacob. "The Clock Paradox." *Scientific American* 208 (February, 1963): 134-144. This most highly recommended article provides an exceptionally good qualitative discussion of the relativistic clock paradox problem in its various aspects. Bronowski demonstrates clearly how the effect depends merely on the Pythagorean theorem. Contains informative illustrations.

Cutnell, John D., and Kenneth W. Johnson. *Physics*. New York: John Wiley & Sons, 1989. An attractive college physics text. Chapter 35 contains a clear discussion of time dilation in special relativity using several examples involving astronauts and space travel.

De Benedetti, Sergio. "The Mössbauer Effect." *Scientific American* 202 (April, 1960): 72-80. A well-written, succinct, and readable article, which was typical of *Scientific American* at the time. The Mössbauer effect is described, and its use as a test of relativistic time dilation is explained.

Einstein, Albert. *Relativity: The Special and General Theory*. New York: Crown,

1961. This short (150-page) book provides a clear insight into the theory of relativity and is written in a straightforward manner which is accessible to high school students. First published in 1916, it has gone through at least fifteen editions, attesting to its success. See chapter 12 about clocks in motion.

Hafele, Joseph C., and R. E. Keating. "Around the World Atomic Clocks: Predicted Relativistic Time Gains." *Science* 177 (July 14, 1972): 166-168.

____________. "Around the World Atomic Clocks: Observed Relativistic Time Gains." *Science* 177 (July 14, 1972): 168-170. In these two articles, the authors summarize their important experimental test of time dilation, which employed clocks flown on commercial jets. The articles are very short; they do contain some technical terminology.

Serway, Raymond A., and Jerry S. Faugn. *College Physics*. 2d ed. Philadelphia: Saunders College Publishing, 1989. One of the more widely used introductory-level, algebra-based (no calculus) physics texts. Chapter 28 provides a good, clear description of special relativity. Time dilation and its applications are thoroughly explained.

George W. Rainey

Cross-References

General Relativity: An Overview, 950; The Lorentz Contraction, 1282; Mesons, 1374; The Michelson-Morley Experiment, 1387; The Mössbauer Effect, 1467; Space-Time: An Overview, 2318; Special Relativity, 2333.

ULTRASONICS

Type of physical science: Classical physics
Field of study: Acoustics

Ultrasound is the region of high-frequency sound above the highest frequencies that can be perceived by humans. Since the 1960's, ultrasonic technology has developed to such an extent that ultrasound is now recognized as a significant branch of physics with numerous important applications.

Principal terms

FREQUENCY: the number of oscillations per second of a vibrating object, defined so that one vibration per second is equal to one hertz; in the ultrasound region, frequencies are measured in kilohertz (thousand hertz), megahertz (million hertz), and gigahertz (billion hertz)

TRANSDUCER: a device that changes energy from one form to another; acoustic transducers are of two types: acoustic generators that produce acoustic waves and acoustic receivers that detect them

ULTRASOUND: literally, beyond sound; vibrations of a material substance similar to audible sound but having frequencies too high to be detected by human ears

WAVELENGTH: the distance between successive crests, or other identical parts, of a wave train; wavelength is inversely related to frequency, thus, the higher the frequency, the shorter the wavelength, and vice versa

Overview

Although the human ear can detect sounds in the frequency range from 16 hertz through 20 kilohertz, the maximum sensitivity lies between 2 and 4 kilohertz. Above this range, the aural sensitivity decreases rapidly to complete insensitivity at 20 kilohertz. Sound waves having frequencies greater than 20 kilohertz are called ultrasonic waves. Although, by definition, ultrasonic sound waves cannot be perceived directly by human ears, the useful applications of ultrasound are as multitudinous as they are varied. Two important properties of sound waves become more pronounced in the ultrasonic region. First, their ability to be focused makes it easier to explore objects and structures of small dimensions. Second, the higher the frequency of a wave, the more rapidly the wave is absorbed. The frequency range of ultrasound is from 20 kilohertz through 1 gigahertz, at which frequency ultrasonic waves can no longer be propagated in air at standard temperatures and pressures. The ultrasonic region may be subdivided into low (20 to 100 kilohertz), medium (100 kilohertz to 10 megahertz), and high (10 megahertz to 0.1 gigahertz) frequencies. Signaling applications, such as range finding, are usually in the low-frequency region, while ultrasonic cleaners and acoustical imaging processors usually operate in the medium-frequency range.

Ultrasonic waves may be generated in one of two ways: by mechanical devices, such as sirens and whistles, or by electromechanical transducers. Although mechanical devices are simple and inexpensive, they are useful only for producing low-frequency ultrasonic waves. Because of their broad spectrum, as well as some frequency and amplitude instability, these devices are seldom employed when accuracy is a criterion. Electromechanical devices use piezoelectric crystals subjected to an alternating voltage to produce medium and high-frequency narrow-band ultrasonic waves of constant amplitude and frequency.

Because ultrasonic waves have wavelengths that are short compared with the objects being irradiated, they can be focused easily. Acoustic lenses and mirrors focus narrow beams within a restricted space. Although X rays, which are highly penetrating electromagnetic waves, have an even shorter wavelength, they are virtually impossible to focus. An X-ray image is thus a shadowy silhouette, while an ultrasonic image shows details and structure.

Other advantages of using ultrasonic image formation over electromagnetic waves (such as light and X rays) are the greater range of wavelengths (frequencies) available with ultrasound, and the fact that many materials opaque to electromagnetic waves appear transparent to ultrasound. The disadvantages of ultrasound are that it cannot be directly seen or heard, it has no effect on photographic film, and, unlike electromagnetic waves, ultrasonic waves do not propagate through a vacuum. Additionally, designing an acoustic lens to focus the waves is a complex task. Achieving good resolving power requires the ratio of the ultrasonic wavelength to the diameter of the lens to be as small as possible. There are practical limits to how large a lens may be and using shorter wavelengths to improve resolution increases the attenuation of ultrasonic waves. Thus, even using an acoustic lens having a 0.50-centimeter diameter with a 0.1-megahertz frequency ultrasonic wave, the resolution is still two orders of magnitude less than for a 0.5-centimeter diameter optical lens system.

Ultrasonic images are made visible to the human eye by an ultrasonic image converter. Ultrasonic waves are produced, focused, and directed through the object to be investigated. The waves strike a metalized surface of a quartz disk, which forms the front plate of a cathode-ray tube (television picture tube). The variation in ultrasonic intensity across the disk surface—resulting from the waves passing through the object—produces a corresponding variation in the (alternating) charge pattern on the disk's rear surface. When this rear surface is scanned by an electron beam, the beam will be modulated in accordance with the surface distribution, which produces a visible image (by the usual means) on the face of the television picture tube. This picture may be considered as an ultrasonic "X ray" of the object.

Applications

The unique properties of ultrasonic waves has led industry and medicine to develop many practical applications for ultrasound. Because these waves have very short wavelengths and can be focused in a manner similar to light waves, they have proved to be a versatile and safe alternative in many situations where electromagnetic waves could not be employed. The applications can be divided into two broad categories: high-intensity ultrasound and low-intensity ultrasound. High-intensity ultrasonics is used to effect physical or chemical changes in some material substance. Low-intensity ultrasonics is concerned with information that can be gathered by means of coherent ultrasonic waves. High-intensity ultrasonic applications include ultrasonic cleaning, mixing, soldering, welding, drilling, and a variety of chemical procedures. The principal applications of low-intensity ultrasonics are the detection of flaws in solid objects, medical sonograms, and acoustical holography.

Ultrasonic cleaning is achieved when the extremely high accelerations of ultrasound waves vibrate small particles of grit off dirty objects. High-intensity ultrasonic waves in a liquid produce regions of intense compression and expansion, which literally tears the liquid apart. The small bubbles, or cavities, formed by this process (known as cavitation) produce shock waves that impinge on the dirty surface and scour it clean. Since cavitation decreases with frequency, ultrasonic cleaners use high-intensity signals of low-ultrasonic frequency, adjusted to produce maximum cavitation in a liquid bath.

Ultrasonic mixing is used for homogenizing and emulsifying substances. When two nonmiscible liquids, such as oil and water, are simultaneously irradiated by a high-intensity ultrasonic signal, an emulsion (a fine dispersion of oil in water, or vice versa) will be formed. Many cosmetics, polishes, paints, and some food products are emulsions. Ultrasonic homogenization is used for condensed milk, baby food, soup, margarine, catsup, mayonnaise, and peanut butter.

Aluminum cannot be soldered by conventional means because it reacts very readily with oxygen, forming a thin film of aluminum oxide on the surface. By propagating ultrasonic waves through a tinning bath, the oxide is torn off the surface by cavitation, thus allowing soldering. In order to weld aluminum, the two metal surfaces are placed in contact and made to vibrate against each other at ultrasonic frequencies. Although no surface preparation is necessary and no heat is required, the two metal surfaces bond together firmly. The ultrasonic vibration removes the dirt and oxide films and produces a localized high temperature at the junction, causing a plastic flow, without melting, of the aluminum. A weld produced by this process is neater, stronger, and requires less power than conventional heat welding.

Another high-intensity ultrasonic application includes ultrasonic drilling, which is used to drill unusually shaped holes in an assortment of materials, many of which are difficult to drill by ordinary means. An example is the drilling of square holes in glass. An ultrasonic transducer is used to vibrate a bit having the desired shape and size. An abrasive slurry, containing small particles of a powdered material, transmits the vibrations to the work piece, which is literally chipped away by the intensely oscillating powder.

Chemical applications of high-intensity ultrasonics include electroplating, atomization of liquids into fogs, and as a catalyst in chemical reactions. When the electrode is irradiated by ultrasound during electroplating processes, the high-frequency waves clean the surface and remove the bubbles, which interfere with the plating process by accumulating on the electrode.

Low-intensity ultrasonic waves in the low- and medium-frequency region are used for the nondestructive testing of materials. A transducer emits a pulse of ultrasonic energy into the solid object to be tested. The boundaries of the material will reflect the pulse, but flaws will be detected as an earlier reflection. Ultrasonic testing is both vital and effective for inspections requiring the highest possible quality control, such as spacecraft components and nuclear pressure vessels. If ultrasonic surveillance were used to reduce further the risk of failure, and ultrasonic testing were used during routine maintenance procedures, nuclear reactors could be made nearly fail-safe. Inspection time would also be shortened, thus reducing reactor downtime, and components difficult to access by other means could be checked for integrity.

It is known that objects subjected to stress will emit ultrasonic waves, which increase in intensity as the body begins to crack. The condition of the stressed object may be evaluated by listening to and interpreting these signals in terms of the deformation processes at work. A related application is to use ultrasound to monitor natural gas stored under pressure in holding tanks of porous rock. The maximum pressure that the impervious rock cap can withstand is known from ultrasonic emission testing. The unique advantage of ultrasonic emission testing is that large structures can be monitored in their entirety in one operation, using a small number of stationary transducers.

Ultrasonics has proved to be an extremely useful tool for medical diagnostics, where it can be employed in two different manners, known as echolocation and imaging. Echolocation is the process by which ultrasonic waves can be passed through body tissues and reflected from internal organs that are moving, such as the heart. When sound bounces off a moving target, the reflected wave frequency is shifted as a result of the Doppler effect. The shifted frequency will interact with the original wave to produce beats (slow variations in amplitude equal to the difference of the two frequencies). By listening to these beats, a trained diagnostician can hear the moving structure. This process is often used during pregnancy because the fluid in the uterus provides an unobstructed path for sound, and as sound waves are not electromagnetic, ultrasound will not harm a fetus as would X rays. The placenta can be located by listening for the sound of blood flowing through it, and the fetus is located by listening to its heart movements.

For the second method, an image of the interior of the body, a sonogram, can be obtained by ultrasonic means. In ultrasonic tomography (from two Greek words meaning "writing a cut"), a complete cross-sectional picture through part of the body is taken by recording a scanning set of ultrasonic waves that are used to modulate an oscilloscope screen. The resulting sonogram—unlike an X ray, which shows only bone and calcified substances—reveals all the tissues in the cross-section. Naturally, this procedure is widely used in obstetrics because the pregnant uterus is an ideal transmitting medium, and the danger of X rays is avoided.

A final interesting application of ultrasonics is in conjunction with holography. Coherent ultrasound used with a laser can produce acoustical holograms in a manner similar to that used to produce optical holograms. An object is irradiated by one ultrasonic beam, which combines with a reference beam to form an interference pattern. One method used for recording an acoustical hologram creates an interference pattern in the form of standing waves on the surface of a tank of water. The resultant ripples can be illuminated by laser light and observed by telescope. By adjusting the focus, one can look at different depths inside the object. As in conventional holography, the object can be recorded on a photographic plate for later reconstruction. Illuminating the plate with laser light re-creates an image of the original object.

Context

The emergence of ultrasound as a subbranch of acoustics can be traced to the invention of the ultrasonic whistle by Francis Galton in 1883. Galton used the whistle to investigate the upper-frequency threshold of human and animal hearing. Another possible application of ultrasonic waves was suggested to Lewis Fry Richardson by the sinking of the *Titanic* in 1912. He proposed that underwater ultrasonic waves could detect submerged obstacles, such as icebergs. This idea was developed by French scientist Paul Langevin during World War I as a means of detecting enemy submarines. By World War II, this primitive system had fully evolved into standard naval equipment for submarine detection.

The advances in electronic technology necessitated by the war led to further refinements in ultrasonic applications, such as sonar systems for underwater sound. Subsequently, the same ultrasonic echolocation techniques were adapted to operate in solids, liquids, and gases, which led to a wide variety of applications.

Considerable potential exists for further developments in both low- and high-energy ultrasound. New materials are being developed for high-intensity generators, and new technology is evolving for better and more sensitive receivers. Although the existence of ultrasonic waves has been known for many years, ultrasonic technology began to develop in the post-World War II period. That this development has been rapid and diverse only exemplifies that scientists are on the threshold of the silent world of ultrasound.

Bibliography

Asimov, Isaac. *Understanding Physics.* Vol 1. New York: New American Library, 1966. This excellent volume includes a short discussion of ultrasound found in the detailed presentation on sound waves and acoustics.

Cracknell, A. P. *Ultrasonics.* London: Wykeham, 1980. In part 1, this important book summarizes the physics of the generation, propagation, attenuation, and detection of ultrasound. Part 2 treats the applications of ultrasonics in science, industry, and medicine, as well as the uses made of ultrasonics by mammals, insects, and birds.

Hewitt, Paul G. *Conceptual Physics.* 6th ed. Scranton, Pa.: HarperCollins, 1989. This amusing and easy-to-read book includes discussions of many aspects of sound as well as a treatment of holography.

Kock, Winston E. *Sound Waves and Light Waves.* Garden City, N.Y.: Doubleday, 1965. An original approach that synthesizes sound and light by emphasizing their similarities and applications.

Strong, W. J., and George R. Plitnik. *Music, Speech, and High Fidelity.* 2d ed. Provo, Utah: Soundprint Press, 1983. A nontechnical treatment of all aspects of sound, including discussions of ultrasonics and infrasonics.

George R. Plitnik

Cross-References

Holography, 1058; Sonics, 2297; Producing and Detecting Sound, 2305; Sound Waves, 2311.

ULTRAVIOLET ASTRONOMY

Type of physical science: Astronomy and astrophysics
Field of study: Observational techniques

Ultraviolet astronomy explores the universe by focusing on wavelengths of the electromagnetic spectrum that are shorter than those of visible light. This portion of the spectrum is particularly important to astronomy, as practically all stars and many of the most abundant elements in the universe emit energy in the ultraviolet range.

Principal terms

ELECTROMAGNETIC RADIATION: waves of electrical and magnetic energy that travel through space at the speed of light

KELVIN: an absolute temperature scale where 0 is set at absolute zero; in this system, a unit of temperature, or Kelvin, is equal to a Celsius degree

NANOMETER: a unit of measurement often used to measure electromagnetic energy; equal to one-billionth of a meter

SPECTROSCOPE: an instrument that spreads electromagnetic radiation into its component wavelengths and photographically records the result

SPECTROSCOPY: the technique of producing spectra and analyzing their constituent wavelengths to determine quantities such as the chemical composition, density, and temperature of the emitting object

SPECTRUM: the entire range of electromagnetic radiation from long-wavelength radio waves to short-wavelength gamma rays; also, a limited range of wavelengths in which an instrument separates the component elements

WAVELENGTH: the distance from crest to crest or trough to trough of a waveform

Overview

Any material with a temperature above absolute zero emits electromagnetic radiation, and that radiation carries with it information about the nature of the event that produced it. An object will emit radiation over a range of wavelengths, with a concentration at a single wavelength. Very hot objects produce shorter wavelengths, while cooler objects emit longer wavelengths. For example, as metal is heated, it first glows red (the longer wavelengths of visible light), then, as its temperature increases, it begins to glow in the shorter-wavelength yellow light. In space, objects that are very cold, perhaps only a few degrees above absolute zero, will emit radiation in the very long infrared and radio wavelengths. At the other extreme, very hot stars give off ultraviolet radiation, X rays, and gamma rays.

Ultraviolet astronomy focuses on the area of the spectrum that is beyond violet light—the shortest wavelengths the eye can see. The ultraviolet portion of the spectrum begins at a wavelength of 390 nanometers, ranges to the extreme ultraviolet at 90 nanometers, and merges into the X-ray portion of the spectrum at 10 nanometers.

While the visible portion of the spectrum can be observed from the surface of the earth, observations at ultraviolet wavelengths must be done outside the earth's atmosphere. Ultraviolet radiation is readily absorbed by gases, both in space and in the earth's atmosphere. Only the longest wavelengths of ultraviolet light penetrate the atmosphere. It is this radiation that is responsible for the tanning (and burning) effects of the sun on the skin. The screening effect of the atmosphere protects life on Earth from the more harmful, shorter-wavelength ultraviolet radiation. At the same time, it prevents astronomers from easily collecting information about the universe from this important range of the spectrum.

Practically every object in the universe emits some radiation at ultraviolet wavelengths. Any material that has a temperature between 10,000 and 1 million Kelvins emits most of its energy in the ultraviolet. This range of the spectrum is important to the knowledge of celestial objects, since the atmospheres of most stars, the surfaces of very massive stars, white dwarf stars, and regions of hot interstellar gas all fall within this temperature range. In addition, the elements that are most abundant in the universe, such as hydrogen, helium, carbon, nitrogen, oxygen, and silicon, all have spectral features that are prominent in the ultraviolet. For this reason, ultraviolet astronomy can provide astronomers with important information about the universe. Although instruments are designed to operate remotely, the actual instruments can be very similar to optical instruments. Ordinary telescope mirrors will focus ultraviolet light. Electronic detectors record the image or, in some cases, the image can be recorded on regular photographic film.

The first ultraviolet telescopes were flown on high-altitude weather balloons. Later, they were launched on rockets, which raised them above the atmosphere for a few minutes at a time. The best way to gain access to ultraviolet information is to place an ultraviolet observatory in orbit. The first satellites to carry ultraviolet instruments were the Orbiting Astronomical Observatory (OAO), a series of four identical satellites that carried different instrument packages to measure ultraviolet radiation from stars and interstellar gas. Only two of the four spacecraft proved functional. The first satellite failed after two days in orbit and another failed to reach orbit. The second OAO achieved a general ultraviolet survey of the sky, discovering ultraviolet sources within the galaxy and measuring ultraviolet light from bright nearby galaxies. The final OAO, Copernicus, was the first to target specific ultraviolet sources. Carrying a 0.8-meter telescope, it was launched on August 21, 1972, and was functional for nine years. Copernicus took the first detailed look at objects in a wide range of the ultraviolet spectrum. Detailed ultraviolet pictures of the Sun came with the OAO mission and the Solar Maximum Mission (SMM) satellite, which splashed into the Indian Ocean in December of 1989. Spectacular ultraviolet solar studies resulted from the American manned space station Skylab, which was launched in

1973. Three three-man crews inhabited the space station for a total of five and one-half months. Throughout this time, they kept a continuous surveillance of the sun with the Apollo Telescope Mount (ATM). The ATM carried eight telescopes, which observed the sun in wavelengths ranging from visible light through the ultraviolet and into the X-ray range.

In 1978, two years before the OAO Copernicus ceased operation, the International Ultraviolet Explorer (IUE) was launched. The IUE was a joint venture by the National Aeronautics and Space Administration (NASA), the European Space Agency (ESA), and the British Science Research Council (SRC). Although the IUE was equipped with a telescope smaller than that of Copernicus (only 45 centimeters), it carried more modern ultraviolet detectors and was able to observe much fainter stars over a broader range of wavelengths. The IUE has been the longest-lived and most productive orbiting observatory. In 1990, it celebrated its twelfth year of operation, with a projected lifetime of sixteen to seventeen years. The IUE carries a 45-centimeter telescope equipped with two spectrographs.

A spectrograph records information by passing light or other radiation through a narrow slit and then through a prism, which separates the radiation into its component wavelengths. The result—a spectrogram—is then recorded on film. Every element emits and absorbs energy according to a characteristic pattern. By analyzing the pattern, or spectrogram, astronomers can determine the chemical composition of very distant objects. A spectrum is governed by the temperature, density, and chemical composition of the object emitting the energy, as well as how the energy has been altered by intervening processes en route to the instruments.

The IUE is run much like a traditional observatory, and, as such, is designed to be used by visiting scientists rather than by a select group of researchers. The IUE remains stationed in an orbit that is geosynchronous (remaining constantly over one area of the earth) about 36,000 kilometers above the Atlantic Ocean. The satellite is in continuous contact with at least one of two ground stations in the United States and Europe. This is an improvement over earlier satellites, which could not remain in continuous contact with ground stations. Their observations were also limited by low orbits, with a large percentage of the field of view blocked by the earth.

Astro-1 was a Spacelab mission launched aboard the space shuttle in early 1990 carrying three ultraviolet telescopes. For the duration of the mission, project scientists conducted observations in the ultraviolet and X-ray regions. Other important ultraviolet satellites have been the European TD-1 and the Soviet Astron satellite. The TD-1, launched in 1972, measured the magnitudes of more than thirty thousand stars in four different spectral regions and gathered ultraviolet spectra from more than one thousand stars. The Soviet Astron satellite, launched in 1983, was similar in size and scope to the OAO Copernicus.

The ultraviolet astronomy of the future lies with instruments such as the Hubble Space Telescope (HST), launched in 1990. The Hubble Space Telescope, although it experienced technical difficulties soon after launch, has the potential to be a powerful ultraviolet instrument, making exciting discoveries in the far ultraviolet, despite its preliminary optical difficulties. The Extreme Ultraviolet Explorer (EUVE) is planned to survey the cosmos for objects emitting very energetic short-wavelength radiation. Astronomers expect the satellite to discover many new objects, including perhaps ten times as many white dwarf stars as are now known.

Other ultraviolet instruments are in the planning stages. The Orbiting and Retrievable Far and Extreme Ultraviolet Spectrometer will carry a 2-meter ultraviolet telescope. It will separate from the shuttle by about 10 kilometers and investigate objects in the far ultraviolet and extreme ultraviolet. NASA's Wind Mission will use multiple instruments, including ultraviolet, to study the solar wind and to sample plasma waves, energetic particles, and electric and magnetic fields. The Lyman Observatory is being planned by the European Space Agency. It will be a high-resolution ultraviolet spectroscopy mission that will examine the dynamics of the Milky Way's halo and will study comets.

Applications

Each energy region in the electromagnetic spectrum allows astronomers to "see" objects in a unique way. The more information that can be discovered about the nature of a celestial object in each of these energy areas, the more the object can be completely understood.

All objects known to exist in the universe—from comets and planets to stars, galaxies, and quasars—can be effectively studied in the ultraviolet range. Ultraviolet telescopes see the hottest stars of all; as a result, they tend to pick out the youngest star groups in the sky. Ultraviolet astronomy can thus focus on the youthful clusters of stars that lie close to regions of star birth. With this particular window, ultraviolet astronomy has been very useful for mapping regions of star formation, both in the Milky Way and in distant galaxies. Other galactic studies have shown that the Milky Way, as well as other galaxies, are surrounded by a hot halo of gas.

There are excellent images of the sun in the ultraviolet. Views of the ultraviolet sun reveal different layers of its chromosphere, transition region, and lower corona. Bright, scintillating points of ultraviolet light in the sun's atmosphere provide a measure of magnetic activity within the sun, with perhaps even more accuracy than the sunspots that are seen on the visible photosphere. By observing other stars in the ultraviolet, astronomers have gained valuable knowledge about the nature of stars, which correlates with what is known about the sun. It has been shown that many stars have hot outer atmospheres similar to the sun. A new class of stars was discovered that had distinguishing characteristics visible only in ultraviolet light.

Ultraviolet astronomy has been very useful in studying binary star systems. A binary system is a pair of stars orbiting around their common center of mass. In a binary system, one star can be much brighter in optical wavelengths, and only a single spectrum can be observed. If the companion star is much hotter, however, it will dominate the spectrum of the system at ultraviolet wavelengths. A binary system gives astronomers a tool with which they can study the nature of these dimmer hot stars. Previously unobserved hot companions have been discovered in stars not suspected earlier of being binaries.

The supernova is another area in which ultraviolet astronomy can contribute significantly. A supernova is a stellar explosion in which all or most of the star's mass is expelled. Astronomers have studied the remains of supernova explosions as well as observing them in the beginning stages of development. Ultraviolet observations can determine the chemical composition of the layers of the star expelled by the explosion. Astronomers discovered that a nova explosion in Cygnus in 1978 produced much nitrogen, while the supernova that created the well-known Crab nebula threw out relatively small amounts of carbon. These facts are important clues to learning how new elements are formed, as well as in understanding the mechanisms behind supernovas.

The IUE observed Comet Kohoutek in 1976, finding a very bright image in ultraviolet wavelengths. Comet IRAS-Iraki-Alcock was observed in 1983 and Comet Halley was observed in 1986. In combination with observations at other wavelengths, it was found that the composition of comets are similar, suggesting that they have a similar origin.

Through ultraviolet observations, astronomers have gained knowledge about the distribution of gas and dust in the galaxy. It has been found that the interstellar "medium" is of similar composition throughout the galaxy but that its distribution is not uniform. Huge bubbles created by the winds of very hot stars are found, giving it a structure resembling swiss cheese.

The sun continually loses matter from its atmosphere through the solar wind, a current of charged particles that streams outward from the sun. Other stars have similar winds. Knowledge of these winds helps astronomers to formulate theories of stellar evolution. Observations in the ultraviolet show that some stars blow off material in bursts, rather than as a steady flow. Astronomers can also determine the speed and direction in which the atoms are traveling. Using this information, it has been discovered that some of the material leaving a star will return to the surface, similar to the dynamics of a solar flare.

Extreme ultraviolet detectors see even hotter objects, those ranging from 100,000 to 1 million Kelvins. Ordinary stars are not this hot, but have extremely hot, tenuous outer atmospheres that can be studied at this range. Other target objects for the extreme ultraviolet are the hot, suddenly exposed cores of old stars, hot streams of gas flowing from a star to its nearby companion, and the gas streams that are flowing into a black hole.

Ultraviolet astronomy has confirmed some long-standing theories that had previously lacked sufficient evidence. The theory of a "gravitational lens" had been predicted by the relativity theories of Albert Einstein but had never been supported by solid evidence. The gravitational lens is a process in which the gravitational field of a very massive object acts as a lens, bending the radiation from a more distant object behind it, distorting its image and often creating a double or multiple image. Observations by the IUE helped to indicate that such gravitational lenses exist. The first such lens system was discovered in 1979.

It is expected that with both ultraviolet and extreme ultraviolet astronomy, new kinds of objects will be discovered that astronomers had not predicted. Observations with the IUE have already discovered a number of ultraviolet sources for which there is no known component in the visible spectrum. The IUE has also found very odd ultraviolet spectra in other known stars that do not match any current theories.

Context

For thousands of years, the human eye was the only astronomical instrument. The eye evolved to be most sensitive to the range of visible light, the most abundant source of radiation at the surface of the earth. Any celestial objects that were dim or emitted radiation at mostly nonoptical wavelengths remained invisible to the eye, which limited the range of information about the universe scientists could study.

The discovery of the telescope radically altered astronomy, not only because of the fainter objects it allowed astronomers to see but also because it opened the possibility that there was more to the universe than what the human eye was able to image. Astronomy improved dramatically over the next few centuries but remained optical. The first sign that there was another way to look at the universe with anything other than optical wavelengths came in 1800, when infrared radiation was discovered by Sir William Herschel, who placed a thermometer just outside the red range of the visible light separated by a prism.

The opening up of the wavelengths of electromagnetic energy got under way with the rapid growth of radio astronomy in the 1950's and 1960's and with the birth of the space program during the same period. The space program allowed ultraviolet astronomy to become an important new area of study. The potential value of ultraviolet observation from space was proposed to the U.S. Air Force in 1946 by the American astrophysicist Lyman Spitzer, Jr. With the establishment of NASA in 1958, the concept of placing orbiting observatories in Earth's orbit became a reality, and a series of orbiting observatories were launched over the next twenty years.

The result of thirty years of observing in all ranges of the spectrum is that astronomers now have a more complete understanding of the processes occurring in the universe. Today's astronomers have taken images of the stars and galaxies which were unimaginable to the ancients, or even to the astronomers of a few decades past.

Ultraviolet astronomy will undoubtedly continue to be an important area of study. It is now at a point where future missions will lose the "frontier" feel of the early missions, with increasingly complex and specialized missions. Even so, exciting new discoveries will continue to be made. In addition to new observations, research using information from the years of observations made by the satellites of the 1960's, 1970's, and 1980's will allow astronomers to gain new insights on virtually every area of the universe.

Bibliography

Cornell, James, and Paul Gorenstein. *Astronomy from Space.* Cambridge, Mass.: MIT Press, 1983. An overview of twenty-five years of astronomical research from space. Summarizes what has been learned in different areas of astronomy. In the form of short articles written by experts in each field, suitable for those with some science background.

Editors of Time-Life Books. *The New Astronomy.* Voyage Through the Universe Series. Alexandria, Va.: Time-Life Books, 1989. One volume of a series examining different aspects of the universe. Comprehensively covers all invisible astronomies, including high-energy astronomy and imaging techniques. Highly illustrated and suitable for the general reader with an interest in astronomy.

Henbest, Nigel. *Mysteries of the Universe.* New York: Van Nostrand Reinhold, 1981. Explores the limits of what is known about the universe. Discusses theories about the origin of the solar system and the universe, exotic astronomy, and astronomy at invisible wavelengths.

Henbest, Nigel, and Michael Marten. *The New Astronomy.* Cambridge, England: Cambridge University Press, 1983. Compares optical, infrared, ultraviolet, radio, and X-ray observations of well-known astronomical objects. Highly visual, and written specifically for general readers.

Marten, Michael, and John Chesterman. *The Radiant Universe.* New York: Macmillan, 1980. An overview of imaging that is not accessible in visible wavelengths. Discusses electronic processing as well as infrared and ultraviolet wavelength imaging. Beautiful pictures, along with easy-to-read and informative text.

Divonna Ogier

Cross-References

Binary Stars, 218; Electromagnetic Waves: An Overview, 735; Interstellar Clouds and the Interstellar Medium, 1167; Main Sequence Stars, 1316; Planetary Atmospheres, 1836; Solar Flares, 2220; Solar Wind, 2228; Star Formation, 2360; Supernovas, 2434; X-Ray and Gamma-Ray Astronomy, 2755.

THE UNCERTAINTY PRINCIPLE

Type of physical science: Atomic physics
Field of study: Nonrelativistic quantum mechanics

The uncertainty principles credited to Werner Heisenberg describe a basic built-in quality of the nature of physical reality. This nature is not often manifested in ordinary human existence, but has extraordinary implications for measurement theory and fundamental understanding of the physical universe.

Principal terms

ENERGY: the physical quantity (E) often regarded as the equivalence to do work; can be transported by particles or waves and be stored in fields

FREQUENCY: number of complete oscillations per unit time

MOMENTUM: the vector physical quantity, conserved under the condition of no net force, defined by the product of the mass and velocity of a particle or system (mv)

PARTICLE: a localizable constituent of matter

PERIOD: the time required for a periodic wave to complete one oscillation

POSITION: the location of a particle or system of particles as measured in a particular coordinate system

WAVE-PARTICLE DUALITY: matter has a dual character; under certain circumstances, it can display a wave behavior and under other circumstances, a particle behavior

WAVELENGTH: the spatial extent covered by a wave during one complete oscillation

Overview

Particles and waves are the two physical entities that transport energy and momentum through space. Consider what is meant by an ideal particle and ideal wave from a classical standpoint. An ideal particle is capable of being completely localized in space. The coordinates of its position at any given time can be precisely known. Ideally, its mass, charge, and all other properties can be measured with absolute precision. Extended objects are composed of particles, but often it is possible in a given problem to regard an extended object as a particle in a larger environment. What can be regarded as a particle depends upon the overall dimension of the system that is under study. For example, in the kinetic theory of gases, even though individual molecules have both finite size and internal structure, it is possible to treat the molecules in a large collection of gas as point mass particles subject to various forces of interaction. At the other end of the size scale, in cosmological dynamics, individual galaxies can be regarded as point masses in a study of the structure and evolution of the universe.

In Newtonian or classical mechanics, the motion of systems of particles or extended objects is analyzed using the basic premise of ideal point masses. Given the forces acting on a system of particles or extended objects and a set of initial conditions—such as position and velocity for the system of particles or extended objects—a complete description of both the future and past motion of the system of particles or extended objects can be obtained. This determinism results because the equation regulating motion—Newton's second law of motion ($F = ma$)—is a second-order ordinary linear differential equation. A unique solution to an ordinary linear differential equation of order n can be obtained so long as n pieces of information, the set of initial or boundary conditions, is available. This solution would then uniquely describe the future time evolution of the system of particles or extended objects.

That which characterizes a wave is its frequency, which is the number of complete oscillations per unit time, or the wavelength, which is the spatial extent of one complete oscillation. A wave is periodic, and its form need not be restricted to a sinusoidal behavior. A wave propagates at a speed v, which is determined by characteristic of the disturbance that created the wave and/or characteristics of the medium through which the wave travels.

An ideal wave is one in which any observer can determine with absolute precision both the frequency and wavelength. Such a requirement forces the observer to have a total lack of knowledge of the wave's location. In other words, if a wave's frequency is known to absolute precision, that wave must be infinite in extent. An investigation regarding an ideal wave leads to the basis of the Heisenberg uncertainty principle. Such an approach is semiclassical because it begins by uncovering a classical uncertainty relation and then applies quantum concepts of energy and momentum to generate the quantum mechanical Heisenberg uncertainty relations.

To outline this approach, consider the following thought experiment. To determine if a wave is monochromatic, meaning it has only one component at a specific frequency, a standard reference must be available to use in comparison. This standard reference could be a mechanical oscillator capable of producing a disturbance and resulting wave in such a way that any desired frequency of oscillation can be available for reference. This could be considered as a standard clock, since a knowledge of the frequency of a reference signal could allow a measurement of time with a knowledge of the wave's frequency and by simply counting the number of wave crests passing by the observer in a given period.

Although no two particles can exist in the same place at the same time, the net effect of two waves passing through a given region of space is the algebraic sum of the two wave displacements. Beats are produced when two waves interfere. This property can be used to ascertain with certainty a given wave's frequency using a standard clock as a reference. Define the frequency of the unknown wave as X and the frequency of the reference wave from the standard clock as Y. If the unknown wave and reference wave are allowed to interfere with each other, the resultant wave pattern can be examined for the production of beats. The beat frequency would be the absolute value of the difference between X and Y, $|X-Y|$. If the reference wave and unknown waves are quite dissimilar in frequency, a high-beat frequency will result. The time between beats will be short as the period of oscillation is the inverse of frequency. In general, if the beat frequency is $|X-Y|$, then the time between successive beats will be $1/|X-Y|$.

If an observer looks at the overall intensity of the interfering waves, the amplitude will vary both spatially and temporally. If X is approximately equal to Y and the observer examines the resultant amplitude of the two interfering waves only over a limited time period, the observer may not notice any significant amplitude change and come to the flawed conclusion that X = Y. To be absolutely certain, on the basis of this experimental method, that the frequency of the reference and unknown waves are identical, an infinite observation time is required to definitely rule out the production of beats. The beat frequency $|X-Y|$ must vanish, thus the time of observation, the time between crests, must approach infinity (∞). During this observation time, the reference wave has been propagating away from the source of the disturbance and as a result has become infinite in spatial extent. The conclusion is inescapable. Precise knowledge of the frequency of a monochromatic wave leaves the position of the wave totally uncertain. This behavior can be summarized in the following classical uncertainty principle. If one observes the behavior of a wave for a time Δt, the knowledge of the wave's frequency is uncertain by at least an amount Δf, given by the inverse of Δt or in equation form, $\Delta f \Delta t \geq 1$.

From this intuitive uncertainty relation, another uncertainty relation can be obtained. This relation involves uncertainty in positional determination of the wave. If a wave is observed to be propagating at speed v for a time Δt, then the wave has moved an extent Δx given by $\Delta x = v\,\Delta t$. Using the previously described frequency-time uncertainty relation ($\Delta f \Delta t \geq 1$) and the dispersion relationship for the wave (the wavespeed v is the product of frequency and wavelength), it can be shown that the product of uncertainty in position and uncertainty in wavelength is at least as large as the square of the wavelength. As previously $\Delta f = 0$ implied $\Delta t = \infty$, now a zero uncertainty in wavelength implies $\Delta x = \infty$. Thus, again a monochromatic wave (one for which a single frequency and wavelength can be specified) must be infinite in extent. Ideal waves lack the localizability in space that ideal particles possess.

Wave pulses are wave disturbances that are confined to some finite spatial extent. Wave pulses propagate through space. (An example of a wave pulse would be the response of a string tied at one end to a single flick of the free end. A pulse would be set up that travels down the string toward the fixed end and then back. The pulse retains its shape as it propagates along the string.) Since a wave pulse has a finite extent in space Δx, the width of the pulse, it cannot have a single wavelength or frequency. Therefore, wave pulses are not monochromatic. By Fourier analysis, any waveform can be shown to be mathematically equivalent to an algebraic sum of superimposed waves of differing amplitudes and frequencies.

Since particles have finite extent in space and some degree of localizability, the matter wave associated with a material particle must be represented as a wave pulse or wave packet. As a result, the matter wave for a particle displaying localizability cannot be monochromatic.

The Heisenberg uncertainty principle states that, if the energy of a wave or particle is E, its frequency f, and momentum components p_x, p_y, and p_z, the uncertainty in energy is related to the uncertainty in time (determined by observational measurement) and the uncertainty in a momentum component is related to the uncertainty in its conjugate positional coordinate. Specifically, these can all be derived directly from the two classical uncertainty relations with the quantum assignment that energy is determined by frequency and momentum by wavelength, the Albert Einstein relation ($E = hf$) and the Louis de Broglie assignment of a matter wavelength (wavelength $= h/p$), respectively. The fundamental constant of nature h is Planck's constant.

If $\Delta f\,\Delta t \geq 1$ and $E = hf$, then $\Delta E = h\,\Delta f$ and $\Delta E \Delta t \geq h$. Similarly, from the de Broglie assignment of wavelength in terms of momentum and the uncertainty relation $\Delta x \Delta$ (wavelength) $\geq$ (wavelength)^2, the second variety of Heisenberg uncertainty relation $\Delta x \Delta p_x \geq h$ is obtained. A similar argument for the y and z components of position yields $\Delta y \Delta p_y \geq h$ and $\Delta z \Delta p_z \geq h$.

These are not the only Heisenberg uncertainty relations. They are the ones most commonly described in elementary textbooks. There are similar uncertainty relationships for angular coordinates and their corresponding angular momenta. In general, any positional coordinate of a system has a conjugate momentum and the uncertainty in that positional coordinate multiplied by the uncertainty in the conjugate momentum is always greater than or equal to the quantum of action h.

Applications

The Heisenberg uncertainty principle provides for the possibility of nonconservation of energy and momentum in cases where classical theory requires conservation of these physical qualities. It also explains much about the nature of wave-particle duality of matter.

Quantum mechanics firmly introduces the wave nature of particles (first postulated by de Broglie using symmetry arguments) and replaces classical deterministic calculations with probabilistic expectation values for physical quantities. The matter wavelength of a particle is determined by its momentum, wavelength being inversely proportional to momentum, and the constant of proportionality is Planck's constant h.

Assigning a wave-behavior to matter establishes a fully symmetric wave-particle duality in nature. Prior to de Broglie's theory, a particle nature had been attributed to electromagnetic waves. The Einstein photon description was necessary to explain fully the photoelectric effect, Compton scattering, pair production/annihilation, and a number of other interactions between matter and electromagnetic radiation that could not be explained satisfactorily purely by classical electromagnetism as expressed by James Clerk Maxwell.

Wave-particle duality has two fundamental attributes. Whenever particles of a given momentum are observed, their physical motion is guided by a matter wave having wavelength determined by the magnitude of momentum. Also, wherever there is a wave of given wavelength, the wave amplitude squared serves as a measure of the probability of observing some type of particle having momentum equal to that carried by the wave and correlated to a particle momentum described by mv. (It must be stressed that masses, m, should always be considered relativistic masses and not rest masses.)

Waves are characterized by frequency and wavelength; particles are characterized by position and momentum. At first glance, wave descriptions and particle descriptions are contradictory. Nevertheless, under quantum mechanical wave-particle duality, they are complementary. When matter displays a particle behavior, the wave description is totally suppressed. The same holds true for waves.

This characteristic is inherent because of the Heisenberg uncertainty principle. Suppose one chooses to characterize electromagnetic radiation using the particle picture in terms of photons. If at any instant of time one knows with absolute precision the photon's location ($\Delta x = 0$), then the photon's wavelength is totally uncertain, meaning a wave-behavior description is impossible. Wave behavior is totally suppressed. On the other hand, if one describes electromagnetic radiation using the wave picture in terms of an absolutely certain frequency ($\Delta f = 0$), then the wave's position is totally uncertain ($\Delta x = \infty$). Particle behavior is totally suppressed.

Another aspect of the Heisenberg uncertainty principle places limits on the measurement process. From the momentum-position uncertainty principle, it is clear that it is impossible to measure simultaneously linear momentum and position with absolute precision. A precise determination of position leaves momentum totally uncertain and vice versa. The same is true of simultaneous measurement of time and energy.

In ordinary human experience, it is difficult to see the effect of the Heisenberg uncertainty principle. For example, if a ball or other object of mass 150 grams is thrown at 24.4 meters per second and the determination of the ball's speed is uncertain by 10 percent (a rather significant uncertainty by human standards), then the uncertainty in momentum of the ball is 0.37 kilogram meter per second. Via the uncertainty principle, this means that a human being attempting to catch this ball would be uncertain about the ball's location by 1.8×10^{-33} meters, an imperceptibly small distance beyond the ability for measurement. A clumsy ball player could never justify his error in ball handling using the Heisenberg uncertainty principle.

Suppose that the universe had a much higher value of Planck's constant, while still retaining the same value for all other fundamental constants of nature. Reevaluation of the previous example would now yield a 0.18 meter uncertainty in ball position if h had a value of 6.63×10^{-2} joule seconds instead of 6.63×10^{-34} joule seconds. Such a universe would make "ordinary" sports virtually impossible. Quantum reality would pervade even the macroscopic world, and even the simple act of catching a ball would prove to be a probability exercise rather than an act of skill.

The uncertainty principle shows itself in the microscopic world. One need not even have to investigate atomic or nuclear physics to find an example of the uncertainty principle at work. Consider a typical virus with mass roughly 7.8×10^{-23} kilograms and a characteristic dimension of 5 nanometers. Such organisms can be imaged using special microscopic systems such as electron microscopes. Getting the virus to remain confined to an area of focus can prove difficult because the virus suffers from a minimum speed Δv determined by the uncertainty principle if it is to be confined to a small volume. If one takes Δx as the characteristic length of the virus and Δp as $m\Delta v$, then the Heisenberg momentum-position uncertainty relation yields a minimum speed Δv of approximately 2 millimeters per second, a significant speed.

Context

Quantum mechanics introduces a number of rather esoteric-appearing qualities to the nature of reality. These properties are merely attributes of reality that are not manifested in "ordinary" existence and as such were not incorporated in classical physics. One of these involves an abandonment of determinism in favor of a probabilistic interpretation. As an example of this, consider electromagnetic radiation. When thought of as a wave phenomenon, electromagnetic radiation is described as a light wave consisting of crossed oscillating electric and magnetic fields propagating at the speed of light. When thought of as a particle phenomenon, electromagnetic radiation is described as a stream of identical massless quanta called photons. An intensity (energy per unit area per unit time) can be defined in terms of wave characteristics; intensity of electromagnetic radiation is high where the wave's electromagnetic field is strong. When the idea of intensity is carried over to the particle description, to specify that the intensity is high is to indicate that the probability of observing a photon is high.

Another example involves a basic uncertainty in the ability to ascertain the precise value of physical quantities. This is not a requirement directly from quantum theory, but instead is fundamental to the measurement process itself. Nevertheless, the last statement should not be interpreted as a problem in the design of measuring instruments. Even if it were possible to devise and construct an absolutely precise measuring device, the basic nature of reality would prevent knowledge of physical quantities with absolute precision. Recognition and acceptance of this quality of nature marked a significant step in the development of quantum mechanical theory.

One curious aspect of the Heisenberg uncertainty principle is the fact that it can be derived semiclassically. Usually that approach is not taken in presenting the uncertainty principle in undergraduate physics texts, which often results in unnecessary confusion for the student. In the semiclassical derivation presented here, the Heisenberg uncertainty principle specifically arises when the quantization of energy (that is, the Einstein photon relation) and the assignment of a de Broglie matter wavelength in terms of a particle's momentum are introduced.

Bibliography

Beiser, Arthur. *Concepts of Modern Physics.* New York: McGraw-Hill, 1987. Excellent text on the development of nonclassical understanding of space and time, matter and energy. Easily accessible to the layperson. Illustrations and examples abound.

Burns, Marshall L. *Modern Physics for Science and Engineering.* San Diego: Harcourt Brace Jovanovich, 1988. Well-written account of modern physics. Although calculus-based and aimed at a college physics audience, qualitative descriptions of basic quantum phenomena are understandable by nontechnical readers.

Halliday, David, and Robert Resnick. *Fundamentals of Physics.* 3d rev. ed. New York: John Wiley & Sons, 1988. Considered by many a classic textbook for investigating elementary calculus-based physics. The section on the uncertainty principle, however, is understandable to those with only algebraic skills. Descriptions of basic quantum reality are quite good and illustrated with insightful examples.

Hawking, Stephen W. *A Brief History of Time: From The Big Bang To Black Holes.* New York: Bantam Books, 1988. A popular book written for the masses by one of the most remarkable scientists of the twentieth century. Details both relativity and quantum physics and stresses their implications for the origin and evolution of the universe. An entire chapter is devoted to the Heisenberg uncertainty principle.

Ohanian, Hans C. *Physics.* New York: W. W. Norton, 1985. Although calculus-based, the text is not mathematically rigorous. Accessible to those with modest mathematics skills. Excellent descriptions of difficult physical concepts. Well illustrated and provides practical examples.

Pagels, Heinz R. *The Cosmic Code: Quantum Physics as the Language of Nature.* New York: Simon & Schuster, 1982. Excellent book (accessible to the serious reader) written as a self-journey of understanding the reality of the quantum nature of the universe. Provides insights into the history of quantum physics and the physicists who developed the theory. Difficult subjects are presented in a meaningful way without the use of advanced mathematics.

Sells, Robert L., and Richard T. Weidner. *Elementary Modern Physics.* Boston: Allyn & Bacon, 1985. Excellent text for the reader with a firm background in elementary physics and good algebraic skills. Qualitative descriptions of the development of quantum mechanical thinking and theory is thorough and explained in an easily understood manner. Detailed diagrams and examples abound.

Wilson, Jerry D., and John Kinard. *College Physics.* Boston: Allyn & Bacon, 1990. Basic text, excellent for the reader unfamiliar with calculus. More qualitative than rigorous. Well-illustrated and thorough description of basic concepts without resorting to advanced mathematics. Excellent for high-school physics instruction.

David G. Fisher

Cross-References

The Big Bang, 210; Cosmology, 589; Electromagnetism, 750; Electrons and Atoms, 778; Mechanics, 1367; Nonrelativistic Quantum Mechanics, 1564; The Interpretation of Quantum Mechanics, 1986; Wave-Particle Duality, 2713.

THE UNIFICATION OF THE WEAK AND ELECTROMAGNETIC INTERACTIONS

Type of physical science: Elementary particle (high-energy) physics
Field of study: Unified theories

The electroweak theory unifies the forces that govern the weak and electromagnetic interactions between elementary particles. This theory also successfully predicts the existence and properties of the particles that transmit the weak force.

Principal terms

ANTIPARTICLE: an elementary particle with the same mass and spin (intrinsic rotation) as another particle, but with opposite charge and magnetic moment; a particle and its antiparticle annihilate into radiation when they come together

BOSONS: a class of elementary particles with integer multiples of the basic quantum unit of spin that mediate forces between other particles; vector bosons have one unit of spin

GAUGE THEORIES: a class of theories for the fundamental forces whose equations remain unchanged in form by certain symmetry transformations at any point in space and time

LEPTONS: a class of elementary particles with a half-unit of spin that are not affected by the strong nuclear force, including electrons, muons, neutrinos, and their antiparticles

NEUTRAL CURRENT PROCESS: a weak interaction mediated by the neutral vector boson (Z^0) that leaves the charge of the interacting particles unchanged

PHOTONS: quantum units of light and other radiation that mediate electromagnetic forces between charged particles

QUARKS: component particles with one-third or two-thirds of the electron charge that join in groups of two or three to form protons, neutrons, and other strongly interacting particles

STRONG FORCE: the strongest of the fundamental forces in nature, responsible for binding the atomic nucleus together through the mediation of mesons (strongly interacting bosons)

WEAK FORCE: the shortest range fundamental interaction between elementary particles, mediated by charged W^+ and W^- bosons and neutral Z^0 bosons, and responsible for radioactivity

Overview

The unified electroweak theory of electromagnetic and weak forces was begun by Sheldon L. Glashow in 1961 and was completed in 1967 by Steven Weinberg, and independently in 1968 by Abdus Salam, for which all three received the 1979 Nobel

Prize in Physics. Together with the more tentative theory of the strong nuclear force, the unified electroweak theory has been remarkably successful in describing elementary particles and their interactions down to one-thousandth of the size of the atomic nucleus, the smallest distance probed by particle accelerators. The electroweak theory is a quantum field theory in which particles interact by means of a field that carries quantum (discrete) units of energy and momentum through space by the exchange of other particles called bosons. It represents the merger of quantum electrodynamics (QED) and quantum weak dynamics in much the same way that James Clerk Maxwell's electromagnetic theory merged classical electricity and magnetism in the nineteenth century.

Like Maxwell's electromagnetic theory, the electroweak theory is a gauge theory in which the equations describing physical processes remain unchanged in form by certain symmetry transformations. The equations for interacting particles can be made gauge-invariant by adding new terms, which correspond to new particles (bosons) that mediate interactions between the original particles. In the equation describing how electric current produces magnetic fields (Ampère's law), Maxwell added a term representing a changing electric field to ensure the local conservation of charge, and thus made the equations linking electric and magnetic fields gauge-symmetric. When these equations were combined, they led to the prediction of electromagnetic waves (the basis for all of optics), identified later with photons as the bosons that mediate electromagnetic forces.

By 1930, QED was formulated by Paul Adrien Maurice Dirac (1902-1984) and refined by Werner Heisenberg (1901-1976) and Wolfgang Pauli (1900-1958). This theory combined classical electromagnetic theory with special relativity and quantum mechanics, yielding the first successful quantum field theory. In the theory, electromagnetic interactions between charged particles are mediated by the emission and absorption of photons (the quantum unit of the field). The theory also predicted the existence of antiparticles such as the positron (positive electron), which would be produced with the electron from radiation of sufficiently high energy (consistent with $E = mc^2$) and would mutually annihilate back into radiation if they came together again.

Because this early version of QED treated the electron as a point charge, it had infinite charge density, and thus infinite interaction energy with its own radiation field, giving it infinite mass. By the late 1940's, Shin'ichirō Tomonaga, Richard P. Feynman, and Julian Schwinger had independently shown how these infinities could be avoided by incorporating infinite corrections in Dirac's equations for charge and mass and then substituting measured values for these quantities. This "renormalization" of QED was possible because of the gauge symmetry of Dirac's equation, giving an incredible precision of about one part in a billion in accounting for phenomena such as the hyperfine splitting of hydrogen spectrum lines and magnetic moments of the hydrogen atom.

Weak interactions were first identified in 1934 by Enrico Fermi (1901-1954) in his theory of radioactive beta decay (electron emission), such as the decay of neutrons into protons and electrons. When the emitted electrons were found to have a continuous distribution of energies, always less than what was needed for conservation of energy, Pauli suggested that the missing energy might be carried off by unobserved neutral particles with little or no mass. Fermi named this supposed particle the neutrino, and he incorporated it into a quantum field theory of beta decay patterned after QED. He was able to account for the electron energy distribution by introducing a new force, much weaker than electromagnetism, acting on the particles at the point where the neutron decayed into a proton, electron, and neutrino.

Although Fermi's theory worked well at most energies, it failed at its high-energy limit because it assumed zero range for the weak force. In 1935, Hideki Yukawa proposed a quantum field for both strong and weak forces in which the weak force was mediated by the exchange of charged bosons (W^+ and W^-, each with one unit of electron charge) rather than acting at a point. Thus, neutrons would change into protons by emitting a W^-, which would then decay into an electron and a neutrino. His theory predicted a mass for bosons that was inversely proportional to the range of the forces they mediate. Thus, photons have zero mass since electromagnetic forces have infinite range, but W bosons would be massive because of the short range of the weak force.

After the discovery of the neutrino in 1953 and the violation of parity (the weak force behaves differently from its mirror image) in 1956, weak-interaction theory was generalized by Feynman, Murray Gell-Mann, and Schwinger in 1958. It was then applied to more complicated processes, such as neutron-neutrino scattering in which boson pairs are exchanged, leading to infinities that could not be renormalized. The problem was compounded by the fact that, unlike the massless photon, massive bosons break the gauge invariance of the field equations for the weak force, which is the symmetry needed for renormalization. Thus, the quantum field theory for massive vector bosons gave a good first approximation for weak interactions in processes such as beta decay, but it broke down when it was applied to more complicated processes requiring higher approximations.

The first attempt to unify electromagnetic and weak interactions was made by Glashow in 1961. He set up an electroweak gauge theory by combining the QED gauge symmetry (the unitary "U(1)" group) with a rotation-type symmetry (the special unitary "SU(2)" group) for the weak force to obtain a natural unification of weak and electromagnetic interactions based on the combined symmetry (the "SU[2] × J[1]" group). This led to the requirement of four bosons: the photon, the charged vector bosons W^+ and W^-, and an unanticipated neutral vector boson Z^0. The symmetry of W^+ and W^- is related to similar symmetries between the electron and its associated neutrino, and between the proton and neutron (or their up and down quarks, u and d), allowing W emission or absorption in transitions between these particles. Unfortunately, this form of the electroweak theory required massless bosons for gauge symmetry, and thus did not allow for the masses required by the short range of the weak force.

The problem of vector boson masses was first solved in 1967 by Weinberg, who was then able to complete the formulation of a successful electroweak theory. He applied an idea suggested by Peter Higgs in 1964 called spontaneous symmetry breaking. Higgs showed that the lowest energy solution (ground state) of a gauge theory may have less symmetry than the equations of motion, and that this asymmetry can be compensated for by a new field. Weinberg introduced this type of Higgs field to restore the symmetry of the weak interactions, and he found that the W and Z bosons can acquire mass by coupling to this field. The success of this Higgs mechanism suggests the possibility of a fifth force in nature and its exchange particle called the Higgs boson. In the resulting unified electroweak theory, the W and Z boson masses can be calculated in the terms of the measured strengths of the weak and electromagnetic interactions, and a mixing angle relating them ($M_W = M_Z \cos\theta_W$). This Weinberg angle can be measured in many different experiments and is an important test of the theory.

The unified electroweak theory attracted more attention after 1972 when Gerard 't Hooft and Martinus Veltman proved that the theory was renormalizable, even with the spontaneous symmetry breaking that gave large masses to the W and Z bosons. It was found that the various multiparticle exchanges of photons, bosons, and other particles add up so as to cancel unresolved infinities. The unanticipated Z^0 particle also suggested the possibility of previously unknown neutral-current processes in which weak interactions between particles might be mediated by the Z^0 without any change in charge of the particles. The observation of neutral-current interactions with accelerators at the Centre Européen de Recherche Nucléaire (CERN) in Geneva and at Fermilab near Chicago in 1973 provided direct evidence to support the theory.

One remaining problem was the possible interaction of Z^0 with the three known quarks (u, d, and s for "strange"), which would permit the neutral-current transition of an s-quark into a d-quark (both with a charge of $-\frac{1}{3}e$) in violation of experimental evidence. A solution to this problem was the 1970 proposal of a fourth c-quark ("charm") linked to the s-quark, which would cancel any transitions from s- to d-quarks. Discovery of the c-quark in 1974 was a further success for the electroweak theory. The final triumph was the 1983 discovery of the W and Z particles with their predicted masses.

Applications

The existence of neutral currents and the discovery of the W and Z particles gave substance to the electroweak theory. The subsequent quantitative tests of the theory have established its basic validity and the underlying unity of electromagnetic and weak interactions. During the 1970's, continuing neutral-current (Z^0 exchange) research concentrated on measuring the Weinberg angle from neutrino scattering experiments, giving remarkably consistent results. Inelastic (multiparticle) scattering from target nuclei of energetic neutrinos produced by the decay of muons (heavy electrons) in high-energy accelerator reactions usually results in charged-current events producing charged muons. The neutral-current signature is the absence of muons in such inelastic neutrino-scattering experiments. The ratio of neutral-current to charged-current events gave the first values for the Weinberg angle. Elastic neutrino-electron scattering (producing no new particles) allows a much cleaner test of the electroweak theory since it requires no corrections for the strong nuclear force. The ratio of neutrino-electron scattering angles to antinuetrino-electron scattering angles gave the same value of the Weinberg angle within the limits of experimental error.

The neutral-current measurements of the Weinberg angle, confirmed by several other types of experiments, led to predictions for the masses of the W and Z particles in the range of eighty-five to one hundred times the mass of the proton. The equivalent energy to produce these particles was far beyond the range of particle accelerators in the early 1980's. In 1982, Carlo Rubbia, Simon van der Meer, and their CERN colleagues recognized that the fixed-target Super Proton Synchrotron (SPS) could be converted to a colliding-beam storage ring with more than a twenty-fold increase in the collision energy. This gave more than enough energy to produce the predicted heavy bosons from proton-antiproton collisions. In 1983, they found all three of the vector bosons predicted by the electroweak theory, for which Rubbia and van der Meer received the 1984 Nobel Prize in Physics.

The measured mass of the W particle is eighty-five proton masses and the mass of the Z particle is ninety-seven proton masses, giving a value for the Weinberg angle that agrees with neutral-current and other measurements. This gives a range for the weak force of about one-thousandth of the size of the atomic nucleus. The W and Z bosons decay rapidly into all possible lepton-antilepton and quark-antiquark pairs. They were detected in the SPS collider at CERN from the W-decay into electrons and neutrinos, and from the Z-decay into electrons and positrons. The detector was the largest ever built for a colliding-beam experiment, with a length of ten meters, a width of five meters, and a mass of 2 million kilograms. Many Z^0 particles have also been produced at Fermilab in a higher energy proton-antiproton collider completed in 1985, and in two electron-positron colliders at Stanford University and at CERN since 1989. Electron-positron colliders produce Z^0 particles from more than 99 percent of the electron-positron annihilations when the collision energy is tuned to the Z^0 mass, whereas proton-antiproton colliders produce less than 1 percent.

Precise measurements of the Z^0 mass and its coupling to leptons and quarks are important not only to test the electroweak theory itself but also to check higher-order corrections of the theory induced by other heavy particles. Because it is a gauge theory, these corrections are not infinite and can be calculated. One such correction shows that boson masses depend on the mass of the predicted "top" quark, which is required to complete the heaviest of the three quark families. The electroweak theory links the u- and d-quarks to the electron and its neutrino, the s- and c-quarks to the muon and its neutrino, and the b- and t-quarks (bottom and top) to the tau lepton (heavier than the muon) and its neutrino. Neutral-current and boson-mass data yield an upper limit of about two hundred proton masses for the top-quark mass, while experiments have set a lower limit of about eighty-five proton masses. The theory is incomplete in one other interesting way: It requires a new force of nature and its associated particles to generate the boson masses. If this force is mediated by Higgs bosons, then it might be observed in the many ongoing Z^0 = decay experiments, or it might induce corrections that would be observed in precision measurements of the Z and W bosons at higher energies.

Context

The success of the electroweak theory is the first step toward a theoretical unification of the fundamental forces of nature and their associated elementary particles. The theory has already contributed to the organization of these particles into family groups, linking the three lepton doublets with the three quark doublets by group symmetry and shedding new light on the c- and t-quarks. The theory may also account for the small differences in mass of strongly interacting particles in family groups, such as the proton and neutron, which lead to infinities from electromagnetic corrections to calculations of the strong nuclear force. Because the intrinsic strengths of electromagnetic and weak forces are similar, they may provide additional corrections that cancel these infinities.

The next major step would be a grand unification of the strong and electroweak forces in a unified gauge-symmetric field theory. The difficulty with this approach is the lack of mathematical methods to cope with the greater strength of the strong interactions. It has been shown, however, that, in some gauge theories, the effective strength of the strong force decreases at sufficiently high energies, so that approximation methods similar to those used in the electroweak theory can be applied to this kind of "asymptotic freedom" at high energies. Calculations of this type agree with experiments but do not yet form a complete theory. If the strong force decreases at high energies and short distances, then it would become larger at low energies and large distances. Thus, ordinary particles cannot be broken up into their component quarks if the strong force increases without limit as the quarks move farther apart (quark "confinement").

Grand unification theories (GUTs) suggest that the intrinsic strength of strong interactions is comparable with that of electroweak interactions, only appearing stronger because of the relatively low energies and large distances of current experiments. At high energies, the strong, electromagnetic, and weak forces would converge to a single unified force. The ultimate goal would be to include the force of gravity in this type of unification, but this would appear to require a quantum theory of general relativity that has yet to be achieved.

The new unified force theories also contribute to an understanding of astrophysics and cosmology. Supernova explosions of giant stars begin with an implosion (collapse), which then reverses and becomes an explosion. Neutrino pressure can account for this reversal, but only because of the electroweak neutral currents that produce "coherent" neutrino interactions with entire nuclei rather than with individual neutrons and protons. Unified force theories also cast light on what took place during the earliest fractions of a second after the creation of the universe, as described by the big bang theory. This theory postulates that the universe began as an expanding fireball from an initial point of infinite energy density. As the temperature falls in the expansion, gravitation and then the strong nuclear force would separate first from an original single unified force, along with the formation of quarks and leptons. This would be followed by separation of the weak force and the decay of massive bosons. At still lower energies, quark confinement would produce protons and neutrons. Finally, expansion would continue until energies were low enough for electromagnetic forces to bind electrons to protons, and for gravity to form stars and galaxies. Radiation and elementary particles left from the big bang may provide the evidence needed for developing a completely unified theory of forces.

Bibliography

Cline, David B., Carlo Rubbia, and Simon van der Meer. "The Search for Intermediate Vector Bosons." *Scientific American* 246 (March, 1982): 48-59. This article contains a readable introduction to the electroweak theory and an extensive description of the experimental equipment that was used in the discovery of the W and Z particles, including photographs, diagrams, and graphs. Rubbia and van der Meer shared the 1984 Nobel Prize in Physics for this 1983 discovery.

Dodd, J. E. *The Ideas of Particle Physics.* Cambridge, England: Cambridge University Press, 1984. The subtitle of this book is "An Introduction for Scientists," but much can be read by the general reader. It has two chapters on weak interaction physics and a chapter on the electroweak theory with diagrams, graphs, and a good glossary.

Eisberg, Robert, and Robert Resnick. *Quantum Physics of Atoms, Molecules, Solids, Nuclei, and Particles.* 2d ed. New York: John Wiley & Sons, 1985. This intermediate-level college physics textbook has good sections on beta decay, Enrico Fermi's theory of weak interactions, quantum electrodynamics (QED), and the electroweak theory. Theoretical results are discussed with equations, graphs, and diagrams, but without lengthy mathematical derivations.

Hughes, I. S. *Elementary Particles.* 2d ed. Cambridge, England: Cambridge University Press, 1985. This undergraduate college text discusses particle physics developments since 1950. Mathematical equations are used, but no lengthy derivations. Three chapters deal with weak forces and the electroweak theory, using many diagrams and graphs.

Langacker, Paul, and Alfred K. Mann. "The Unification of Electromagnetism with the Weak Force." *Physics Today* 42 (December, 1989): 22-31. A good discussion of the history of the electroweak theory with particular emphasis on the successful experimental tests of the theory. References to original articles are also given.

Trefil, James S. *From Atoms to Quarks.* New York: Charles Scribner's Sons, 1980. This is a very readable introduction to elementary particle physics with several chapters on quarks and two chapters on weak interactions and unified force theories. Many diagrams and a glossary are included, but no index.

Weinberg, Steven. "Unified Theories of Elementary-Particle Interaction." *Scientific American* 23 (July, 1974): 50-51. An excellent and readable discussion of the electroweak theory, including its background and applications. The article was written by one of the formulaters of the theory who shared the 1979 Nobel Prize in Physics for its development.

Joseph L. Spradley

Cross-References

Antimatter, 123; The Big Bang, 210; Bosons, 252; The Dirac Equation, 685; Gauge Theories, 942; Grand Unification Theories and Supersymmetry, 985; Group Theory and Elementary Particles, 1022; Leptons and the Weak Interaction, 1242; Quantum Elecrodynamics, 1965; Quarks and the Strong Interaction, 2006; The Standard Model, 2353.

THE EVOLUTION OF THE UNIVERSE

Type of physical science: Astronomy and astrophysics
Field of study: Cosmology

In the past 10 billion to 20 billion years, the universe has expanded from a single point in space to a vast area containing billions of galaxies. In the future, the universe will either continue to expand or will slow down and contract to a single point again, depending on how much mass it contains.

Principal terms

ANTHROPIC PRINCIPLE: the idea that the universe is "programmed" by physical laws to evolve complex life

BIG BANG: the unimaginable explosion that created the universe (space and time); it is thought to have occurred sometime between 13 and 20 billion years ago

CLOSED UNIVERSE: a universe where all matter will eventually recompress into a tiny volume of space

COSMIC BACKGROUND RADIATION: the low-level radiation, created by the big bang, that bombards the earth from all directions of space

COSMIC CRUNCH: the eventual recompression of all matter in the universe that may occur if the universe is closed

HEAT DEATH: the eventual loss of all usable energy in the universe that will occur if life exists in an open universe

MISSING MASS: the invisible mass, such as black holes, black dwarfs, or neutrinos, that cannot be found by direct observation of galaxies

OPEN UNIVERSE: a universe that expands forever and will eventually lose all usable energy

OSCILLATING UNIVERSE: a type of closed universe where a big bang and subsequent recompression of all matter periodically occur forever

Overview

The evolution of the universe can be conveniently discussed in two parts: the evolution of the past and the evolution of the future. The past evolution begins with the origin of the universe and continues to the present. Most cosmologists (scientists who study the origin and fate of the universe) believe that the universe originated sometime between 13 billion and 20 billion years ago. The so-called big bang occurred at that time, creating the universe from a tiny point of nothingness. This event created not only space but also time, because both are linked in the space-time continuum as shown by Albert Einstein's special theory of relativity. There are two basic lines of evidence for the big bang. One is the visible expansion of the present universe. Astronomers have known since the mid-twentieth century that nearly everywhere in the universe, galaxies are moving away from one another at high speeds.

Measurement of the rate of expansion allows astronomers to work backward, estimating when the galaxies were all together at a single point. The second line of evidence is the observation of cosmic microwave background radiation that is a "fossil" of the tremendous energies released at the time of the big bang. This radiation is observed as a weak radio "hiss" that bombards the earth from all directions of space.

Events following the big bang can be reconstructed in great detail by using known physical laws to predict the behavior of matter and energy under various conditions. Thus, as all the matter in the known universe is compressed to smaller and smaller areas of space, calculations show that temperature and pressure become increasingly large. As compression continues to a tiny point in space, the enormous temperatures and pressures that existed during and shortly after the big bang can be re-created. Under these conditions, the big bang was found to have consisted of the release of unimaginable amounts of energy. As space expanded, temperature and pressure began to decline and some energy began to form matter.

Matter and energy can be converted into each other from Einstein's famous equation: $E = mc^2$, where E is energy, m is mass, and c is the speed of light. By the first microsecond following the big bang, subatomic particles, such as protons and neutrons, had begun to form. Then electrons formed as the huge fireball continued to cool. As cooling continued, the electrons, neutrons, and protons began to be attracted by electromagnetic and nuclear forces to form atoms. One hundred thousand years after the big bang, atoms were common, and matter became more abundant than energy. These early atoms consisted of the elements hydrogen and helium, which are the two simplest atoms (having only one and two protons, respectively). Even today, these atoms compose about 98 percent of the universe.

A billion years after the big bang, gravity caused the hydrogen and helium atoms to cluster into huge clouds that became "galaxies." Galaxies contain billions of stars; therefore, these clouds were of tremendous size. Soon after, the stars began to form within the galaxies. Stars form when local clouds of hydrogen and helium begin to condense through gravitational attraction of the atoms. As matter accumulates, the atoms at the core of the cloud come under increasing temperature and pressure until they begin to unite, or "fuse," nuclei. This "nuclear fusion" is what causes the star to emit energy. Most important for the evolution of the universe, such fusion of the simple hydrogen and helium atoms is how atoms more complex than hydrogen and helium were created. Thus, the present universe is composed of more than one hundred elements, each composed of its own kind of atom. Without nuclear fusion by stars, there would be no silicon, oxygen, carbon, or any of the many other elements that form a planet and its life.

The past 10 billion years of the universe's evolution have seen all the galaxies form, even though new stars continue to form today. The universe has continued to cool until it has reached the very low average temperature of only 3 degrees above absolute zero. Many of the newer stars, such as the sun, are "second-generation" stars, in that they formed from the debris of older stars that exploded. Only these second-generation stars can have planetary systems, since the more complex atoms created by the older stars are needed to form the planets. The Milky Way galaxy formed about 12 billion years ago and the sun and its nine known planets began to condense from a debris cloud between 5 billion and 4.5 billion years ago.

The future evolution of the universe is considerably more controversial than the past. Nevertheless, the vast majority of cosmologists believe that there are only two possible ultimate fates for the universe: The universe either will continue to expand forever (an "open universe") or will cease to expand and will contract to recompress to another pinpoint in space (a "closed universe"). The crucial determining factor is whether there is sufficient mass in the universe to cause the separating galaxies to be attracted gravitationally enough to slow down and recondense eventually. There are three lines of evidence that cosmologists have used to infer whether humans live in an open or closed universe. Unfortunately, this evidence gives conflicting results, although most of it superficially seems to favor an open universe.

The first line of evidence on the future of the universe is the direct measurement of the amount of mass it contains. The most obvious way of measuring the mass is to estimate the number of galaxies in the universe and multiply it by the average mass of a galaxy. This measurement has been done by many astronomers, and the general conclusion supports an open universe. The amount of mass in the universe, when estimated this way, is only about 10 percent of that needed to cause the galaxies to recondense eventually. A major flaw of this method, however, is that it takes only visible objects into account. The average mass of galaxies is calculated on the assumption that stars and other luminous objects that are seen form all (or nearly all) of each galaxy. Yet measurements of galactic mass based on other methods indicate that much of the matter is invisible. For example, when mass is estimated by how much a galaxy's gravity affects surrounding galaxies or how fast it spins, the estimated mass becomes considerably greater. Scientists are researching what forms the "missing mass." Many objects have been suggested, but there is little proof for any of the suggestions. Among the objects most often cited are black holes and black dwarfs, both of which are the remnants of dead stars. Another possibility is that vast unseen halos of gas and dust surround galaxies. Finally, some scientists have suggested that tiny particles called "neutrinos," traditionally thought to be massless, may have mass. Because neutrinos are so common, this would give the universe a much greater mass.

The second line of evidence is to measure the speed at which the galaxies are separating to see if the expansion of the universe is slowing down. This is easiest on the galaxies closest to Earth, but, unfortunately, these galaxies are also the least useful. This results from the fact that more distant galaxies emitted light a longer time ago and therefore give more information on past rates of separation. Measurement of the nearer galaxies indicates that the rate of expansion is not slowing down, so that an open universe is once again implied. Nevertheless, when information on the motion of distant galaxies can be obtained, the opposite conclusion is reached. Unfortunately, this measurement on more distant galaxies is so difficult that the information is of dubious quality.

The third line of evidence on the future of the universe comes from measurements of the number of certain atoms now present in the universe. For example, certain isotopes, such as "heavy hydrogen," could have formed only in certain conditions present early in the formation of the universe. Calculations show that if the amount of matter in the early universe were high enough for a closed universe, then virtually no heavy hydrogen would have survived to the present time. Yet, if less matter were present so that an open universe exists, then the calculations show that heavy hydrogen would be fairly common (about 1 in 100,000 hydrogen atoms). The observed amount of heavy hydrogen in today's universe is close to 1 in 100,000 so that, once again, there seems to be support for the open universe. Nevertheless, the calculations on which these estimates are based rely on assumptions of the early universe that are very uncertain. Therefore, the results provided by the third line of evidence on the future of the universe, like the others, are also highly questionable.

Applications

Every thinking human being wonders about the origin and fate of many things. This includes the origin and fate of the earth, its life, and even human life. These thoughts are dwarfed by the origin and fate of the universe itself, which includes everything that exists throughout time and space. Therefore, while it might seem that the evolution of the universe is not especially relevant to daily life, it has profound implications for giving human life meaning: understanding how humans evolved and where the human species will one day find itself. Indeed, the origin and evolution of the universe has been called the major question of all time. Science cannot answer all such questions, but it does provide many key facts that are highly relevant to major questions that humans have asked for thousands of years.

Perhaps the most crucial of these questions is why the universe exists. With current knowledge, instead of simply speculating about the origin of the universe, specific questions can be asked about what happened to cause the big bang. Some scientists have suggested that the universe (space and time) was created from nothingness as a kind of statistical "hiccup." That is, physical systems often show natural fluctuations in energy even when the average energy present is zero. For example, physicists have observed that elementary particles literally appear out of nowhere during certain subnuclear reactions. This does not violate the laws of physics because a new particle of matter is counterbalanced by the production of a particle of antimatter. If these two particles collide, they both disappear and there is no net change. Thus, the current universe is composed of matter, and if it also contains antimatter (as some scientists suspect), then it could have erupted from nothingness as do new particles today, without violating known laws.

The knowledge of the universe's evolution has also been directly applied to religious thoughts about God. Some cosmologists have pointed out that the physical laws of the universe that were created by the big bang are "programmed" to create life: stars to provide energy, habitable planets, and chemical bonding that promotes complex, self-reproducing molecules. Cosmologists point out that if many of the physical laws, such as the electromagnetic force attracting protons to electrons, were changed only the slightest bit, then neither stars, planets, nor life would form. This idea—that the universe is programmed (perhaps by God) to create complex intelligent life—is called the "anthropic principle" (*anthropos*, meaning humankind). Many other scientists dispute this notion, pointing out that it involves circular logic: If the universe did not create intelligent life, humans would not be here to speculate about it. In other words, there may be (or may have been) many other universes where no stars or life arose.

In addition to the origin of the universe, yet another profound philosophical and religious application of the universe's evolution is in its ultimate fate: What will eventually become of the human species and civilization? This fate depends entirely on whether the universe is open or closed. If it is open, then the stars will eventually burn out and the universe will have no usable energy. Life, including humans, cannot possibly survive this so-called heat death, because all life needs energy to maintain itself. On the other hand, if the universe is closed, then it seems impossible that any life could survive the eventual compression of all the matter in the galaxy into a tiny point in space. This has been called the "cosmic crunch." If this occurs, many cosmologists have shown evidence that the resulting pinpoint of matter will then re-explode into another big bang. Thus, the current universe may be merely one of many cycles in an "oscillating universe" that periodically explodes and collapses. This kind of universe holds out more hope for life than the open universe because it can at least re-evolve when destroyed, whereas in the open universe the "heat death" is the eternal end of all life.

There may also be some practical applications for the growing knowledge of the evolution of the universe. There are four basic forces that determine everything that occurs in the universe: the electromagnetic force, the gravitational force, the weak nuclear force, and the strong nuclear force. Physicists have tried for many years to interrelate these forces into a single force by a grand unified theory (GUT). If this is possible, a better understanding of the forces will result that may allow incredible control over nature, such as nearly unlimited sources of energy. It would be even greater than that unleashed by the discovery of nuclear forces. Calculations show that the best chance for creating a GUT is to be found in reconstructing the very early history of the universe, because the four forces did not become separate until sometime after the big bang. Thus, by reconstructing how they separated, only then can the forces' interrelationship be understood.

Context

Because of the human need to understand their origins, nearly every human society that has ever existed has had some kind of "creation myth." In Western culture, many religious and philosophical beliefs about the origin of the universe can be traced back many thousands of years to the creation myths of societies in the Near East. The most familiar of these to members of the Judeo-Christian tradition is the

Book of Genesis in the Bible. It was not until the science of astronomy was maturing during the Renaissance of Europe, however, that accurate scientific evidence about the universe was gathered. Once this revealed the true nature of the universe, especially its vastness, beliefs about its origin became progressively modified to account for the new facts. For example, Nicolaus Copernicus and others showed that Earth was not at the center of the known universe. By the 1800's, Sir Charles Lyell and other geologists had shown that Earth was millions of years old.

While these earlier discoveries led humans to question older beliefs, it was not until the 1920's that the most direct and important evidence about the evolution of the universe was discovered. Edwin Hubble showed that the universe is expanding. By analyzing the light from distant and nearby galaxies, Hubble noticed that the most distant galaxies are moving faster than those nearby. By plotting the speed and direction of the galaxies' motion, it was soon evident that this expansion occurred from an explosion that caused matter to be dispersed long ago in many directions. This explosion eventually came to be called the big bang. The second major line of evidence for the big bang was discovered in the 1960's. Researchers trying to eliminate "static" from radio and other communications transmissions found that they could not isolate the source of a persistent hiss. They soon found that it was not an earth-bound source (such as electrical storms) but came from all directions of outer space. By analyzing its frequency, wavelength, and other features, this so-called cosmic background radiation is now known to be the "fossil" radiation pervading space that was created by the big bang explosion.

The understanding of the evolution of the universe will greatly improve as advances in astronomical instrumentation occur. A major advance is the Hubble Telescope (named for the discoverer of the expanding universe), which was put in orbit around Earth in 1990. Once the unfortunate technical flaws are corrected, this telescope will gather information about more distant galaxies and therefore see even further back in time.

Bibliography

Chaisson, Eric. *The Life Era*. New York: Atlantic Monthly Press, 1987. A popular account of the evolution of the universe, written in an informal style by a well-known astronomer. One of the main points is the argument that life is an inevitable product of the physical laws of the universe.

____________. *Universe: An Evolutionary Approach to Astronomy*. Englewood Cliffs, N.J.: Prentice-Hall, 1988. An introductory text on astronomy with a strong evolutionary theme. Part 3 is devoted entirely to the evolution of the universe.

Harrison, Edward R. *Cosmology*. Cambridge, England: Cambridge University Press, 1981. An introductory text on the origin, evolution, and fate of the universe. Suitable for the highly motivated college student.

Hartmann, William K. *Astronomy: The Cosmic Journey*. Belmont, Calif.: Wadsworth, 1987. An excellent basic text with a good discussion of the origin, evolution, and fate of the universe. Written by a respected, well-known astronomer.

Sagan, Carl. *Cosmos.* New York: Random House, 1980. One of the most popular science books of all time. It is superbly written and illustrated by a famous astronomer and is based on the popular television series of the same name. Fun and easy reading.

Seielstad, George A. *At the Heart of the Web.* Orlando, Fla.: Harcourt Brace Jovanovich, 1989. Another popular account of the evolution of the universe that argues that the evolution of life is inevitable. This book devotes more to biology and is less philosophical than the popular book by Chaisson (cited above).

Time-Life Books. *Voyage Through the Universe.* Alexandria, Va.: Time-Life Books, 1988. A beautifully illustrated series that covers the evolution of the universe from the big bang to the late twentieth century.

Michael L. McKinney

Cross-References

The Big Bang, 210; Cosmology, 589; Planet Formation, 1828; Star Formation, 2360; The Expansion of the Universe, 2650; Large-Scale Structure in the Universe, 2657.

THE EXPANSION OF THE UNIVERSE

Type of physical science: Astronomy and astrophysics
Field of study: Galaxies

The galaxies of the universe are traveling away from one another, with the most distant galaxies receding at the greatest velocities. Understanding the expansion of the universe will lead to an understanding of its beginning and of its ultimate fate.

Principal terms

CEPHEID VARIABLE STAR: a variable star whose period of variation is related to its absolute brightness (luminosity); used to measure distances to star clusters and galaxies

DOPPLER SHIFT: a change in the observed frequency, or pitch, of sound and electromagnetic waves because of the relative motion of the source and observer; applies to the shift in the wavelength of light caused by the motion of source and observer

HUBBLE'S LAW: the relation that the spectral redshift of light from a distant galaxy is a measure of its velocity, which is dependent on the galaxy's distance

LIGHT-YEAR: an astronomical unit of distance equal to the distance light travels in one year, 9,460 billion kilometers

MILKY WAY: the bright, dense band of stars seen overhead on a summer's night; the name for the galaxy in which Earth is located

PARALLAX: an apparent shift in the position of a star in the sky because of a change in the location of its observer; used to measure the distance to nearby stars

REDSHIFT: the shift in the spectrum of light from a star or galaxy toward the lower-frequency, or red, end of the spectrum, caused by the velocity of the galaxy moving away from the observer

SPECTRUM: the component wavelengths of energy generated by a star, generally referring to the visible wavelengths (violet to red); includes the distinct wavelengths (colors) absorbed by heated gases at a star's surface

Overview

The universe is composed of stars, much like Earth's sun, clumped together by the billions in groups called galaxies. Billions of galaxies are scattered like islands throughout the sea of space. In 1929, it was discovered that the galaxies were not fixed in place but were speeding away from one another; the more distant the galaxy, the greater its speed. If one represents galaxies by raisins in raisin bread, each raisin (galaxy) moves away from all the other raisins as the bread rises. The galactic motion says that the universe is expanding. It also implies that the universe had a definite beginning. This one observation revolutionized the ideas on the nature of the universe, of its beginning, and of its ultimate fate.

Looking up at the night sky, the expansion cannot be seen. Although the Moon, stars, and planets rotate overhead with the hours and the seasons, their patterns have remained unchanged throughout human history. It has taken a combination of theory and observation to lead from the idea of a static universe, as explained by the Greeks, to the dynamic expanding universe of the twentieth century.

In 1610, Galileo Galilei turned the newly invented telescope toward the heavens and discovered not only that millions of stars formed the Milky Way but also that some objects were not starlike. These objects were called nebulas, from the Latin for "cloud." These spiral and spiderlike shapes were thought to be clouds of dust and gas. In the mid-1700's, philosopher Immanuel Kant suggested that the spiral nebulas were composed of individual stars, like the Milky Way, and were actually "island universes" scattered about the vast void of space. Although Kant's idea, partly based on a work by English philosopher Thomas Wright, was reasoned to be intuition, not observation, it inspired later astronomers.

In the 1800's, advancements in observational astronomy provided the techniques needed to uncover the nature of the nebulas. Physicist William Hyde Wollaston discovered the presence of dark lines in the spectrum of colors that make up sunlight. These lines represent the absorption of certain frequencies (colors) characteristic of elements in the sun's outer layers. Using this information, astronomer Sir William Huggins used the spectroscope (an instrument to observe spectra) to study the spectra of nebulas. He discovered two kinds of nebulas: those with the spectra of heated gas and those with the spectra of stars. Throughout the nineteenth century, astronomers sought to improve the resolution of their telescopes as they cataloged stars and nebulas and collected their spectra.

The main stumbling block of the time was the size of the Milky Way. Astronomers simply did not know the distance to most objects they saw. Distances to a few hundred stars were known, worked out by the parallax method. "Parallax" is the apparent motion of an object, compared to distant ones, because of the motion of the observer. This apparent motion can be seen by looking at one's finger on an outstretched arm first with one eye and then with the other. The finger appears to move, compared to distant buildings or trees. Using the motion of Earth around the sun, the distance to nearby stars can be found. Yet, this works only for stars a few tens of light-years away. (A light-year is the distance light travels in one year, 9,460 billion kilometers.) In the 1890's, Henrietta Swan Leavitt, at the Harvard College Observatory, discovered a relation between brightness (luminosity) and period for Cepheid variable stars. Variable stars change their brightness with time. The time for one cycle of change is the period. For Cepheid variables, the brighter the star is, the longer its period. The observed brightness of any star decreases regularly with distance. Measuring the period of a Cepheid determines its brightness. Therefore, by comparing it to a star of known brightness and distance, the distance to the Cepheid can be found.

In the early twentieth century, Harlow Shapley, at Mount Wilson Observatory, used the method to show the Milky Way as a disk-shaped collection of stars and gas surrounded by a sphere of globular star clusters. Beginning in 1919, Edwin Powell Hubble began to study spiral nebulas, sorting them into distinct categories. In 1923, using the newly completed 254-centimeter reflector telescope, Hubble found individual stars—Cepheid variables—in the Andromeda nebula. Over the next two years, Hubble discovered fifty more Cepheids in nebula NGC 6822. Using Shapley's distance method, Hubble found these nebulas to be several hundred thousand light-years away. He also photographed nebula M33 with new, sensitive film, resolving it into distinct stars. Thus, as Kant had proposed, nebulas were galaxies, independent of the Milky Way.

As Hubble looked for Cepheid variables, he also collected the nebular spectra. Although the spectra resembled that of the stars, they also differed. The dark absorption lines for the elements were found in the same order as in stellar spectra but shifted from the expected frequencies (or colors). This phenomenon was discovered by Christian Johann Doppler in the nineteenth century as he studied the change in pitch (frequency) of moving sources of sound. The pitch of a train whistle sounds higher as the train approaches the listener and lower as the train departs. This change of the pitch, or frequency, of sound from a moving source is called the Doppler effect, or shift. The spectra of light from the galaxies is "Doppler shifted" if the galaxy is moving toward or away from an observer. Hubble found a few galactic spectra shifted toward the shorter-wavelength, or blue, end of the spectrum, meaning they were traveling toward the Milky Way. Doppler shifts for the vast majority of galaxies were toward the red wavelengths, meaning the galaxies were traveling away from the Milky Way.

In 1929, Hubble and his assistant, Vesto Melvin Slipher, had forty-six redshift measurements for eighteen galaxies and the Virgo star cluster. By plotting the data, Hubble concluded that redshift (and velocity) were proportional to distance. The relation is now known as Hubble's law. Thus, an object's velocity depends on its distance; the greater the velocity, the farther away the object is. Further confirmation was provided by Milton Humason at Mount Wilson. By 1935, Humason had added 150 redshifts to the list, with velocities in excess of one-eighth the speed of light. With the completion of the 508-centimeter telescope, Humason found, by the late 1950's, redshifted galaxies traveling at one-third the speed of light, at a distance of several billion light-years. Hubble and Humason found that all but the nearest galaxies were rushing away from the Milky Way at tremendous speeds. They discovered that the universe was expanding; with the distance between galaxies steadily growing larger.

An expanding universe had been predicted independently before Hubble's discovery by Soviet mathematician Aleksandr Aleksandrovich Friedmann and Dutch astronomer Willem de Sitter, as a solution to Albert Einstein's general theory of relativity. Nevertheless, the idea of such a universe was difficult to accept. Even Einstein rejected his own solution for an expanding universe because of the lack of observational data. The observations of Hubble and Humason, however, forced astronomers to accept a new view of the universe.

Two basic models evolved to explain the observed expansion. In the steady-state theory, as the galaxies expand, new matter is created to fill the space between the galaxies. Such a universe has neither beginning nor end; it is expanding but unchanging. The big bang theory proposed a universe expanding outward from an initial explosion at some time in the past. Both models had supporters throughout the 1930's and 1940's. Nevertheless, as astronomers tested the models against more and better observations, evidence mounted to show that the universe was expanding from an early high-density state. The time since this explosion can be found from Hubble's law. If a galaxy's velocity is proportional to its distance, then at some time in the past all the galaxies were close together. The slope of the redshift (velocity) versus distance relation is called the Hubble constant. Taking the reciprocal of the Hubble's constant, one can calculate how long the galaxies have been moving apart. Hubble's constant, about 15 kilometers per second per million light-years, gives a maximum time of 20 billion years. Because of uncertainties in the data and revision of distance estimates, most astronomers agree that the universe began with an initial explosion between 10 and 20 billion years ago.

Astronomers also realized that Hubble's law is not strictly linear. The expansion must slow with time because of the gravitational attraction of the mass of the universe. Scientists building on Friedmann's work have discovered a family of expanding models for the universe. The difference in the models depends, in part, on the total mass of the universe. If the mass is great enough, the universe will slow to a stop and eventually collapse; otherwise, the universe will continue to expand forever. Astronomers need to determine the rate of expansion in order to learn the future of the universe.

Applications

The discovery of the expansion of the universe forced a revolutionary change in scientists' ideas about the universe. Before then, the question of the origin of the universe was a subject for metaphysics or theology. With Hubble and Humason's discovery, questions regarding the beginning of the universe moved into the realm of science.

The redshifted spectra of galaxies showed that the universe was expanding. The relation between redshift and distance, Hubble's law, allowed astronomers to estimate the size and age of the universe. To explain these astronomical observations, scientists had to develop new models of the universe. Before any model can be accepted, however, it must be tested against observations. It also must predict phenomena to be searched for. These expanding universe models also promoted technological development of new techniques to study distant stars and galaxies. Astronomers look into space not only in the visible part of the spectrum but also in the infrared, ultraviolet, X-ray, and radio wavelengths. Yet they are limited in what they can see by the speed of light.

The speed of light is constant at 300,000 kilometers per second. The light seen from a star or galaxy has been traveling for some time before it can be observed. The more distant the object, the longer the time the light has been traveling. Therefore, light from distant galaxies is light from the distant past. Hubble's law for redshift and distance is related to a galaxy's velocity at some past time. Thus, observations of galactic spectra reveal the universe as it was when the light left the stars. Detailed study of the expansion of the universe has prompted the need for larger telescopes with better resolution in visible and nonvisible wavelengths. Seeing the fainter, more distant galaxies gives information farther back in time. This has a twofold use: It provides scientists with information on the early history of the universe and will allow them to predict the universe's ultimate fate more accurately.

The new models of the universe are generally concerned with whether the universe will continue to expand or will eventually collapse. For example, consider the universe to be somewhat like a ball tossed up in the air. The ball travels upward, slows, comes to a stop, and then falls downward because of the gravitational attraction of the earth. If the ball is thrown hard enough (with enough velocity), it can overcome the force of gravity and escape the earth. (Spacecraft to the Moon and planets are given enough velocity to escape Earth's attraction.) If the ball is thrown upward on the Moon, which has less mass than Earth, the ball can escape the attraction of the Moon with less velocity than on Earth. Whether the universe continues to move outward, like the ball, depends on the gravitational attraction of the total mass of the universe. If the mass of the universe is large enough, the universe (like the ball) will expand, slow, come to a stop, and eventually collapse. If the mass of the universe is not large enough, however, the universe will continue to expand. Hubble's relation for these different models is not linear but slightly curved. Nevertheless, the difference in curvature between the models is small. Astronomers need more observations from more distant galaxies to determine how fast the expansion of the universe is slowing. This will enable astronomers to learn which of the models of the universe best explains what is seen. It may be that the universe will continue to expand or it may eventually collapse. Whatever new observations are made, theories and models of the universe will need to explain them.

The increased understanding of stars and galaxies as one looks farther away in distance and time have forced scientists to look more deeply into the two basic theories of the universe: the general theory of relativity and quantum mechanics. Humans yearn to understand their world. Therefore, scientists seek to combine these theories into one unified explanation of the universe. Astronomical observations of the expanding universe will help them in their search. It is also possible that new astronomical techniques will uncover quite unexpected observations, forcing scientists to form new ideas again about the universe.

Context

Since first looking up into the sky, human beings have sought to understand the universe. The Greeks placed Earth at the center of their universe, with all the heavenly bodies moving on crystal spheres around Earth. In 1543, Nicolaus Copernicus proposed a universe where the planets, including Earth, moved about the sun. Galileo confirmed the idea by observation in 1610. Yet, the basic notion of the stars fixed on a crystal sphere just past the orbits of the planets remained unchanged.

In the mid-eighteenth century, new ideas about the universe began to emerge. They dealt primarily with the size and shape of the universe. Wright suggested that the sun was part of a great swarm of stars forming a flat disk, which explained the bright band of stars seen overhead on a summer's night. Kant proposed the Milky Way to be one of many such galaxies scattered at vast distances about space. Yet, the universe was still seen as unchanging with time.

The twentieth century idea of a dynamic universe has been a difficult one to accept. The Milky Way appears as a stable collection of stars, though some stars might die as others are born. Even Einstein had difficulty with the idea of a dynamic universe. His first application of his general theory of relativity to equations describing the universe gave solutions saying the universe must be either contracting or expanding, not static. At the time, observations did not support the theory, so Einstein added a term to give a solution for a universe that neither expanded nor contracted. Improved astronomical tools allowed Hubble and Humason to discover the expanding universe. This implies a universe that evolved with time from some definite beginning. Even then, the steady-state theory sought to avoid the implication of Hubble's law. By creating matter in the voids between galaxies, the universe could expand while remaining unchanged. This universe would have neither a beginning nor an end. Nevertheless, predictions by this model could not be supported by observations, but the explosive big bang theory could. Later models predict an expanding universe, but one that will either continue to expand or that will eventually contract. Only future observations will decide which model will be best.

As did the Greeks, modern astronomers seek to explain the heavens. Scientists continue to expand the knowledge of the universe as they look further back in time, hoping to reach the time of the universe's birth. By seeing and understanding the early history of the universe, astronomers hope to predict the universe's ultimate fate.

Bibliography

Caes, Charles J. *Cosmology: The Search for the Order of the Universe*. Blue Ridge Summit, Pa.: Tab Books, 1986. Suitable for the general reader on the evolution of the universe. Chapter 6, "Origin and Expansion of the Universe," discusses the basic models for an expanding universe. Includes glossary and bibliography.

Eddington, Arthur. *The Expanding Universe*. Reprint. Cambridge, England: Cambridge University Press, 1987. Reissue of a classic early work supporting the theory of an expanding universe by a foremost scientist of the time. Contains a foreword by William McCrea, placing the work in its historical perspective.

Ferris, Timothy. *The Red Limit*. New York: William Morrow, 1983. A well-written history of modern astronomy focusing on the people involved. Chapter 1 explains how the expansion of the universe was discovered. Includes glossary, photographs, and bibliography.

Gribbin, John. *In Search of the Big Bang*. New York: Bantam Books, 1986. Explains the search for an understanding of the nature of the universe. "Part One: Einstein's Universe" gives a historical understanding of events leading to the discovery of the expanding universe. Well-written book for the general reader. Third in a series on important scientific achievements of the twentieth century.

Hawking, Stephen W. *A Brief History of Time*. New York: Bantam Books, 1988. Geared for the general reader by one of the leading theoretical physicists of the twentieth century. Provides a nonmathematical step-by-step explanation of the expanding universe and why the universe must expand.

Hodge, Paul W. *Galaxies*. Cambridge, Mass.: Harvard University Press, 1986. Well-illustrated introduction to galaxies by a professor of astronomy. Chapter 10, "The Distance Scale," explains the method used to determine galactic distances.

Moore, Patrick. *Men of the Stars*. New York: Gallery Books, 1986. Well-illustrated book providing a short biography on fifty men who made significant advances in the science of astronomy. Includes a brief explanation of each scientist's discovery.

Pasachoff, Jay M. *Astronomy: From the Earth to the Universe*. Philadelphia: Saunders College Publishing, 1991. A well-illustrated introductory text for the college student. Chapters such as "The Early History of Astronomy" and "Cosmology" provide a good understanding of the evolution of ideas about the universe. "Galaxies" discusses Hubble's law and the expansion of the universe.

Weinberg, Steven. *The First Three Minutes*. New York: Basic Books, 1977. Suitable for the general reader. By a well-known physicist on the origin of the universe. Chapter 2, "The Expansion of the Universe," discusses the discovery of the expansion and explains the differences between the various expanding models for the universe. Includes glossary, mathematical supplement, and suggested readings.

Pamela R. Justice

Cross-References

The Big Bang, 210; Cepheid Variables, 345; Cosmology, 589; General Relativity: An Overview, 950; Optical Astronomy, 1650; Optical Detectors, 1657; Optical Telescopes, 1665; The Evolution of the Universe, 2643.

LARGE-SCALE STRUCTURE IN THE UNIVERSE

Type of physical science: Astronomy and astrophysics
Field of study: Cosmology

There is evidence of connective patterns of galaxy clusters across dimensions at least as large as many hundreds of millions of light-years. Explaining the observed features of this large-scale structure is a challenge to cosmologists that severely limits acceptable models of the origin and evolution of the universe.

Principal terms

BUBBLE: the possible structuring of galaxies, galaxy filaments, and galaxy clusters around a void

COSMIC STRING: a hypothetical early concentration of energy that initiated formation of galaxy filaments

DARK MATTER: mass in the universe that does not give off any form of electromagnetic radiation and is thus invisible, but is known by its gravitational influence

HOMOGENEITY PROBLEM: the difficulty of reconciling observations of the extreme uniformity of the cosmic background radiation with the early inhomogeneity required to account for large-scale structure

VIRIAL THEOREM: the result of statistical mechanics used to estimate masses of galaxies in a cluster, assuming their energies of motion and gravitation are balanced

VOID: the region of space hundreds of millions of light-years across, which is relatively free of galaxies

Overview

The hundred thousand light-years which span the diameter of the Milky Way galaxy enclose complex organizations of matter and energy, including stars emitting visible light, dense molecular clouds radiating at infrared wavelengths, and dark matter known only through its gravitational influence. To the contemporary cosmologist, all of these are small-scale structures when viewed from the perspective of attempting to construct a comprehensive model of the principal features of the composition and evolution of the entire physically observable universe. In this context, a large-scale structure means a feature still distinguishable when comparing the average contents of volumes each at least several million light-years across. One such volume encloses the Local Group of galaxies, whose dynamical behavior is determined by its gravitational interaction with other such volumes.

Understanding the structure and evolution of the universe is thus dependent upon a theory of gravitation and evidence about the distribution of matter and energy on the largest scales. As soon as Albert Einstein's theory of space-time and gravitation was completed, in 1916, it was immediately recognized that this theory would have a profound impact on cosmological models. The most extensively studied cosmological solutions of the general theory of relativity are those that also satisfy the cosmological principle. This is the assumption that at each moment in the history of the universe, an observer at any place would find the large-scale structure of surrounding matter the same as an observer at any other place (homogeneity) and appearing the same in all directions (isotropy).

In 1922, Aleksandr Aleksandrovich Friedmann derived the only three possible spatial geometries that can be solutions of Einstein's gravitational field equations under these restrictions. One of these is a space of uniform positive curvature and finite volume though without boundary. In two dimensions, the surface (not including the interior volume) of a sphere is such a space: Motion on it is never blocked by a barrier (perimeter line), but its area is finite. Another, and particularly simple, possibility is a flat space satisfying all the assumptions (and thus displaying all the derived features) of Euclidean geometry. In two dimensions, the surface of a plane without a boundary must be infinite. The only other homogeneous and isotropic space possible in Einstein's theory is a negatively curved one, also of infinite extent. In two dimensions, such a surface is termed "hyperboloid": A saddle is an example of a part of such a surface, which has unlimited area, if lacking a boundary curve. The common feature of these models is that their scale factors (distance between representative points, such as clusters of galaxies) are required to change with time: The universe must be either expanding or contracting. Before 1930, Edwin Powell Hubble published the first evidence that the universe is indeed expanding. If the scale factor of the universe was smaller in the past, its local densities of matter and radiation were higher. Since the mass density of matter varies inversely as the scale factor cubed, while the energy density of radiation varies inversely as the scale factor to the fourth power, at sufficiently early times, the pressure driving the expansion of the universe came from radiation. This is the basis of the "hot big bang" concept of the origin of the universe.

In modern years, there has been increasing interest in the study of inhomogeneous and anisotropic cosmological models. Initially, the motivation came primarily from the desire to understand better the implications of Einstein's theory of gravity. Besides the fundamental intellectual challenge of discovering and analyzing new solutions of the gravitational field equations, this work has been driven by the need to understand which features of the standard models are robust—that is, not sensitively dependent on the assumptions of exact homogeneity and isotropy. After all, this is a universe which has different small features in various places and presents slightly different views as one gazes outward in various directions.

Concurrently, observational and theoretical evidence has been accumulating, which challenges the old assumption that large-scale structure is not important in formulating a model that explains the general features of the universe. Pioneering work by George Abell and others, beginning after World War II, suggested that the distribution of rich clusters (each containing at least a thousand members) of galaxies included many of irregular shape and diverse contents. Yet, since the late 1970's, it was believed that the largest inhomogeneities in the distribution of matter in the universe were less than 300 million light-years across. Since then, new data and new analysis of older data have disclosed substantially larger structures. For example, a three-dimensional map of the tens of thousands of galaxies within a sphere more than 1 billion light-years in diameter, centered on the Milky Way, has been constructed by John Huchra and Margaret Geller. Based on more than five years of data gathered by a telescope, on the surface of the earth, of a size comparable to the Hubble Space Telescope, this map shows voids and filamentary structures on scales greater than 400 million light-years. Understanding the origin of structure in the universe on such large scales challenges the creativity of cosmologists.

Applications

First images, unfortunately of disappointing quality, were returned by the Hubble Space Telescope in May of 1990. Scientists expect that, when its optical problems are corrected, it will produce data of comparable quality to those described above throughout the remainder of its fifteen-year estimated life. It will be looking substantially deeper into space, perhaps ultimately surveying a volume a thousand times greater than heretofore. The motivation for having such data is that there are millions of times as many volumes that could be mapped as have been, and looking out is also looking back in time (because of the finite speed of light) toward the beginning of the universe. In the region already surveyed in detail, the density of galaxies in voids is typically a factor of ten less than average, and the density in the narrow but long filaments is typically a factor of ten thousand greater than average. Scientists are confronted with the major challenge of explaining the origin of such variations in density from an early universe that was amazingly homogeneous. The experimental evidence for this last claim is the extreme uniformity of the cosmic microwave background radiation, as observed by the Cosmic Background Explorer (COBE) satellite, launched in 1990. This radiation, now known to have a temperature of 2.735 Kelvins (absolute degrees) in all directions, is a relic of the primeval fireball, informing scientists of conditions throughout the universe at a time no more than a few hundred million years after its beginning. The difficulty of understanding how a universe that was so smooth at early times became so lumpy on such large scales in later times is called the homogeneity problem.

The solution of the homogeneity problem may involve what has been, somewhat misleadingly, often termed the "missing-mass problem." It is better to call this problem that of the identity of dark matter, since it is light rather than mass that appears to be missing. There are two distinct ways to estimate the total mass of the galaxies in clusters. One is based on assuming that a statistical mechanics result known as the virial theorem, relating kinetic and gravitational potential energies, applies to their observed speeds and separations. Another method is based on applying ratios of mass to light and is derived from knowledge of the composition of the light-producing objects within galaxies. These two methods are in serious conflict: The estimate of mass from light is typically less than that from mechanics by a factor of ten or more.

Most astrophysicists consider the arguments for the estimate from mechanics more firmly established, and thus are challenged to explain the nature of the dark matter, which apparently interacts very little (if at all) with electromagnetic radiation. Though this challenge has not been resolved in a manner satisfactory to all the experts in this field, the mere existence of dark matter, which is much more abundant than the familiar matter through its interaction with photons, provides a possible solution to the homogeneity problem. The idea is that the distribution of dark matter may be much more nearly uniform than that of the forms seen. If so, both matter and radiation are quite homogeneous now and were quite homogeneous at early times. It is only where densities of matter are slightly higher than average for the universe that the kinds that can be seen are strongly concentrated. An analogy for this "biased" formation of visible large-scale structure in the universe would be the view of a passenger aboard a large oceangoing vessel who gazes just above the horizon. Such an observer would see only unusually high wave peaks, while all the time being at a great and only slightly fluctuating height above the ocean floor. Unfortunately, it has not yet been possible to gather convincing evidence in support of quantitative tests of biased structure formation scenarios.

Another relevant concept to the understanding of large-scale structure in the universe may be the activity of cosmic strings. These hypothetical concentrations of energy are lines formed during spontaneous symmetry breaking at the end of the era of the grand unification of strong, electromagnetic, and weak interactions. They may stretch across the entire extent of the very early universe or join their ends to form loops. In either case, they could be the "seeds" for the formation of concentrations of matter, like droplets condensing on a wire, which later evolved into filamentary groups of galaxies. The absence of cosmic strings in the later universe (which has been observed) is conveniently explained through the generation of gravitational waves by their accelerations in the early expanding universe. There is hope that gravitational wave astronomy and observations of gravitationally lensed distant quasars may provide direct experimental evidence for the existence and behavior of cosmic strings in the not too distant future.

Finally, going back even closer to the beginning of the space-time and mass-energy of the universe leads one to the era of inflation. There are speculations that the high energy ("elementary particle") quantum processes during these exceedingly brief times after the birth of the universe may have been associated with inhomogeneous inflation. This suggests that the earliest structures, and therefore those now of the largest scales, would be bubbles. If such concepts could be confirmed by any relic evidence surviving from those extraordinarily early times, it could be said that not only the stuff of the cosmos but also its grandest patterns of organization are determined in its earliest moments. The desire to predict what such relics might be and then to find them drives a cutting edge of research in cosmology.

Context

The history of the quest to understand the large-scale structure of the universe has been, at least for the past few centuries, an uneven progression toward the recognition of ever more subtle organization at ever-larger distances.

Perhaps some early observers of the "patterns" that the mind can "recognize" among the stars visible to the unaided human eye believed these constellations to be real large-scale structures in the universe. Nevertheless, in one of the earliest applications of telescopes in astronomy, Galileo's observations from Italy at the start of the seventeenth century revealed many previously unseen stars in the Milky Way. Careful reasoning supported by this new evidence supplanted the reigning paradigm of a sphere of fixed stars centered on the earth with that of a universe of indefinitely large extent, where the earth is at a position of no particular importance. Thus, the constellations came to be seen as merely convenient direction indicators from a vantage point, not necessarily physical associations.

Further progress in understanding the distribution of matter in space was dependent on actual measurements of the distances to the stars. A direct approach through trigonometric parallax (changes in the apparent angular positions of nearby stars in relation to more distant stars as seen from Earth at various points in its orbit around the sun), became possible only when the precision of angular location surpassed the level of 1 second of arc. (Recall that there are 60 seconds in one minute, 60 minutes in one degree, and 90 degrees in the right angle between two perpendicular lines.) The first reports of success, by Friedrich Wilhelm Bessel in Germany, Thomas Henderson in South Africa, and Friedrich Georg Wilhelm von Struve in Russia, came in 1838 and 1839. By 1890, distances were known for nearly one hundred stars in the immediate neighborhood of the sun, but such observations were limited by having to look through the earth's atmosphere at visible light wavelengths. It took slightly more than a century for a variety of indirect distance indicators to be developed and calibrated before astronomers had the data from which to derive a substantially correct view of the size and shape of the Milky Way.

At the start of the twentieth century, most considered the Milky Way the observable universe. By the late 1920's, studies of Cepheid variable stars and observations of spectroscopic redshifts had established that the "spiral nebulas" were structures comparable to the Milky Way, far outside it and on average moving away at speeds proportional to their distances. Thus, the expanding universe once again was perceived to be immensely larger and richer in structure than had been believed only a short time before. Ever since, central goals of observational and theoretical cosmology have been to produce accurate maps of the distribution of these galaxies and explanations of the origin of the features in this distribution.

Bibliography

Bahcall, Neta A. "Large Scale Structure in the Universe Indicated by Galaxy Clusters." In *Annual Reviews of Astronomy and Astrophysics*. Vol. 26. Palo Alto, Calif.: Annual Reviews, 1988. An extended, somewhat technical, article with a formidable specialist bibliography, this paper is a masterful survey of evidence up to the late 1980's presented largely through an illuminating collection of maps, diagrams, and graphs.

Cohen, Nathan. *Gravity's Lens: Views of the New Cosmology.* New York: John Wiley & Sons, 1988. A book for the general reader by a researcher in general relativity and cosmology, this clear and well-illustrated volume features extensive discussion of the evidence for large-scale structure, and prospects for refined and more extensive observations in the future.

Ferington, Esther, et al. *The Cosmos.* Alexandria, Va.: Time-Life Books, 1988. This profusely illustrated large-format book presents a readable and surprisingly comprehensive brief introduction to the field of modern cosmology. The final third concentrates on the connection between high-energy ("elementary particle") physics and the very early universe as the unifying element in understanding the origins of large-scale structure.

Hartwick, F. D. A., and David Schade. "The Space Distribution of Quasars." In *Annual Reviews of Astronomy and Astrophysics*. Vol. 28. Palo Alto, Calif.: Annual Reviews, 1990. A rather technical article with much cited literature summarizing knowledge of the clustering of these most distant discrete objects yet observed. Includes a discussion of their importance in understanding the development of large-scale structure in the universe.

Peebles, P. J. E. *The Large-Scale Structure of the Universe*. Princeton, N.J.: Princeton University Press, 1980. This authoritative text, written at an advanced level, contains many accessible discussions of early evidence for and historical developments in the physical understanding of large-scale structure. Much larger and more complex structures have been found since this volume was published.

Trefil, James. *The Dark Side of the Universe*. New York: Charles Scribner's Sons, 1988. A lucid account of contemporary cosmology by a physicist for the general reader, this well-written volume concentrates on the evidence for dark matter in the universe, emphasizing its role in explaining large-scale structure.

John J. Dykla

Cross-References

The Big Bang, 210; Cosmology, 589; Types of Galaxies and Galactic Clusters, 928; General Relativity: An Overview, 950; Grand Unification Theories and Supersymmetry, 985; Gravitational Waves, 1000; Space-Time: An Overview, 2318.

VACUUM TUBES

Type of physical science: Classical physics
Field of study: Electromagnetism

Vacuum tubes, despite being replaced by transistors in many applications, are still in use today in high-power applications. With the recent success in developing vacuum tube microelectronic circuits, there is a need to study vacuum tubes in great detail.

Principal terms

EMITTER: also referred to as a cathode; in vacuum tubes, it is an electrode that emits electrons when heated, a phenomenon often referred to as thermionic emission

GRIDS: all electrodes other than the emitter and the plate in a vacuum tube; they are generally in the form of mesh, and depending on their functions, they are classified as control grid, screen grid, suppressor grid, and the like

MICROWAVE FREQUENCY SIGNALS: alternating voltages whose frequencies are in the range of 300 megahertz to 30 gigahertz

PLATE: also called an anode; in vacuum tubes, it collects electrons, emitted by the cathode as it is held at a voltage that is positive with respect to an emitter

RADIO FREQUENCY SIGNALS: alternating voltages whose frequencies are in the range of a few kilohertz to 30 megahertz

Overview

Vacuum tubes are devices in which electrons flow in a complete or partial vacuum, and the flow is controlled for a variety of radio, microwave, and optical applications. Historically, the oldest vacuum tube is a diode. It consists of a glass or metallic tube with two electrodes in complete vacuum. One of the two electrodes is called the emitter, or cathode, and consists of a tungsten wire or an oxide-coated plate, which releases electrons when heated to a temperature greater than 1,900 degrees Celsius by passing a suitable current either through it or through a filament insulated from it. These electrons then flow toward the other electrode, called a plate or an anode, when a positive voltage is applied between the plate and the emitter. The tube is completely sealed, but pins are available externally for connecting proper voltages to the plate and the emitter and for supplying filament current to heat the emitter. A diode is useful for converting alternating currents to direct currents.

To study the current-voltage characteristics of a diode, its emitter is connected to the negative terminal of a direct current supply (battery), and its plate is connected to the positive terminal of the supply through a resistance. When the plate supply voltage is low, current flow is also low, because most of the electrons released by the emitter stay close to the emitter and constitute a space charge cloud. Only a few electrons will be attracted as a result of the positive charge on the plate. As the positive voltage on the plate is increased, current in the circuit will also increase as more electrons are attracted toward the plate. Eventually, there will be a plate voltage called "saturation voltage" at which all electrons released from the cathode will be pulled toward the plate, and further increase in the plate voltage does not lead to any increase in plate current. On the other hand, if the plate is at a negative voltage with respect to the emitter, all the electrons released from the emitter will be repelled by the plate, and there will be no current flow.

A great advance took place when a metallic mesh was inserted between the emitter and the plate of a diode tube. This mesh is called a "control grid," and the new device is called a triode, indicating it has three electrodes. In general, a negative direct voltage is applied between the control grid and the emitter. The control grid repels the electrons moving toward the plate and is more efficient in doing so because it is closer to the emitter than the plate is. Yet, since it is only a mesh, many electrons will slip through the mesh and reach the plate. To study the current voltage characteristics of a triode, the emitter is connected to the positive terminal of a fixed-voltage battery and the negative terminal is connected to control the grid. The plate is connected to the positive terminal of another battery through a resistance and the negative terminal of this battery is connected to the emitter. Now, plate current (current in the circuit connecting the plate and the cathode) in the triode circuit is a function of the voltage between the plate and the emitter (the plate voltage) and voltage between the control grid and the emitter (the grid voltage). For a given control-grid voltage, the higher the plate voltage, the higher the plate current will be. As in the case of the diode circuit, there is a saturation current, which now depends on the control-grid voltage. The more negative the control-grid voltage is, the less the corresponding saturation current will be. Plate current versus plate voltage for different values of control-grid voltage are plotted on a graph for use in designing circuits. On the other hand, for a given plate voltage, the more negative the grid voltage is, the less the plate current will be. Therefore, another choice is to plot plate current versus control-grid voltage (negative) for different values of plate voltage.

A triode is more versatile than a diode. It can amplify small voltage signals applied between the control grid and the emitter. It can be used to design oscillators, wave-shaping circuits, modulators, and demodulators. The important characteristics of a triode are its plate resistance, amplification factor, and transconductance. The plate resistance of a vacuum tube is defined as the ratio of a small change in plate voltage to a corresponding small change in plate current and is, in general, on the order of 100,000 ohms. The amplification factor is given by the ratio of a change in plate voltage to a corresponding change in grid voltage to produce equal changes in plate current. It can range anywhere from 5 to 100 ohms or higher, and it is so important a design criterion that special physical constructions of tubes are adopted to achieve high, low, or medium amplification factors. The transconductance is defined as the ratio of change in plate current to a small change in the grid voltage. Its value may range anywhere from hundreds to thousands of micromhos. A micromho is a millionth of a mho (the reciprocal of the ohm). For a given tube, plate and control-grid voltages will also play a role in determining exact values of the amplification factor and the transconductance. The physical construction of the tube depends on the desired application. For example, when the tube is used in a receiver circuit, requirements are low power handling, large amplification factor, and least amount of distortion.

There are three capacitances in a triode. One capacitance is between the plate and the control grid, another is between the emitter and the grid, and the third capacitance is between the plate and the emitter. Of these three capacitances, at radio frequencies the capacitance between the grid and the plate feeds back energy from the plate to the grid, leading to oscillations and instability of circuits at times. Thus, a need arose to reduce this capacitance, which is achieved by putting another mesh electrode between the grid and the plate. This is called a "screen grid" and is maintained at a positive potential with respect to the emitter, but at a voltage less than the plate voltage. Such a device is called a tetrode and the device characteristics now depend on three voltages: those on the plate, the grid, and the screen grid. The screen grid is also a mesh and thus electrons can pass through it easily to the plate. A finite amount of current will flow in the circuit, connecting the emitter and the screen grid. A bypass capacitor, which offers a low impedance at the desired frequency of operation, is connected across the screen grid and the emitter to reduce an unwanted screen current.

At high plate voltages, electrons in the tube travel at very high speeds. When they strike the plate, electrons are emitted from the surface of the plate. These are called secondary electrons, and some of these electrons are collected by the screen grid in a tetrode, as it is at a positive voltage. This leads to a reverse current flow between the screen and the plate. Another grid in the form of mesh is inserted between the plate and the screen and is maintained at the potential of the cathode. This is called a "suppressor grid," which then shields the secondary electrons from reaching the screen. This five-electrode tube is called a pentode. Such a device has a very high plate resistance, as plate voltage now does not have much influence on current because of the presence of the screen grid. Nevertheless, the control grid has substantial influence on plate current, and this leads to the same level of transconductance as in an equivalent triode. A combination of both factors leads to a high amplification factor.

Most practical emitters in vacuum tubes are made of tungsten, thoriated tungsten, and oxide-coated materials. Tungsten is generally operated at 2,270 degrees Celsius to give maximum life for the filament. Thoriated tungsten gives good electron emission at 1,900 degrees Celsius. The oxide-coated emitters consist of a mixture of barium and strontium, coated on nickel or a nickel alloy and are typically operated at 900 degrees Celsius. The thorium layer on the surface of the thoriated tungsten emitter gets stripped off as a result of bombardment by positive ions from any residual gas in the tube. Oxide-coated emitters are most efficient in that they give maximum emission per watt of heating power, but they deteriorate more rapidly. They can give high instantaneous emission for a few microseconds, which is used for generating pulses for radar. They are generally used in small receiver tubes. For power tubes operated at extremely high voltages, tungsten or thoriated tungsten is preferable. In all cases, the cathode can be heated either directly when the filament current passes through the cathode, or indirectly when an insulated tungsten filament heater inside a thin nickel cylinder cathode with an external oxide coating heats it.

Major difficulties with vacuum tubes are the deteriorating effects of residual gas in the tube, transit time effects caused by finite electron velocities and capacitances between the electrodes. The residual gas gets ionized in the presence of heat and the electric field. The resulting positive ions may strike the emitting surface and then either mechanically or chemically degrade it. So, operating the tube at high voltages reduces the lifetime of the device. Note, however, that the tungsten filament cathode is immune to the gas effects. To reduce the gas effects, substances such as barium or magnesium are introduced into the tube. These materials, called getters, help in obtaining a high vacuum initially and combine with any gas that is released by the cathode materials or left as residue.

Electrons take a finite amount of time to travel from the cathode to the plate. At low frequencies up to a few megahertz, transit time of the electrons in a vacuum tube under normal operating voltages is small compared with the time period of the applied frequency. This time period is defined as one divided by the frequency in hertz. At low frequencies, electron flow, which constitutes the plate current, will be able to follow the fluctuations of the applied alternating voltage on the control grid or the plate. Yet, at very high frequencies, the applied voltages will change their magnitudes too rapidly and the current caused by electron flow will not be able to follow the voltages. This leads to distortion and reduced amplification of the signal. Further degradation of the basic characteristics of the vacuum tubes results if applied signals are large in magnitude. Some of these effects can be reduced by using closely spaced small electrodes, which reduce capacitance and leads with reduced inductance.

The failure of triodes, tetrodes, and pentodes at high frequencies caused by transit time effects led to alternate strategies, such as to making use of transit time in achieving high power and high-frequency operation. These are the high-power microwave tubes, such as klystron, magnetron, traveling wave tube, and backward wave oscillator. For example, in a traveling wave tube, a radio-frequency voltage to be amplified is applied near the cathode end of a long and loose metallic helix and guided by it. An electron beam is shot from the cathode end along the axis of the helix and collected by the anode at the other end of the helix. Under suitable conditions, an amplified signal of the applied microwave voltage appears at the other end of the helix.

There are other types of tubes, such as television picture tubes and photomultiplier tubes. The photomultiplier tube consists of a photocathode, several dynodes, and an anode. The photocathode, which is made up of a material such as gallium arsenide or indium phosphide, generates electrons when light falls on it. These electrons are accelerated by the voltage on a secondary electrode so that when the electrons strike it, more electrons are released by it. By using several secondary electrodes, good amplification of electrons can be achieved. These are collected by an anode to produce current in the plate circuit. A photomultiplier tube is capable of detecting extremely small optical signals. In television and cathode-ray tubes, the screen consists of phosphor materials such as zinc orthosilicate, which emits light when electrons strike it.

Applications

With the advent of vacuum tubes, radio engineering was born. A host of circuits for generation, amplification, and detection of radio waves were designed, and these gave birth to radio, television, radar, and modern communication engineering.

A simple diode circuit can be used as a detector, a rectifier, or a mixer. In a simple diode rectifier circuit, an emitter is connected to an alternating sinusoidal signal, and the other terminal of the source is connected to the plate along with a series resistance. In this configuration, the voltage of the plate with respect to the emitter is changing sinusoidally. For half of the sinusoidal cycle, the voltage on the plate is positive with respect to the cathode, and there will be a current flow in the circuit. There will be no current flow for the other half of the cycle, as the plate is negative with respect to the emitter for that period. This is called the half-wave rectifier, and the resulting voltage across the resistance contains a direct voltage, an alternate voltage of the same frequency as that of the applied voltage and harmonics, which are alternating voltages at frequencies that are integer multiples of the frequency of original applied voltage. There are other variations of this simple rectifier circuit, such as a full-wave rectifier, which is more efficient in converting alternating currents to direct currents and to currents at higher harmonics. Thus, in a radio receiver, a simple diode can be used as a detector to recover the applied signal. Also, the rectification property can be used to detect and demodulate the transmitted signal.

When diodes are used for detection, the resulting detected signal is very low. Yet, a triode can amplify low power signals for clarity and further uses. It is found that when an alternating sinusoidal signal is applied between the control grid and the emitter, a sinusoidal signal of the same frequency at a much larger voltage level is observed across resistance, connected between the emitter and the plate. This results from the fact that the control grid is more efficient in repelling electrons from the emitter than the plate is in collecting the electrons. Thus, the signal applied between the control grid and the emitter has been amplified and appears across the plate and the emitter. This led to great advances in radio engineering, as very small signals can now be amplified and studied in detail. Other tubes are merely improvements of triodes to overcome some of the defects in their performance. The often-used design criteria for amplifiers are gain, bandwidth, center frequency, power, and distortion.

By proper choice of bias supply and alternating voltage applied to the control grid, it is possible to control the amount of time the plate current is nonzero. This control can be used for linear amplification, power amplification, and harmonic generation. Several vacuum tube amplifier circuits were described in *Electronic and Radio Engineering* (1948) by Frederick Emmons Terman.

When a part of the output signal (voltage between the plate and the cathode) of an amplifier is fed back to the input (voltage between the control grid and the cathode) with proper circuit design, it leads to oscillations. In actual practice, no input signal is needed. Depending on the actual details of feedback network and device, the oscillator circuit chooses a frequency from noise power of the device and keeps on amplifying until it reaches stable oscillations. Thus, oscillators, which can generate stable output voltages, are designed using feedback techniques.

When a sum of more than two alternating signals of different frequencies are used as a plate voltage, the rectification property of the diode generates frequencies that are the sum and difference of integer multiples of these frequencies. This property is called mixing, and it has several useful applications in instrument design, such as spectrum analyzers and intermediate frequency generation. Triode circuits can also be used for mixer operation. Special-purpose vacuum tubes, such as hexodes and octodes, are efficient mixers.

In some communication systems, amplitude or frequency of a sinusoidal wave, called a carrier, is varied according to the signal being transmitted; this is called modulation. For modulation, a carrier signal in series with the modulating voltage and a fixed bias is connected between the control grid to the cathode. This will produce a modulating plate current that carries information around both the carrier wave and the modulating voltage. By mixing it with a local oscillator at the carrier frequency, the modulating voltage can be reconstructed; this is called demodulation.

Digital vacuum-tube circuits played a major role in the development of the first large-scale computers. Also, photomultiplier tubes continue to be used for amplifying extremely small optical signals. Microwave tubes, such as magnetron and traveling wave tubes, are still in use today for generating large amounts of high power. Cathode-ray tubes, and its variation picture tube, continue to be used in television receivers and electronic instruments.

Since 1985, efforts have been under way to develop miniaturized vacuum-tube electronic circuits, using the concept of field emission, whereby electrons can be pulled from the tip of a metal by applying a strong electrical field. Using several such tips together, called field emitter arrays, engineers are developing large-scale integrated circuits. One of the motivations behind this goal is that the speed with which electrons can move in a transistor is limited by properties of semiconductor materials. No such limits exist for electron flow in a vacuum, and space-charge problems can be circumvented by proper arrangement of cathodes. Also, these devices can be switched very rapidly. Several prototype field emitter arrays and microminiature integrated circuits are being reported. This may revolutionize the design of flat-screen television sets, superfast computers, optically controlled computers, and communication equipment, besides several military applications such as electronic counter measures and light-weight radars.

Context

Sir William Crookes was the first scientist to study the flow of charges from one electrode to another electrode through a vacuum. This led to discoveries of X rays by Wilhelm Conrad Röntgen and electrons by Sir Joseph John Thomson. The Braun tube, introduced by Karl Ferdinard Braun in 1897, was the earliest vacuum tube and later evolved into the cathode-ray tube used in oscilloscopes. In 1883, the American inventor Thomas Alva Edison noticed an unexplained current flow between the two filaments of his electric lamp under certain conditions. This current flow was in a direction opposite to that of the lamp current. This phenomenon was studied and identified as a thermionic emission. Sir John Ambrose Fleming got the idea of using the Edison effect to study radio waves and develop the vacuum-tube diode for detecting radio waves in 1904. It detected radio waves very well, but the resulting signals were weak. In 1907, an American engineer Lee de Forest and a little later, van der Bijl discovered the triode. This was a singularly important contribution to the entire development of radio engineering. The triode amplified extremely weak signals by converting the bias direct current power into alternating current power of the signal. Walter Hans Schottky invented the tetrode in 1919, and Jobst and B. D. H. Tellegen designed the pentode in 1926 to improve the performance of the triode. The iconoscope, which converted an optical signal into an electrical signal, was invented by Vladimir Zworykin in the 1920's. In later years, the iconoscope was to become the television tube.

The study of electron emission from various materials was also under way simultaneously. Heinrich Rudolph Hertz discovered the photoelectric effect in 1887. Studies on oxide emission began in 1903 and culminated in the famous thermionic emission equation given by Sir Owen Willians Richardson and S. Dushman in 1923. Schottky introduced electron emission under intense fields in 1918. These resulted in good cathodes for vacuum tubes. Also, a theoretical understanding of noise in vacuum tubes such as Schottky noise, flicker noise, and thermal noise was achieved.

World War II gave great impetus to electronic and radio engineering. Many reliable and excellent electronic circuits for a vast variety of applications were designed during this time. The war needs such as radar, electronic guidance for weapons, aerospace and naval communication equipment, and general communication equipment required the use of microwave frequencies. At microwave frequencies, the conventional vacuum tubes, such as the triode and the pentode, did not work well, and this led to the development of klystrons, magnetrons, and traveling wave tubes.

When the transistor was invented in the 1950's, the use of vacuum tubes in electronic circuitry rapidly diminished as the transistor was much smaller in size, lighter in weight, and required less power consumption. It also led to miniaturization and better integration of large electronic circuits. No cooling circuits were needed. Yet the transistor could not deliver high powers. Therefore, vacuum tubes are still being used today for high-power applications, such as radar, microwave ovens, and broadcasting stations.

It is fair to say that almost all major applications of electronics and principles of design were first conceived in the vacuum-tube era and later reused for transistor technology. Even the first all-electronics computer was designed using vacuum tubes.

Efforts to develop vacuum microelectronic circuits, which use the field-emission concept, are meeting with success. The future study of vacuum tubes will be directly related to the physics and engineering of vacuum microelectronics. Some of the topics that will be pursued are switching speeds of the devices, their ability to operate at high and variable temperatures, field configurations at such short distances, resistance to radiation, and noise mechanisms.

Bibliography

DeMaw, Doug, ed. *The Radio Amateur's Handbook.* Newington, Conn.: American Radio Relay League, 1970. This is the standard manual of amateur radio communication. Provides information on radio engineering. The material can be used for practical training and also as a quick reference. The treatment of several radio topics is essentially nonmathematical.

Fink, Donald G., ed. *Electronics Engineers' Handbook.* New York: McGraw-Hill, 1975. This book covers a variety of topics in electronics engineering. Of particular interest are chapters on amplifiers, oscillators, modulators, demodulators, and frequency converters. Also, the chapter on ultrahigh-frequency and microwave devices is valuable. The mathematical treatment is easy to follow. Examines vacuum tube technology with reference to other aspects of electronics in detail.

Millman, Jacob, and Christos C. Halkias. *Electronic Devices and Circuits.* New York: McGraw-Hill, 1967. This can be used as a first course in electronics for serious students. Millman and Halkias provide extensive physics-based discussion for each topic and also provide a good mathematical treatment. A popular book for undergraduates in physics and engineering. Many chapters are relevant.

Port, Otis. "Taming Space and Time to Make Tomorrow's Chips." *Business Week*, March 13, 1989, 68-74. This is a popular article on vacuum microelectronics and quantum chips with regard to the semiconductor industry and is written in a style that any layperson can understand. Covers the current state of work at various research laboratories in the United States and discusses the implications for semiconductor industry in the future.

Spangenberg, Karl R. *Vacuum Tubes.* New York: McGraw-Hill, 1948. Spangenberg drew upon his extensive experience as a teacher at Stanford University to provide a comprehensive treatment of vacuum tube physics and engineering. It is perhaps the most referenced book on vacuum tubes. There are extensive discussions, which cannot be found elsewhere, on topics such as electron emission, determination of fields in the tubes, space-charge effects, noise in the tubes, and special tubes such as octodes, heptodes, and hexodes.

Terman, Frederick Emmons. *Electronic and Radio Engineering.* New York: McGraw-

Hill, 1948. An important book on vacuum-tube techniques that emphasizes circuit design. Written by one of the influential figures in electronics education and industry, it contains extensive information on designing vacuum-tube circuits. Easily understandable and gives many principles and techniques of radio engineering for students and practicing engineers. The mathematical treatment of the topics is kept to a minimum.

V. S. Rao Gudimetla

Cross-References

Electrical Circuits, 706; Electrical Test Equipment, 715; Generating and Detecting Electromagnetic Waves, 743; Electron Emission from Surfaces, 764; Radar, 2020; Radio and Television, 2042; Rectifiers, 2086; Transistors, 2572.

VARIATIONAL CALCULUS

Type of physical science: Mathematical methods
Field of study: Calculus

The calculus of variations addresses the problem of finding extreme values of expressions called functionals. It is an extension of the differential calculus, which seeks the extreme values of functions. Many problems in classical mechanics, including the motion of pendulums and the shape of soap bubbles, can be considered by means of the variational calculus, as can problems in electromagnetic theory, quantum mechanics, and relativity.

Principal terms

ADMISSIBLE FUNCTION: a curve that meets certain requirements, such as being unbroken or having no corners

BRACHISTOCHRONE PROBLEM: the problem of finding the path that minimizes the time of descent for a particle under the influence of gravity alone

CONSTRAINT: an external condition placed on a solution curve

DERIVATIVE: the rate of change of a function, or of one variable with respect to another

EULER-LAGRANGE EQUATION: an equation that places a necessary restriction on the solution of a variational calculus problem

FUNCTION: a relationship between two sets of numbers, graphically represented as a curve or surface

FUNCTIONAL: a relationship between a function and a real number, such as the length of its graph or the difference between its starting and its ending value

VARIATION: an approximate measure of the change in the value of a functional when the underlying function is changed

Overview

The calculus of variations is an extension of the ideas of the differential calculus. While the differential calculus is primarily concerned with the determination of extreme values (either maxima or minima) of functions, the variational calculus addresses the question of finding the extreme values of functionals. In its simplest form, a function can be viewed as a set of points that can be represented graphically as a curve or a surface. The words "function" and "curve" are here used interchangeably, although there are technical differences between the two. Similarly, a functional, in its simplest form, is an association between a function and a real number. For example, if one considers two points, A and B, in the plane, one can draw a continuous, unbroken curve that connects them. This curve is the graph of a function. A measure of the length of this curve is a functional; that is, associated with any such curve is a real number that represents its length.

It is reasonable to ask which, of all possible curves joining A and B, has the shortest length. The obvious answer is the straight line segment joining the two points, but a change in the functional under scrutiny may give no hint about the required optimal value. An example is to find the path that a particle would follow from point A to point B so that the elapsed time of descent is minimal when the only external force is gravity. The calculus of variations provides a framework for solving problems such as these.

The basic tool of the differential calculus is the derivative, which gives a measure of the rate of change of a function. The extreme values of a function can be determined by examining these points where the function either is not changing (the derivative is zero there) or is undergoing a possible abrupt change (the derivative does not exist there). A fundamental result in the calculus says that if a function has an extreme value, then it must occur at these critical values. Critical values where the derivative is zero are called stationary points, and the terminology originates with the idea that if the function represents distance as a function of time, then the derivative is a velocity (rate of change of distance with respect to time). When the velocity is zero, the particle is motionless, or stationary. An example is a ball thrown vertically, which has zero velocity at the peak of its trajectory. More complicated trajectories require more sophisticated analysis, but the candidates for extreme values are always obtained by means of the derivative.

The calculus of variations has an analogue of the derivative that is called the first variation of the functional. The concept is based upon the notion that although the explicit evaluation of the functional for many, if not all, of the admissible functions may not be possible, a comparison between the optimal solution and others can be made. An attempt to minimize the difference or variation between the possible candidate functions and the optimal solution can be made, and the ultimate goal is to make this difference zero. This optimal candidate function is said to be a stationary value for the functional, in analogy with calculus. An ingenious use of a technique from the integral calculus, called integration by parts, is then employed to shift the emphasis from evaluating the functional to obtaining a necessary condition for this variation to be zero. This latter condition is the Euler-Lagrange differential equation, which, when solved, identifies a candidate for the optimal solution for the functional under study. Solving the Euler-Lagrange equation is usually very difficult by itself, but there are at least three factors that complicate the situation.

The first of these is the variety of external conditions that must be satisfied. These usually arise not so much from the mathematics of the problem as from physical restrictions. These are sometimes called constraints, and they can be as simple as a specification of where the endpoints must be or as complicated as allowing the endpoints to move about freely. Other constraints may even involve a limitation on the solution that itself takes the form of a functional.

A second factor is the identification of which type of functions are admissible candidates for the solution. If one places *a priori* requirements that are too severe on the candidates, solutions may not even exist. For example, the admissible functions must always be continuous (unbroken) curves, but it may be desirable to have solution curves that have no corners. A case in point is a hanging rope suspended from two points, which physically must be a smooth curve. A restriction like this is equivalent to prohibiting abrupt changes in the velocity of a particle. Higher dimensions, complicated geometries, and systems of functionals pose other problems.

The third factor—usually the most difficult problem—is showing that the candidate yields an optimal value for the functional under study. The analogue here is that of using the second derivative test in the calculus to distinguish maximum and minimum values. In the calculus of variations, there are tests that can identify weak extrema (simply continuous functions) and strong extrema (smooth functions). These are quite sophisticated, and research on this and other problems in the calculus of variations continues.

Applications

The variational calculus provides a powerful and elegant framework for analyzing many of the physical problems in optics, electromagnetic theory, and mechanics, both classical and modern. The foundation for these applications is Pierre de Fermat's principle that nature behaves in ways that are "easiest and fastest." Easiest implies that a minimum amount of energy is expended, and fastest indicates a minimal time problem. A few examples of how the variational calculus is used to explain physical phenomena follow.

The simplest minimal distance problem is to find the path that joins two points A and B in the plane so that the total length of the path is as small as possible. The arc length of a curve can be represented as an integral (essentially, a sum of infinitely many small quantities) of an expression involving square roots and the derivative of the curve. For plane curves, the minimal path is a straight line. If the points are on some surface, the minimal paths are called geodesics. If the surface is a sphere, the arcs of minimal length are great circle routes. As expected, the functional representing the distance in these latter cases will be more complicated, to accommodate the geometry of the surface.

An extension of this problem is obtained if one considers the rotation of a curve around some fixed line and seeks the curve that minimizes the surface area of the resulting solid of revolution. The shape of a soap bubble and the shape of a red blood cell can be ascertained from the solution of such minimal surface problems.

Other applications arise when the effects of external physical forces, such as gravity, are included in specifying the functional. Examples here include the minimum time problem for a particle in the plane descending from point A to point B under the influence of gravity alone. This is the famous brachistochrone problem, and the solution curve is part of a cycloid. Other examples include the path that a ray of light follows through media with different indices of refraction. The crucial observation here is that the quantity that is minimized is not the total length of the path, but the time of transit. As a consequence, Snell's law for refraction of light can be obtained.

Further examples in physics can be found when the functional represents the energy of a system. A hanging, flexible chain, suspended from two points A and B in the plane influenced by gravity alone, takes the shape of a catenary. The quantity minimized here is the potential energy of the chain, or equivalently, the center of gravity is found at the lowest possible point. It is also possible to ascertain motions when external requirements of the system are given in advance. For example, if a pendulum is required to swing in an arc so that the elapsed time of each cycle is the same, then one has the solution to the tautochrone problem, which was actually used to design many large-amplitude pendulum clocks.

All the previous examples illustrate the elegance of explaining physical phenomena by means of minimizing an important parameter such as distance, time, or energy. One of the crowning achievements of mathematical physics is a much more sophisticated application of the ideas discussed above. Let the functional take the form of an integral (or summing up) of an expression called the action of the system. The action is defined to be the difference between the kinetic and the potential energy, and it is usually referred to as the Lagrangian function for the system. Fermat's idea of "easicst and fastest" becomes the problem of minimizing this action integral. In order to obtain such a minimum, the Euler-Lagrange differential equation must hold, and this dictates the form of the Lagrangian, which identifies the motion of the system. The resulting Euler-Lagrange equation is referred to as Lagrange's equation of motion.

In the absence of any external constraints, Lagrange's equations reduce to Sir Isaac Newton's laws of motion, and the analysis by the action integral does not present a new theory, since each set of equations can be derived from the other. It is in the presence of external constraints and in complicated geometries that the action integral method proves to be superior. A significant impediment occurs if frictional forces are present, since these are not accounted for in the action integral method. Problems in physics (but not engineering) adapt easily to this approach.

In many problems in physics, it is advantageous to change the coordinate system to suit the geometry of the problem. For example, a change from rectangular to polar or cylindrical coordinates when studying fluid flow in a circular pipe is desirable, and a change to spherical coordinates in gravitational, electromagnetic, and atomic physics simplifies the mathematics. Each problem generally has a natural setting in which the mathematics is most transparent, and the preferred coordinate system is called the set of canonical variables for the system. The surprising feature is that spatial coordinates are not always the best choice for describing the motion of a system. Sometimes it is more reasonable to use "generalized" coordinates, such as the components of momentum (which is a product of the mass times a velocity component). While Newton's equations of motion depend on the choice of the coordinate system, the value of the Lagrangian, which is found from the action integral, is the same in all coordinate systems. In fact, the form of the Euler-Lagrange equation remains coordinate-free. This is of crucial importance in quantum mechanics and in relativity, where the usual rectangular coordinates do not even exist.

If the generalized coordinates include those of momentum, it is possible to define a new function, the Hamiltonian, which is a combination of the Lagrangian and these coordinates and which represents the total energy of the system. When considering the Euler-Lagrange equation in minimizing the action integral in this form, it can be shown that certain expressions must remain constant. These are mathematical results, yet the quantities that remain fixed are the physical quantities of energy, momentum, and angular momentum. Thus, the conservation laws that form the basis for much of Newtonian mechanics follow logically from the calculus of variations. In fact, there are many other conserved quantities that do not have direct physical meaning.

A stable equilibrium point occurs for a physical system when the potential energy is minimized, and the variational calculus provides a mechanism for identifying motions near such equilibria. The motions of vibrating strings, rods, and membranes, as well as buckling beams with the subsidiary conditions of how the ends of edges are fixed, are suited to analysis by the calculus of variations. The ability to incorporate conservation laws, external constraints originating from the geometry, and energy restrictions allows one to formulate and solve problems in elasticity and electromagnetism in a natural way. Relativity problems can be discussed by making appropriate changes in the masses when velocities approach the speed of light. Using complex variables and probability theory, it is possible to minimize an energy functional in the sense of quantum mechanics and obtain the Schrödinger equation for wave mechanics.

Finally, Newton's laws of motion can be derived from the Lagrangian and Hamiltonian functions with the calculus of variations, and vice versa. The latter are more general, since the coordinates need not be lengths, and they are much better adapted to complicated systems, even in the presence of external constraints.

Context

The earliest appearance of a problem in the variational calculus was the isoperimetric problem of Dido in Greek mythology, but the first mathematical treatment was most likely by Galileo in 1638, when he conjectured (incorrectly) the solution of the brachistochrone problem. Pierre de Fermat in 1662 made the first attempt to use the calculus to solve a problem in optics, stating the principle of least action. Later, in 1696, Johann Bernoulli challenged others to solve the brachistochrone problem, not mentioning that he had already found a solution that used a variational approach. Solutions were submitted by Newton, Gottfried Wilhelm Leibniz, Johann's brother Jakob Bernoulli, and Guillaume F. A. L'Hôpital. All except L'Hôpital gave the correct solution as a cycloid.

It was the Bernoullis' solutions, however, that essentially initiated the subject, giving a general method for solving problems of this type. Prior to this, questions had always focused on finding the equations of motion for the trajectory of some object, such as the earth around the sun or the flight of a cannonball. Once the unique trajectory was found, questions concerning maximum height, maximum range, or total elapsed time of transit were considered. The difference now was in the nature of the problem. The task was to choose, from the infinitely many possible trajectories, the one that minimized (or maximized) a quantity such as the time of travel, the length of the path, or the total energy used.

Leonhard Euler, a student of Johann Bernoulli, realized the mathematical significance of a problem of this type and, in 1744, gave a necessary condition that a solution must satisfy. Later in 1775, Joseph Lagrange introduced the concept of "variations," and Euler deferred to Lagrange's more elegant approach and named the subject the calculus of variations. Lagrange, however, saw more than the mathematical importance of this problem; he realized its physical significance as well. His monumental work, *Mécanique analytique*, introduced in 1788 the notion of generalized coordinates and exploited the concept that nature acts in such a way that the action of a system is minimized.

In 1835, William R. Hamilton extended Lagrange's ideas even further and developed Hamiltonian mechanics. By minimizing, or at least making stationary, an expression for the total energy, he derived the equations of motion for complicated systems. Both Lagrange and Hamilton made significant contributions because of their abilities to see intimate interconnections between theoretical physics and the mathematics in the variational calculus.

In the late 1800's, Karl Weierstrass gave conditions that guaranteed an extreme value for functionals, which gives a structure that is somewhat parallel to the differential calculus. In the twentieth century, David Hilbert essentially completed the analytical structure of the variational calculus and opened the way for more extensions using geometric approaches. One direction for such extension was by L. S. Pontryagin and his coworkers, who let the constraints take the form of differential equations. This area is known as control theory, and Pontryagin gave constructive methods for finding optimal solutions.

Bibliography

Bliss, Gilbert Ames. *Calculus of Variations.* The Carus Mathematical Monographs. Chicago: Mathematical Association of America, 1925. A readable account of the variational calculus with emphasis on minimal distance, time, and area. The discussions of these problems and extensions are accessible to the layperson.

Courant, Richard, and Herbert Robbins. *What Is Mathematics?* London: Oxford University Press, 1967. Chapter 7 is a particularly well-written account of the minimal-surface and soap-bubble problems, along with other extreme problems in mathematics.

Gelfand, I. M., and S. V. Fomin. *Calculus of Variations.* Englewood Cliffs, N.J.: Prentice-Hall, 1963. An introductory textbook on the subject with an explanation of problems along with physical motivations concerning Lagrangian mechanics.

Goldstine, Herman H. *A History of the Calculus of Variations from the Seventeenth Through the Nineteenth Century.* New York: Springer-Verlag, 1980. The definitive account of the subject from its earliest conception to the twentieth century. Original discussions, interconnections, and historical anecdotes make this the best source for the subject. Extensive bibliography.

Struik, D. J. *A Source Book in Mathematics, 1200-1800.* Princeton, N.J.: Princeton University Press, 1986. Originals and translations of the earliest problems of the subject and fascinating footnotes that contain excellent references. (See, especially, pages 391f.).

Eugene P. Schlereth

Cross-References

Calculus: An Overview, 281; Conservation Laws, 546; Differential Calculus, 650; Differential Equations, 658; Integral Calculus, 1146; Newton's Laws, 1509; Nonrelativistic Quantum Mechanics, 1564; Schrödinger's Equation, 2156; Trajectories, 2559.

VECTORS

Type of physical science: Mathematical methods
Field of study: Algebra

A vector is a quantity that has both magnitude and direction and adheres to the basic rules of algebra with respect to the operations of addition, subtraction, and multiplication. This quantity can be used to represent the displacement of a point through time and space.

Principal terms

COORDINATE SYSTEM: a grouping of magnitudes that are used to define the positions of points in space

INITIAL POINT: the beginning point of a vector

ORIGIN: the reference point at which a coordinate system is centered

SCALAR: a quantity that has magnitude but not a direction, such as the nondimensional measurements of time or temperature

TERMINAL POINT: the final or end point of a vector; it is the end at which the head or arrow of the vector is affixed

VELOCITY: the rate of change of the position of a body, measured in terms of distance per unit time (for example, meters per second)

Overview

A vector is defined as a quantity that has both magnitude and direction. Vectors are used to describe the position and motion of a body in space. An abstract concept (for example, an airplane flying from Chicago to San Francisco, an automobile driving on a highway, a person walking down the street) possesses some amount of velocity which is pointed in a specific direction. In other words, each object has a vector associated with it that can define their respective positions and motion. Sometimes the object (which can be thought of as a point) is large, such as an airplane; perhaps it is small, such as a person walking, or perhaps it is infinitesimally small, as in the case of an electron moving around the nucleus of an atom.

The magnitude of the motion or displacement in each of the previous examples is described in terms of their respective velocities and directions, which can be measured as an angle with respect to an arbitrary origin. This origin is part of a coordinate system that is used to describe the position of a point in space. Many different types of coordinate systems exist; some have very special purposes in solving advanced mathematical problems. Vectors can exist in any one of these different systems, and also in any dimension (from one-dimensional to n-dimensional, the latter being one with limitless coordinates). Although vectors can exist in any number of dimensions, they are usually discussed, and most clearly visualized, in two or three dimensions. Each of the earlier examples is operating in an assumed level plane, so these are thought of as lying in a two-dimensional plane. Of these examples, the airplane is capable of moving literally into a third dimension, which it does when it is climbing or descending along its flight path.

Coordinate systems allow one to establish a one-to-one correspondence between points in a plane or surface and a series of numbers that actually define the positions of the points. The coordinate system most commonly used is the Cartesian coordinate system. This system, which was devised by the French mathematician René Descartes (1596-1650), provides a means of referencing points in either two or three dimensions. This discussion will be limited to Cartesian coordinate systems and to two or three dimensions.

Suppose a person is in a room that has a linoleum tile floor along with four walls and a ceiling, all being perpendicular to one another. The walls and floor define a coordinate system, which is thought of as being orthogonal—that is, one in which all axes are mutually perpendicular to one another. If the person is standing in one corner of the room facing the openness of the room, the corner at the person's feet can be defined as the origin. The two lines along the floor where the walls adjacent to the person intersect the floor form the abscissa (or x-direction, which is to the right) and ordinate (or y-direction, which is to the left). Thus, one can use the x-axis and y-axis to measure distances from the origin. This relationship of having the x-axis to the right of the y-axis establishes a right-handed coordinate system (the more common one).

In measuring distances on the floor, let each linoleum tile represent one unit of distance. If there exists a point on the floor (two-dimensional space) which is three units away from the origin in the x-direction and four units away from the origin in the y-direction, then this point can be represented in terms of the point (x, y), in this case (3, 4). A vector describing this point would extend from the origin (0, 0) to the point (3, 4). The length, or magnitude, of the vector is found by using the Pythagorean theorem; hence, the magnitude is the length of the hypotenuse of the triangle whose sides are 3 and 4 units each in length. The magnitude of the vector is thus 5 units. The direction of the vector is determined by finding the angle between the vector and the x-axis. For this vector, the angle is 53 degrees (arcsine of $4/5$).

If the coordinate system is expanded to include the vertical direction by using the two walls, the number of dimensions is increased from two to three. In doing so, another axis will need to be incorporated, that of the vertical (z-axis). By doing so, a point in space can now be represented, such as defining the position of a light bulb suspended from the ceiling. The vector describing the position of the bulb has an initial point (in this case, located at the origin) and a terminus point (located at the bulb).

For a three-dimensional case, points in space can be described by using three numbers (x, y, z), which then gives the position of the point from the origin point. Consider a room that is 5 meters in length, 5 meters in width, and 3.5 meters in height. If a light bulb were suspended in the center of the room 0.5 meter below the ceiling, then the coordinates of the bulb would be 2.5, 2.5, and 3, where the unit length along each axis is in terms of meters.

If the light bulb has to be replaced, prior to doing so, it is placed in the corner; this is defined as the origin. The light bulb is then moved in a straight line to the fixture suspended from the ceiling. This line defining the movement (or displacement) of the bulb has now produced a three-dimensional vector, because the bulb moved at a particular velocity and in a specific direction. If this movement were photographed using time-lapse photography and some type of strobe light, one could actually see the vector in space. (This record of the motion of an object in space can also be seen by following the condensation trail left by a jet plane when it flies at a high altitude.)

Vectors can be compared to one another in terms of their magnitudes and directions. Two vectors are termed "equal" if both their magnitudes and directions are equal; this is true irrespective of the positions of their initial points. Also, one vector can be the exact opposite of another (sometimes referred to as an antivector) if the magnitudes of the two vectors are equal but their respective directions are exactly opposite (that is, the vectors are parallel to one another but point in opposite directions).

The operations of addition, subtraction, and multiplication can be extended to develop an algebra for vectors. These operations are necessary in order to aid in the description of how objects (or points) respond to motion in space. The end result of applying a series of forces to the object can be determined by an additive process. After the second force is applied following the first one, the object will come to rest (its new terminus point) in a position found by adding the two vectors.

The addition of two vectors produces their sum or resultant. In figure 1a, vector **A** has an initial point *P* and a terminus *Q*. If vector **B**, which has *Q* as its initial point and terminus *R*, is added to **A**, then the resultant vector is vector **C**, which has *P* as its initial point and *R* as its terminus. The sum is written as $\mathbf{C} = \mathbf{A} + \mathbf{B}$. Notice that the resultant **C** is the main diagonal of a parallelogram having vectors **A** and **B** as sides. This summing process can be applied to more than two vectors. The procedure simply involves placing the initial point of one vector at the terminus of another until all the vectors are connected together.

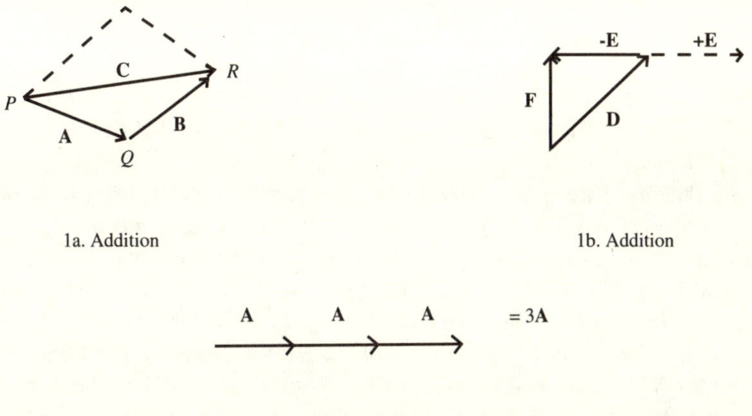

1a. Addition

1b. Addition

1c. Addition

Figure 1. Basic Algebra of Vectors

The subtraction of vectors is performed in a similar manner. The difference of vectors **D** and **E** is written as **D** − **E**. This is the same as adding the negative of **E** to **D** [**D** + (−**E**)]. Therefore, the resultant **F** is the one which, when added to **E**, produces **D** (see figure 1b). Notice that if two vectors are equal, then their difference produces the null, or zero, vector. This has a magnitude of 0 but no direction is attached to it.

Vectors can be multiplied by a scalar, thus changing the value of the original vector. If a scalar *c* is multiplied times **A**, then the result is *c***A**. If *c* is positive, then the magnitude of **A** is increased *c* times in the same direction as **A** (see figure 1c). If *c* is negative, then the direction of the product *c***A** is opposite that of **A**. Should *c* be zero, then the product *c***A** produces the null vector.

Unit vectors have two primary properties. The first property is that a unit vector is a dimensionless value defined as having a length of unity. The second property is that a unit vector is parallel to (actually coincident with) one of the coordinate axes (see figure 2). Thus, for a three-dimensional Cartesian coordinate system, there are three unit vectors, each parallel to one of the three axes, respectively. Standard nomenclature is used to represent the unit vector in the *x*-direction as **i**. The unit vector in the *y*-direction is **j**, while the unit vector in the *z*-direction is **k**.

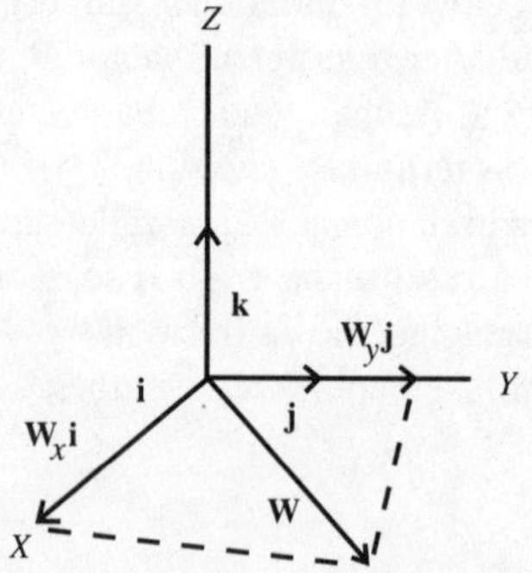

Figure 2. Resolution of a Vector (W) in the *X-Y* Plane

Vectors that have more than one dimension can be broken down into various parts. This process is termed "resolving a vector into its components." For a two-dimensional vector in the *x*-*y* plane, the components of the vector can be determined in the *x* and *y* directions. This is done by projecting the vector onto each of the two axes. In actuality, this procedure can work for any *n*-dimensional vector in any coordinate system. Therefore, *n* components of the vector would be produced. In figure 2, vector W has been resolved into two components: $\mathbf{W}_x\mathbf{i}$ is the *x*-direction component and $\mathbf{W}_y\mathbf{j}$ is the component in the *y*-direction. Note that for two-dimensional vectors, no reference is made to the third direction. Hence, a two-dimensional vector in the *x*-*y* plane has no *z* component.

Applications

In the physical world, vectors can be used to represent such quantities as velocity, displacement, force, acceleration, and magnetic or gravity fields, all of which have a magnitude and direction associated with them. In physics, if an object is affected by a force, then that object might move in response to the force. This motion can be described in terms of distance and rate. Thus, a vector can be produced that would characterize the motion and describe the new position of the object.

Vectors are used in determining the paths of rockets when they are launched to place satellites into space. The velocity and acceleration of a rocket is expressed in terms of vectors, as well as the gravitational forces acting on the rocket to pull it back to Earth. Once a satellite is placed into orbit following the rocket launch, vectors are crucial in helping define the path and motion of the satellite. Scientists must be able to calculate the gravitational forces that are being exerted on the satellite in order to determine the proper orbit so it stays in orbit a maximum amount of time. Vectors are used to represent both the gravitational forces and the actual motion of the satellite around the earth. In this example, the coordinate systems are much more complicated than the simple Cartesian system. The fact that the earth is moving (both by rotation and revolution), the Moon is moving, and the satellite is in motion produces a very complex set of coordinates. Vectors, however, can serve as a means of describing the motion of each of these bodies in space.

Vectors are used by navigators to determine their future positions. In the case of a ship (which is essentially operating in a two-dimensional system), the navigator considers the forces exerted on the ship by its engines, the wind, and any ocean currents. Each of these forces can be represented by a vector and their sum that will produce the resultant motion of the vessel. A similar set of calculations can be applied to an airplane, which must consider the effects of wind on the airplane.

Context

Vectors were developed in the late nineteenth century as another method to describe the mathematics associated with quaternions, a mathematical system that involves the combination of real and complex numbers. Vectors are an important means whereby physicists and mathematicians can describe the motion of a body in space. They provide a type of shorthand that simplifies the description of the magnitude and direction a body possesses. Because vectors can be written in general terms, they can be applied to any of the numerous coordinate systems scientists must use in describing motion.

Since vectors adhere to the basic algebraic operations of addition, subtraction, and multiplication, they can be used and understood in relatively simple terms. These characteristics allow them to reduce the complexity of many problems physicists and mathematicians encounter in the physical world and thus enable them to arrive at solutions to seemingly impossible problems. The need to describe motion which occurs in a very complex problem can be handled with vectors and their associated operations. Most are familar with fixed coordinate systems (ones that do not have any apparent motion). Consider the need to describe the displacement in a series of coordinate systems, which are both moving through space and rotating at the same time, such as that experienced in satellite motion or in outer space with other solar systems. Vectors are crucial in allowing scientists to make an analysis of these types of motions.

Bibliography

Besancon, R. M., ed. *The Encyclopedia of Physics.* 3d ed. New York: Van Nostrand Reinhold, 1985. An expansive volume that provides relatively short, but occasionally advanced explanations of terms in physics. The level of explanation is variable. The discussion of vectors is concise, but somewhat advanced. The articles that discuss the use of vectors are related to specific problems in physics. Geared for high school students through college graduates.

Gellert, Walter S., et al., eds. *The VNR Concise Encyclopedia of Mathematics.* 2d ed. New York: Van Nostrand Reinhold, 1989. This book provides a succinct presentation of vector space, including vector algebra. The word *concise* in this text's title is quite appropriate. Suitable for advanced high school students and college-level mathematics students.

Halliday, David, and Robert Resnick. *Fundamentals of Physics.* 3d ed. New York: John Wiley & Sons, 1988. This widely used college-level text provides numerous examples of specific uses of vectors in solving classical problems in physics.

Lerner, R. G., and G. L. Trigg. *Encyclopedia of Physics.* Reading, Mass.: Addison-Wesley, 1981. This text provides brief explanations of various concepts and terms in physics. Each article lists brief citations of related references at the advanced level. Suitable for college students and above.

Pearson, Carl E., ed. *Handbook of Applied Mathematics: Selected Results and Methods.* 2d ed. New York: Van Nostrand Reinhold, 1983. An extensive volume that covers all areas within mathematics. A well-written and clearly illustrated text that can be of assistance to the advanced high-school student and beyond.

David M. Best

Cross-References

Differential Calculus, 650; Differential Geometry, 663; Electric and Magnetic Fields, 699; Forces on Charges and Currents, 907; Mechanics, 1367; Newton's Laws, 1509; Orthogonal Functions and Expansions, 1704; Trajectories, 2589.

VISCOSITY

Type of physical science: Classical physics
Field of study: Fluids

The internal friction of fluids is called viscosity. Viscosity causes resistance to flow and is important in every problem involving fluid motion from painting to the curve ball in baseball.

Principal terms

BOUNDARY LAYER: a layer of viscous fluid immediately adjacent to a solid; in it, the velocity of fluid rapidly approaches zero relative to the solid

NEWTONIAN FLUID: a fluid in which the viscous stress is proportional to the velocity gradient

POISEUILLE FLOW: the steady flow of viscous fluid in a tube driven by an external pressure difference

VELOCITY GRADIENT: the rate at which the flow velocity changes spatially

VISCOUS STRESS: the internal frictional force per unit contact area between two parts of a fluid in nonuniform flow

Overview

Viscosity is a property of fluids (gases and liquids) by which the flow motion is gradually damped (slowed) and dissipated into heat. Viscosity is a familiar phenomenon in daily life. An opened bottle of wine can be easily poured; the wine flows easily under the influence of gravity into a glass. Maple syrup, on the other hand, cannot be poured so easily; under the action of gravity, it flows sluggishly. The syrup has a higher viscosity than the wine. Almost all fluids have viscosity; the only exceptions are the two isotopes of helium, helium 3 and helium 4, which at extremely low temperatures exhibit no viscosity at all. (In such a state, they are known as superfluids; quantum mechanics is required to understand superfluids.)

A viscous fluid has internal friction. This frictional force exists between the flowing fluid and the surface of the container, such as a tube (the boundary effect), as well as between different parts of the fluid (the bulk effect). Quantitatively, the bulk effect can be illustrated in the following way. Imagine a fluid flowing east. If the flow is uniform, that is, if different parts of the fluid flow with the same velocity, then there can be no frictional force between different parts of the fluid. This results from the fact that if the flow is uniform, then the fluid looks static from the point of view of an observer moving with the fluid. Sir Isaac Newton's first law of motion states that it should stay that way. Consequently, the flow will continue and no viscous damping is possible, at least in bulk.

To observe viscous effect in the bulk, it is essential that the flow be nonuniform. If the easterly flow is assumed to be nonuniform, the magnitude of the flow velocity increases in the northern direction; the flow is more rapid in the north than in the south. To determine how nonuniform the flow is, it is measured in terms of a velocity gradient, which measures the change in flow velocity (meter per second) per unit distance (meter) in the direction of that change (north in the example). In a viscous fluid, such a nonuniform flow generally implies a nonzero viscous force, which in the example is exerted by the relatively slow fluid to the south on the relatively fast fluid to the north to slow down the latter. Following Newton's third law of motion, there is a force equal in magnitude and opposite in direction, which is exerted by the fast fluid to the north to speed up the fluid to the south. This force is proportional to the area of contact between the two parts of the fluid. The frictional force per unit area is known as the viscous stress, which causes the flow to be more uniform. In the example, it slows down the fast flowing fluid and speeds up the slower flowing fluid.

The boundary effect of viscosity forces the fluid that is in direct contact with the boundary to have the same velocity as the boundary. If one considers the flow of a fluid in a tube, and if the tube is stationary, then the viscosity has the effect of slowing down the flow. In the absence of any externally applied pressure (gravity, pump), this will stop the flow completely; with applied pressure the flow may continue, but the viscosity causes resistance to the flow.

Exactly how the viscous stress varies as a function of the velocity gradient depends on the fluid. In many simple fluids, such as water and air, the viscous stress is simply proportional to the velocity gradient. The proportionality coefficient is called the viscosity of the fluid. Thus, viscous stress equals viscosity times velocity gradient. Such a fluid is known as a Newtonian fluid. Viscosity is measured in units of poise, in honor of the French physician, Jean-Louis-Marie Poiseuille. In a fluid with a viscosity of 1 poise, if the velocity gradient is 1 (meter/second/meter), the viscous stress will be 0.1 newton per square meter. Under normal conditions, water has a viscosity of 0.01 poise and air has about 0.00017 poise.

Many polymers are non-Newtonian fluids in that their viscosity depends on the velocity gradient (so that the viscous stress is not simply proportional to the velocity gradient—the functional relationship is nonlinear) or even how long the velocity gradient has been maintained, among other things. Paint, for example, is a non-Newtonian fluid whose viscosity decreases (the fluid becomes thinner) after it has been stirred.

Usually, the viscosity of a fluid depends sensitively on the ambient temperature. The viscosity of gases generally increases with increasing temperature. In contrast, most liquids become less viscous at high temperatures. The viscosity of some substances can vary by a factor of many billions as a function of temperature. Pitch, for example, has a viscosity in excess of 10 billion poises at room temperature but is poured quite easily at elevated temperatures. In the process of glass formation, the viscosity increases from less than several hundred poises at high temperatures to more than 10^{14} poises below the glass transition temperature, which is the temperature at which the viscosity is equal to 10^{13} poises.

One can get a sense of where viscosity comes from by considering the following microscopic model. Imagine many molecules forming a gas. Initially, it is assumed that the gas is at rest. Although the macroscopic velocity of the gas is zero, the molecules are in constant random motion at any nonvanishing temperature. The velocity is zero only on average. Imagine such a gas is set up in a flow. Microscopically, the molecules are still undergoing random motion. The difference is that now the average velocity is no longer zero but rather equals the flow velocity. One can now assume that the flow is nonuniform, with the flow velocity being higher on the left. The molecules on the left have a higher average velocity than the molecules on the right. At this point, the random motion of the molecules will tend to mix the fast and the slow. Some of the fast-moving molecules on the left may be carried by random motion to the right, where they will collide with the molecules already there and that originally had a lower average velocity. In this process, the molecules exchange momentum; the fast ones give up some of their momentum to the slow ones, causing the latter to speed up. A similar process, in which the slow molecules on the right are carried by random motion to the left, causes the fast ones to slow down. This manifests itself on the macroscopic scale as viscosity.

The viscosity of a gas can be understood on the basis of this model. The thermal random motion of the molecules mixes the velocity of molecules in a nonuniform flow. The farther a molecule can travel between two collisions, the more effective this mixing will be. At higher temperatures, the random motion of molecules is faster and can carry a molecule farther between two collisions. This is the reason that the viscosity of a gas increases with temperature.

A simple model is not possible for liquids. In a liquid, the molecules are very close together and the motions of individual molecules are highly correlated. This presents a formidable theoretical problem. The microscopic calculation of the viscosity of a liquid has met with only limited success.

Applications

With the exception of superfluids, all fluids have viscosity. It is impossible to understand the flow of fluids without a proper understanding of viscous effects. Because of the viscous effect, the velocity of fluid in direct contact with the wall is zero. In a tube, the flow velocity is not uniform but, rather, has a parabolic distribution, going to zero at the boundary and reaching a maximum at the center. The steady flow of viscous fluids in a tube is known as Poiseuille flow. The total flux of fluid is proportional to the fourth power of the tube radius. All valves depend on viscosity. Closing the valve reduces the opening and increases the resistance, which slows down the flow. Without viscosity, devices ranging from a water faucet to the gas pedal in a car would fail.

To drive the flow in a tube, the viscous resistance of the fluid must be balanced by a pressure difference on the two ends of the tube, whether it is the Alaskan oil pipeline or the human artery system. The viscosity slows down the falling raindrops, causing them to reach some terminal velocity. Without viscosity, the rain drops would shoot down from the sky with the velocity of a bullet.

The thin layer of fluid near a solid object where the flow velocity rapidly decreases to zero relative to the solid is known as the boundary layer. In the boundary layer, the fluid is largely dragged by the object. This is important in a number of sports. A spinning ball drags the air around it and causes the air to circulate. If such a ball is thrown through the air, the circulation and the flow of the air rushing past the ball will be in the same direction on one side of the ball and in opposite directions on the other side. Consequently, the net air velocity on the two sides of the ball will be different. Using Bernoulli's principle, this leads to a difference in pressure on the two sides, which drives the ball to one side of the trajectory. This phenomenon is familiar in baseball and volleyball. To pitch a curve ball, the pitcher gives the ball a spin around a vertical axis. The pressure difference will therefore be between the left and the right side of the trajectory, causing the ball to be deflected to one side. The surface of a golf ball is intentionally roughened with dimples. This enhances the dragging effect and gives the ball a significant lift to an upward spinning ball, greatly increasing its range.

Non-Newtonian fluids are important technologically. For example, paint must be easy to spread with a brush and, at the same time, sufficiently viscous once it is applied to avoid dripping. These two conditions can both be satisfied because of its non-Newtonian viscous properties. The viscosity of a paint depends on how fast the paint is stirred. It decreases when the paint is stirred and sheared by the brush but increases once the stir ceases.

Context

Viscosity can be studied macroscopically and microscopically. The macroscopic study of viscosity belongs to hydrodynamics. In hydrodynamics, the value of the viscosity is assumed to be known; the theory concentrates on working out the dynamical consequences of viscous flow. The significance of viscosity was known in the early days of hydrodynamics. Not until the twentieth century, however, were effective mathematical tools developed to treat viscous flow problems of any complexity. The complicated problem of turbulent flow with viscosity is only beginning to be understood in the last quarter of the twentieth century. The application of hydrodynamics to the viscous flow of unconventional fluids, such as water and oil, in sediments and in biological systems such as human organs and the body, will continue to be of great interest for many years.

The microscopic study of viscosity belongs to the field of statistical mechanics. Here, one is interested in understanding the origin of viscosity, the non-Newtonian behavior, and viscoelastic properties of a fluid on the basis of its molecular properties, such as the intermolecular forces and the deformation of the molecules under stress. Extensive work was done in the last one hundred years. The understanding of the viscosity of gases under normal conditions is essentially complete. One can now calculate the viscosity quantitatively with considerable confidence. The theory for the viscosity of liquids is at a more primitive stage; quantitative results can be expected only for simple liquids. Still less well understood are the viscous properties of concentrated polymer systems. These fluids are often non-Newtonian. Despite their technological importance, no fundamental theory is available. It is likely that qualitatively new ideas are needed before any progress can be made on this problem.

The fascinating behavior of fluids, the ubiquity of viscosity, its technological significance, and the theoretical challenge of its understanding from a microscopic point of view will continue to be a major driving force in statistical mechanics and hydrodynamics.

Bibliography

Feynman, Richard P., Robert B. Leighton, and Matthew Sands. *The Feynman Lectures on Physics.* Vol. 2. Reading, Mass.: Addison-Wesley, 1963. Feynman's lectures are insightful and entertaining. Traditional textbooks on hydrodynamics tend to emphasize the elegant mathematical theory of highly idealized models of fluids (which Feynman called "dry water"). Feynman, on the other hand, emphasizes the basic physics.

National Committee for Fluid Mechanics Films. *Illustrated Experiment in Fluid Mechanics.* Cambridge, Mass.: MIT Press, 1972. Contains pictures of fluids in motion, with accompanying text.

Shapiro, A. H. *Shape and Flow.* Garden City, N.Y.: Doubleday, 1961. An introduction to hydrodynamics with illustrations. No mathematics.

Smith, N. F. "Bernoulli and Newton in Fluid Mechanics." *Physics Teacher* 10 (1972): 451. Explains the seemingly strange trajectory of a spinning ball.

Trevena, D. H. *The Liquid Phase.* New York: Springer-Verlag, 1975. Gives a well-balanced treatment of liquids at the beginning college level. Chapters 9 and 10, dealing respectively with Newtonian and non-Newtonian viscous fluids, are of particular interest.

Yaotian Fu

Cross-References

Glasses, 965; Polymers, 1920; Superfluids, 2425.

THE VON NEUMANN MACHINE

Type of physical science: Computation
Field of study: Computers

The flexibility and speed of computers are derived, in part, from the fact that a program can be stored in the computer's high-speed memory. This idea of a stored program is the fundamental identifying characteristic of a von Neumann machine.

Principal terms

CENTRAL PROCESSING UNIT (CPU): the part of an electronic computer that performs the calculations and that controls the execution of the program

MEMORY: the part of the computer used to store or "remember" data or instructions

PROGRAM: a series of instructions given to a computer so it will do some desired task

PROGRAM COUNTER: a special-purpose register that contains the address of the next instruction to be executed

REGISTER: a storage location within the central processing unit of a computer, used for results or addresses

STORED-PROGRAM CONCEPT: the idea that program commands and data can be together in memory

Overview

A von Neumann machine is a general-purpose, stored-program computer with a single central processing unit (CPU). This type of computer, which includes almost all computers developed since 1945, takes its name from the mathematician John von Neumann, who was involved in the development of the concept at the Moore School of the University of Pennsylvania during the 1940's.

First, a general-purpose computer is one that is designed to do many different tasks, depending on how it is programmed. A computer program is simply a set of instructions. To have a general-purpose computer, it must be possible to change the program so that the computer can change the task it does. This makes it possible for the same computer to process the payroll, forecast the weather, and play a game of chess. Before the invention of the general-purpose computer, there were some attempts to develop special-purpose electronic computers, which could solve only one problem or class of problems. For example, John V. Atanasoff designed, but did not complete, an electric computer to solve systems of simultaneous linear equations.

The computer needs memory for whatever problem or problems it is designed or programmed to solve. Memory is the part of the computer that is used for storage. The memory can be used to store information that is inputted from the outside world or from the results of computations. A good way to think about computer memory is as a set of many slots, each of which can contain one piece of data. In addition to the datum it contains, each slot has an address, a number that is used to reference the slot. The number of slots and the form of the datum that can fit in each slot vary from computer to computer.

The key to developing a fast, flexible general-purpose computer was the invention of the stored-program concept. In a stored-program computer, the memory is used to store both data and the program itself—hence "stored program." In other words, some of the slots in memory will contain data and some of them will contain instructions to the computer, telling it what to do with the data. This makes the computer extremely powerful, since it can be reprogrammed by changing the contents of the slots that contain its instructions. This should be contrasted with earlier designs for general-purpose computers. In these, it was possible to change the program, but only by either changing the physical configuration of the computer (usually by changing the arrangement of wires or electronic plugs of some kind) or by changing the external paper tapes or cards that controlled the device. Although such a computer is theoretically as powerful as one with a stored program, the time-consuming task of reconfiguring it prevents easy use of this capability.

The idea of storing commands and data in the same slots, however, raises interesting questions. How does the computer tell the difference between data and programs? How does the computer find and follow the program? There are two possible solutions to the first question. First, it is possible to mark the commands (or the data) in some way so that they are distinguished. If this is done, it is possible to tell from the form of the contents of a slot whether that slot contains data or program. This solution is rarely used in general-purpose computers. The second solution is simpler. The data and the program commands are not distinguishable. A slot is assumed to contain data if the computer executes a command that tells it to use the slot as data; it is assumed to contain a program command if the computer is told to use the contents of the slot as a command.

Now, how is a computer with a stored program told what to do and where to find data or programs? One needs to look at the structure of the CPU. The CPU is the part of the computer that actually does the computations. It contains the circuitry that allows the computer to do arithmetic and to perform logical functions. It also contains the circuitry that controls the way the computer does its tasks under instruction from the program. The CPU contains a few memory slots, which are called registers. They are used for the operands of commands and for memory addresses. (An operand is a piece of data used in the execution of a command. For example, if the command is to add two numbers, the addends and the sum are the operands.) Registers that contain addresses of memory slots are used to keep track of where the computer is looking in its memory. It is these address registers that allow the smooth execution of a program.

When a program is placed into the computer—a process called loading—the program instructions are usually placed into consecutive slots in memory, so that each command is preceded by the command executed before it and followed by the command that should be executed after it. At the end of the loading process, the address of the first command in the program is placed in a special register, called the program counter. The program counter always contains the address of the slot that contains the next instruction to be executed.

Starting from this situation, the execution of a program is simple. The computer repeats the following tasks over and over. First, the instruction in the slot whose address is currently in the program counter is "fetched"—that is, it is transferred from memory into the CPU. Second, the program counter is "incremented" (one is added to it) so that it contains the address of the next slot, which should contain the next instruction. Finally, the instruction is "executed." This may involve changing the registers in the CPU, accessing data in memory, and other operations.

If one imagines a program being followed according to the tasks above, one can see that after an instruction is fetched to the CPU, the program counter is incremented in the second step. If the third task does not change the program counter, the next time the first step is reached the instruction physically following the current instruction will be fetched to be executed. The computer follows the instructions in the order in which they are stored in memory.

It is not always adequate to execute instructions in this order. Sometimes, programs must "loop," or execute a set of instructions repeatedly. For example, if a program is to print a check for each employee, it will need to execute the instructions that print the check once for each employee. This requires backing up in the program to the beginning of the set of instructions. A program may also need to be able to choose between alternative actions. For example, a program that runs an automated teller machine must do one thing if there is money in the account and another thing if there is no money in the account. This will require skipping some instructions. In all such cases, the program will need to go from one instruction to an instruction that does not follow it in memory; it will need to jump. The use of a program counter makes this easy to do. Suppose that the execution of the current instruction in the third step reveals that the next instruction to be executed is not the instruction in physical sequence. Part of the execution of the current instruction will be the placement into the program counter of the address of the next instruction to be executed. Because the first step looks at the program counter to determine the location of the next instruction, the jump is automatically taken.

This is the same process that occurs in the loading process. The last step of loading is to put the address of the beginning of the program in the program counter. In fact, the computer is always following the fetch-increment-execute cycle. When an application is being loaded into memory, a program is running that does the loading. This program is usually part of a large controlling system called the operating system.

An example of another strength of the von Neumann machine design is that the instructions of one program can be the data of another program. The loader treats the program it is putting into memory as data. Since data and commands are stored together in the same memory, this is simple. This process also simplifies the development of programs to translate from a high-level programming language, which is easy to understand, into the numerical code that is used to run computers. Without such programs, called interpreters and compilers, programs would have to be written in the numerical machine language of the computer, which would be difficult for users to learn. The ability to treat programs as data is taken one step further in artificial intelligence programming, in which programs write other programs and then run them.

Another characteristic of the von Neumann machine is that it has a single CPU that acts as a centralized controller. Instructions are executed one at a time, sequentially, and such a computer is sometimes called a sequential machine.

Applications

Although computers rarely use an explicit data/program distinction, the idea can be illustrated as it is used in one kind of common program: the spreadsheet. A spreadsheet program turns the screen of a computer into a "snapshot" of part of the computer memory and allows the user to manipulate it. Spreadsheets are usually used to do simple analysis of numerical data. The screen appears as a grid of numbers arranged in columns and rows. The numbers are contained in cells, which roughly correspond to slots. Although numbers appear on the screen, a spreadsheet cell can contain either a piece of data (for example, 186, the regional sales for July) or a formula explaining how to do a calculation. As an example of the formula, a cell might contain an instruction to add up the twelve cells above it. If the twelve cells above it contain the regional sales for each month, then the cell that contains the instruction will contain the yearly total. In other words, a cell can contain a little piece of a program. The user can modify this program, but it is the result of the instruction that appears on the screen. To distinguish between data and instructions, many spreadsheet programs require that if a cell contains a formula, the data in the cell must start with an equals sign (=). A cell that contains data does not start with an equals sign. This means that it is possible to tell by inspection whether a cell is an instruction or data. If it is an instruction, the computer evaluates the instructions and displays the result of the computation; if it is not, the computer displays the data itself.

The von Neumann machine is the basis for almost all real computers. For example, one can see how the basic ideas are reflected in the design of a real computer, using the popular minicomputer line of VAX computers designed by Digital Equipment Corporation. In a real computer, each memory location can hold a number of some fixed maximum size. The actual size varies from computer to computer. In a VAX, each memory location can contain integer numbers that range from 0 to 255, a size that is called a byte. When it is necessary to store larger numbers, several adjacent memory locations are used. The VAX has facilities for treating one, two, four, or eight adjacent bytes as though they were one larger slot to make it easier to use data of various sizes.

Real CPUs also differ from machine to machine in several ways. First, the number and size of the registers vary widely from one machine to another. Some machines have only one register for use in computation; some have many. The VAX computers have fifteen of these general-purpose registers. Each of them can hold numbers that range from 0 to 4,294,967,295. This results from the fact that each of the registers is designed to hold four bytes. The program counter register is the same size, which means that the VAX can distinguish more than 4 billion different addresses. VAX computers do not actually have as many physical address slots, but most do have several million bytes of memory.

The commands for a stored-program computer are stored in memory, so they must be written in a numerical code, called machine code. One of the main tasks forced on the designer of a computer by the von Neumann architecture is the design of the instruction set. How many instructions should there be? What should they do? How much memory should they use? For various reasons, the size of instructions may vary even on one computer; this complicates the design. Another consideration in the design of a machine is that instructions must access data; for example, an instruction may transfer data from memory to a register in the CPU. Such an instruction must contain, implicitly or explicitly, the address of the data to be transferred.

The machine language of the VAX consists of between two hundred and three hundred instructions. There are approximately fifteen different ways to specify the address of an operand in an instruction. One of them, called the program counter relative mode, is an interesting extension of the use of the program counter. In program counter relative mode, the address of the operand of an instruction is stored in the locations immediately after the instruction itself. For example, an instruction to increment a memory location (add one to it) will be stored in memory as the numeric code for increment, followed by the address of the operand. Yet, the operand's memory location is not stored explicitly. Instead, the difference between the address of the operand and the address of the increment instruction is stored. This number is called the offset. To find out where the operand is located, the CPU control circuitry adds the offset to the value of the program counter, which should contain the address of the instruction. (Actually, this is not quite right, since when the instruction is executed, the program counter has been incremented. The CPU compensates for this.) At first glance, this seems unusual. The actual address is not stored because there is an advantage to doing it with the offset. If the entire program is moved to start at a different location, the offset does not change, the actual address would change. By extending this idea, almost everything in a VAX program is independent of where the program is loaded. This makes the task of the operating system loader simpler than it would be if the actual address of each operand had to be recalculated based on the starting address.

Context

In the nineteenth century, Charles Babbage developed the idea of a programmable calculating machine. He designed plans for two of these, which he called the difference and analytical engines. He did not build the machines because of the primitive state of manufacturing. Ada Lovelace wrote several programs that were designed to solve problems using the analytical engine; therefore, she is credited with being the first programmer. These programs were not stored. The device was controlled by external punch cards. This method of programming was actually taken from programmable weaving looms developed by Joseph-Marie Jacquard in 1801.

External means were also used to program the first large operational electronic computer, the ENIAC (Electronic Numerical Integrator and Computer). In this case, reprogramming was done by physical reconfiguration. The ENIAC showed that it was practical to build a computer completely from electronic devices; it was built at the University of Pennsylvania in the 1940's by John Presper Eckert and John William Mauchly. The largest problem with the ENIAC was the difficulty of externally setting up a new program. After ENIAC, researchers developed the ideas that came to be embodied in the von Neumann Machine. Many researchers contributed to this accomplishment. Despite the name, it is not the exclusive invention of von Neumann. The Pennsylvania computer based on this design, called EDVAC (Electronic Discrete Variable Automatic Computer), was completed in 1952. The stored-program concept had been previously used in the EDSAC (Electronic Delay Storage Automatic Calculator), completed at the University of Cambridge in 1949.

It is difficult to overestimate the influence of the stored-program concept on the history of computers. There are no electronic digital computers that do not use stored programs. The stored-program concept guarantees that faster memory leads to faster program execution. Cheaper memory allows larger programs to be resident in memory. Programming computers would be a tedious task without stored programs.

The other hallmark of the von Neumann machine is that it is a single CPU machine; that is, it is a sequential computer. Interestingly enough, this was not the case in EDVAC's predecessor ENIAC, which incorporated some parallel task execution. The motivation for restricting to sequential operation was apparently the technical difficulty of building the appropriate electronics. One problem with sequential machines is the so-called von Neumann bottleneck. The computer does only one thing at a time. All the information that is used in the CPU must be brought, byte by byte, across the connection between the memory and the CPU. The speed at which this can be done often limits the overall speed of a program. In the late 1960's and early 1970's, computers began to use limited numbers of multiple CPUs and multiple data transfer paths to speed operation. In the 1980's, more progress was made on the programming and design of decentralized computers, called parallel machines. Computers were designed that contained thousands of processors. After their dominance of early computing, the importance of single CPU computers has gradually diminished. There is no reason to believe, however, that the stored-program concept will diminish in importance.

Bibliography

Aspray, William. *John von Neumann and the Origins of Modern Computing.* Cambridge, Mass.: MIT Press, 1990. Using the archival record, the author provides a broad and detailed account of von Neumann's contributions to computing. These extend far beyond his well-known work in the design and construction of computer systems to include important scientific applications, the revival of numerical analysis, and the creation of a theory of computing.

Bell, C. Gordon, and Allen Newell. *Computer Structures: Readings and Examples.* New York: McGraw-Hill, 1971. A collection of papers about computer architecture, only a few of which are of interest to a general audience. Includes "Preliminary Discussion of the Logical Design of an Electronic Computing Instrument," a paper by Arthur Burks, Herman Goldstine, and John von Neumann, which should be read by everyone interested in the subject.

Biermann, Alan W. *Great Ideas in Computer Science: A Gentle Introduction.* Cambridge, Mass.: MIT Press, 1990. A successful attempt to explain the intellectual achievements of computer science for a nonspecialist audience. In particular, the concepts in this article are further developed in chapter 9, "Machine Architecture."

Goldstine, Herman H. *The Computer, from Pascal to von Neumann.* Princeton, N.J.: Princeton University Press, 1972. A nontechnical history of the computer, ending shortly after the development of the stored-program concept and the von Neumann machine. The concepts are nicely developed in their historical context. Goldstine was a colleague of von Neumann and was a major figure in the development of early computers.

Langholz, Gideon, Joan Francioni, and Abraham Kandel. *Elements of Computer Organization.* Englewood Cliffs, N.J.: Prentice-Hall, 1989. A well-written introductory text that discusses the way computers are designed. Chapter 1, "The Evolution of Computers," and chapter 4, "Basic Computer Organization," combine to give an extended discussion of the material in this article.

Randell, Brian, ed. *The Origins of Digital Computers: Selected Papers.* New York: Springer-Verlag, 1973. A collection of the papers that made the history of computers. The reader will find works by Babbage, von Neumann, Atanasoff, and many others. In particular, the editor has included part of von Neumann's "First Draft of a Report on EDVAC," which is readable and remarkable. It contains a description of a computer that is almost a completely correct prediction of what subsequently developed. The editor has provided well-written introductions that provide background and place the papers in context.

John E. Rager

Cross-References

Compilers, 455; Computer Architecture, 477; Parallel Computer Architecture, 485; Computer Operating Systems, 502; Computer Programming Languages, 510; Computer Programs and Subroutines, 518; Programming Languages for Artificial Intelligence, 1943.

THE CHEMISTRY OF WATER POLLUTION

Type of physical science: Chemistry
Field of study: Environmental chemistry

Water pollution occurs when physical and chemical change has been made to water which limits its usefulness to humans, animals, and plants. A variety of physical and chemical substances are considered water pollutants. Anthropogenic activities are usually the cause of polluted water.

Principal terms

CONCENTRATION: the amount of a substance present in water or soil, usually expressed in terms of weight per volume or mass of water

CONTAMINANT: an unwanted chemical, biological, or physical substance that degrades the quality of water

DOSE: the concentration and amount of a substance to which an organism is exposed

EUTROPHICATION: the condition caused by overfertilization of a lake or river, which involves the addition of large amounts of nutrients to the water

HYDROPHOBIC: refers to a material that does not readily associate with water

HYDROSPHERE: the reservoirs of the earth that contain water, including the atmosphere, oceans, surface waters, and groundwater

NONPOINT SOURCE: a widespread source of pollution that may be hard to isolate and identify; examples include runoff from agricultural lands and automobile exhaust

PERSISTENT CHEMICAL: a chemical that is not readily decomposed or neutralized in the environment

POINT SOURCE: a single isolated source of pollution (such as a sewage outfall) that is often easily identified

WATER POLLUTION: the degradation of a water mass because of physical and chemical changes

Overview

Water pollution describes the condition where water is made undesirable or unfit for human, animal, and plant consumption because it contains unwanted chemicals or biological substances called contaminants. Water is called the universal solvent because, to some degree, it readily dissolves almost all materials. Thus, water quality is easily degraded by liquid, gaseous, and solid materials, which are dissolved during the migration of water through the earth. In addition, even the most insoluble solids and hydrophobic liquids may be transported by surface or ground waters, either as solid particles carried along by moving water or as liquids transported on the surface of the water. Most water pollution is the result of human activities involving disposal of industrial products, chemicals, or wastes, but in certain instances, pollution may result from natural processes. Pollution resulting from direct or indirect human activity is termed "anthropogenic pollution." Most pollutants are inorganic or organic chemicals and compounds; however, they can also be physical substances such as litter and sediment. If the contaminant is from a single source or location of input, such as the outfall from a chemical plant, a waste treatment plant, or a storm sewer, it is termed "point-source pollution." Nonpoint source pollution results from more widely distributed sources. Runoff of pesticides and herbicides from agricultural fields or lawns or from atmospheric deposition are examples of nonpoint source pollution.

Regardless of the contamination source or the water reservoir that is affected, there are four important aspects to consider concerning chemical pollution of water: the type of chemical contaminant, the danger a contaminant presents because of its toxicity, the concentration or amount of the contaminant present, and the behavior of contaminants in the environment. A variety of chemicals may cause water pollution. These include inorganic elements such as the heavy metals lead, cadmium, and mercury, and inorganic nonmetals such as arsenic and selenium. A large number of organic compounds are also identified as pollutants, including insecticides, pesticides, industrial wastes, by-products of fuel and waste combustion, and hydrocarbon fuels. Nutrients and sediments can also cause water pollution, albeit indirectly, by causing overfertilization of plants and other organisms, thereby resulting in water "unfit" for consumption.

The danger a chemical pollutant presents is discussed in terms of contaminant toxicity. For example, lead (Pb) is often referred to as a dangerous "toxic" metal. Toxicity refers to a chemical's ability to disrupt cellular and tissue functions in organisms. If the disruption is severe, cells may stop functioning and, if severe enough, the organism may eventually die. It is the exposure or dose that describes the degree an organism has interacted with a pollutant. The dose is a function of the amount or concentration and the length of time over which exposure occurs. Generally, the larger the concentration an organism is exposed, the shorter the length of time before the organism is affected. The chemical form, or speciation, of a pollutant also affects its toxicity. For example, mercury (Hg) is highly toxic and dangerous when it is the form of methyl-mercury but is less of a problem when it is present as metallic mercury.

Pollutants, and, in fact, most chemicals, are toxic when some critical dose or exposure is surpassed. Particularly good examples are the trace metals, many of which, such as copper and zinc, are essential for proper cell function when present at low concentrations. At elevated doses, however, these same metals can become very toxic to organisms. The same concept applies to many organic compounds; however, as a general rule-of-thumb, organic compounds are usually toxic at lower doses. There will be some level at which all chemicals will be dangerous. These levels are different for each chemical pollutant and for each organism and are related to the dose or length of time of exposure. Based on doses from laboratory studies of animals and case studies of humans, government health agencies have prepared lists of recommended safe drinking water concentrations for many potential chemical pollutants. These concentrations are designed to minimize the dose, and therefore, the health risks of polluted water.

Because it is exposure or dose that makes chemicals dangerous, it is very important to consider both the chemical type and abundance when water pollution is evaluated for the danger it presents. The amount or concentration of a chemical in water is expressed in terms of weight per unit volume or mass of water. Units of concentration commonly used are milligrams per liter of water, parts per million of water, and parts per billion of water. As an example of how small a total contaminant amount these concentrations represent, 1 part per billion represented slightly greater than 4 centimeters of the total circumference of the earth (40,077 kilometers). Very small concentrations of extremely toxic contaminants, at levels such as several parts per billion, may cause water to be unfit for use, while less toxic pollutants may be dangerous when they are present in the part per million range or higher. Even at these low concentrations, contaminants are often thousands or millions of times greater in polluted water than in natural, noncontaminated water. As a result, a number of problems are faced when remedial actions are taken to clean up contaminated water, mainly because of the amounts of toxic pollutants that can be dispersed in large volumes of lake, river, or groundwater. Many times, polluted water will contain elevated concentrations of compounds that are so toxic that their effects cannot be easily removed by simply diluting the polluted water with clean water. This case illustrates that, in some cases, a very small total amount, and concentration, of a toxic contaminant may cause water to be unsafe.

Any discussion of water pollution and its causes must focus on the different, interconnected reservoirs of water that make up the earth's hydrosphere. These reservoirs are the atmosphere, the oceans, surface waters (lakes, streams, rivers, swamps), glaciers, and the underground reservoir termed "groundwater." The hydrologic cycle refers to the pathways of water movement between these reservoirs, as well as the volumes of water in each reservoir and the rate of transfer between reservoirs. Because each of these reservoirs may be affected by different types or classes of chemical and physical pollutants, the study of water pollution generally follows the hydrologic cycle.

Applications

The major sources of pollution to atmospheric water are gases and dust particles that enter the atmosphere. These pollutants are generated from the exhaust and particulates emitted during the burning of coal, gasoline, and solid wastes. In some areas, atmospheric pollution results from wind-borne dust from mining activities. The gases react with water in the atmosphere, while the dust particles act as nuclei for rain drops, which carry both types of pollutants to the earth's surface.

The two groups of gases that are most important as pollution sources are nitrous oxide compounds and sulfur oxide compounds. Nitrous oxide compounds are formed and emitted during the burning of fuel in internal combustion engines. Sulfur oxide compounds are generated primarily during the burning of coal as the result of oxidation of sulfide minerals contained in the coal. Both gases dissolve into rain and snow during cloud condensation and can also be washed out of the air by falling precipitation. Once dissolved, they quickly convert to sulfuric acid (H_2SO_4) and nitric acid (HNO_3), respectively, which are the main agents of acid rain. These acids may migrate hundreds of kilometers in the atmosphere before being swept to the ground by precipitation. In many areas far from ground-based sources of pollution, atmospheric precipitation of acids, nonburnt hydrocarbons, and minute dust and ash particles are the primary sources of water pollution.

Because of the ease with which they are accessed by humans, surface waters such as lakes, streams, and the ocean are perhaps the most easily polluted reservoirs in the hydrosphere. Traditionally, lakes and rivers were used as disposal sites for untreated anthropogenic waste products. As a result, the most abundant contaminants of surface water pollution are the chemicals found in industrial and municipal wastes. These wastes supply contaminants such as heavy metals and organic compounds. Waste contamination is often introduced from point sources such as municipal waste treatment plants, sewer outfalls, and industrial outfalls.

Surface waters are also affected by nonpoint sources of pollution. The most important of these are direct atmospheric deposition of particles, along with overland runoff of precipitation. Large inputs of dust, ash, and other particles may occur in lake basins in arid regions or near heavily populated regions where fuels or other materials are burned. Overland flow of precipitation draining off the land also delivers dissolved contaminants to streams and rivers. This source, along with groundwater, transports to surface water bodies many pollutants such as fertilizers, pesticides, and herbicides, which are applied to crops, fields, and lawns. Perhaps the most recognized example of this type of pollutant is the once widely used pesticide DDT (dichloro-diphenyl-trichloromethane), which was banned after having been linked to declines in populations of fish-feeding birds. These types of chemicals are often termed "persistent" because they are very slow to break down, degrade, and dissipate from the environment. Thus, these pollutants may remain active in the environment many years after application, still moving through the hydrologic cycle far from the site of their use, and still having deleterious effects on plants and animals for many years. Another important group of chemical pollutants delivered to surface water by overland flow are the essential nutrients phosphorus (P) and nitrogen (N). These chemicals originate from animal and human waste and are also major components of fertilizer. Nutrient pollution occurs from nonpoint sources, including fields and animal feedlots, as well as from point sources, such as waste treatment outlets. The result of excess nutrient input to surface water is a condition termed "eutrophication": the rapid growth of plants and algae in lakes resulting from the oversupply of essential nutrients. Another type of surface water pollution is thermal pollution, which results from inputs of hot water from such sources as electrical and steam generation plants.

An important concept concerning the fate of pollutants in surface water is the ability of these reservoirs to cleanse themselves naturally of pollutants once the pollution source is removed. This ability results from two natural processes occurring in lakes and rivers: The first of these processes involves the continual replenishment and flushing of water through lakes and rivers. As a result, clean water replaces polluted water relatively rapidly, but only if the source of pollution is eliminated. This flushing is fastest in rivers and slowest in large lakes or lakes that do not have rivers flowing from them. The other mechanism occurs through the deposition and burial of sediment particles on river and lake beds and on the ocean floor. Many contaminants are hydrophobic; that is, they readily associate with sediment and solid organic debris rather than remaining dissolved in the water. As they settle to the bottom, solid particles will scavenge contaminants from water, thereby removing them from the water. While the overall effect is to cleanse dissolved contamination from surface water bodies, the sediment and organic debris may, in fact, act as a long-term source of pollution to the overlying water. This is caused by a number of processes, among which are changes in the oxidation state of the sediment after burial. These changes may cause the sediment to dissolve and release contaminants back to the overlying water column. Resuspension of contaminated sediment by floods, storm waves, or human activity can also reintroduce the contaminant to the overlying water. Additionally, some plants may extract contaminants directly from sediment and water. Finally, ingestion of the contaminants by worms and other sediment feeds can introduce the contaminant into the food chain, including fish eaten by humans.

Major surface water contamination also occurs by oil spills such as the 1989 accident of the *Exxon Valdez* in Alaska or the actions in the Persian Gulf in 1991. After a spill, the petroleum rapidly changes through evaporation of the most toxic, lower-molecular-weight components. The heavier components tend to spread and are altered through microbial activity and oxidation. The chemical evolution of a spill depends upon the type of petroleum involved, the temperature of the air and water, ocean turbulence, and any actions taken to control or contain the effects of the spill.

Groundwater is the portion of the hydrosphere that lies below the ground surface; it is the water that moves by gravity through the open void space of consolidated and nonconsolidated geologic materials. Geologic materials that are saturated and supply water are called aquifers. Layers that cannot supply water and act as barriers to water flow are called aquicludes or aquitards. Aquifers that are surrounded by aquitards or aquicludes are called confined. Confined aquifers are usually deeper and are protected by aquitards from surface sources of pollution. Pollution of confined aquifers can occur if contaminants are introduced through such activities as drilling or in areas of soil excavation. Aquifers having an upper water surface, termed the "water table," which is connected to the atmosphere through the soil, are called unconfined aquifers. These aquifers are particularly sensitive to pollution from spilled liquids and polluted surface water that infiltrates into the ground. Polluted water may directly enter an unconfined aquifer, and clean water, infiltrating through contaminated soil or wastes, can pick up chemicals and transfer them into underlying aquifers as well.

A variety of soluble and insoluble or hydrophobic chemicals may pollute groundwater. Inorganic chemicals that are soluble in water may be introduced from concentrated solutions, such as saline oil-field brines, or heavy metals concentrated in waste waters. Organic chemicals of particular importance are the persistent organic compounds, especially aromatic organic compounds, which contain chlorinated benzene derivatives. Examples of these compounds include benzene, toluene, ethylbenzene, and xylene, referred to as BTEX, a suite of compounds often used as indicators for the presence of other aromatic compounds in groundwater. Another group of these compounds include polychlorinated biphenyls (PCBs). Many organic compounds are naturally degraded with time by the actions of bacteria, sunlight, and oxygen penetrating into soils. Nevertheless, many classes of chlorobenzene compounds are toxic to bacteria and are not readily degraded; therefore, they often last for many years in slow-moving groundwater. The density of organic compounds may either be lighter than or heavier than water, depending on their atomic makeup. This group of chemicals may therefore either float on the groundwater table or sink to the bottom of aquifers if they are more dense than water. Sources for organic chemical pollution include leakage from chemical or gasoline storage tanks, broken pipelines, waste storage ponds, and landfills.

Solid materials and solid wastes are also cited as contamination sources. These include road salt, solid waste piles (both municipal and industrial wastes), ash generated from coal or refuse combustion, and any other solid material that can be dissolved by precipitation or washed off the surface of a solid material.

Unlike surface waters, which may cleanse themselves quickly if the source of contamination is eliminated, groundwater will remain polluted long after the contamination source is removed. This results from the fact that the slow flow rates typical of groundwater eliminate the rapid flushing of contaminants from aquifers, and because of the lack of rapid means for removing contaminants by processes such as sedimentation. Reactions between the dissolved chemicals and the clays and other minerals in the aquifer may be the only means by which contaminants are removed from the water. If the solid materials in an aquifer become contaminated, however, they may act as an almost permanent source of recontamination of fresh groundwater entering the aquifer.

Context

The study of water pollution grew out of the awareness of the importance of maintaining a healthy environment. Perhaps the first major book recognizing the problem was *Silent Spring* (1962) by Rachel Carson, who documented the threat of environmental pollution. The problem of widespread pollution first surfaced in the 1950's and 1960's after a series of incidents of human illness and obvious degradation of the environment were brought to light. Major examples of these incidents include mercury poisoning, which occurred in Minamata Bay, Japan; the disappearance of pelicans along the California coast; and the extreme eutrophication and metal pollution that has occurred in the Great Lakes, especially Lake Erie.

In Minamata Bay, Japan, an outburst of illness and disability, affecting more than a hundred people and leading to the deaths of many, was related to inorganic mercury used in an industry along the bay. Some of the mercury, lost during processing, was flushed into the bay and poisoned the food chain and ultimately the people who ate seafood along the bay. In California, the rapid decrease in the number of seabirds was linked to the movement of the persistent pesticide DDT off the land and into the ocean food chain. This chemical affected the ability of the birds to reproduce. The resulting ban on DDT gradually has allowed a return of pelicans in the area. Pollution of the Great Lakes, once recognized, has led to local, state, and federal actions to reduce the sources of contamination and overfertilization. This is gradually allowing the cleanup of these important waterways to occur.

Although the importance of maintaining a clean environment is now recognized and major steps have been taken to return the earth's water to an unpolluted state, many problems remain. Perhaps the most important is the existence of toxic sites, where both soil and water have been polluted through industrial activity or waste disposal. Even where the industrial and disposal activities were done legally and with consideration for environmental needs as they were then understood at the time, advances in environmental chemistry now show these disposal sites are "time bombs" that threaten the health of the population and the usefulness of the land for hundreds of years to come. Major strides have been made in understanding the nature of water pollution and the interaction of the chemicals with different parts of the hydrosphere. Much remains to be done to develop efficient means of detecting, confining, and neutralizing pollution of the environment.

The study of water pollution, its sources, behavior, migration, and effects on organisms is of extreme importance, since without clean, safe, pure drinking water, humans, plants, and animals cannot survive. Because water plays such a vital role in many different aspects of life and Earth processes, aspects of water pollution are studied by many sciences such as chemistry, geology, geochemistry, biology, forestry, zoology, and civil, mechanical, and chemical engineering. In addition, because of the vital economic role that water has in human endeavors, water pollution is also studied in fields such as resource management and utilization.

Bibliography

Berner, E. K., and R. A. Berner. *The Global Water Cycle: Geochemistry and Environment.* Englewood Cliffs, N.J.: Prentice-Hall, 1987. A review of the global water cycle that discusses the important natural and anthropogenic inputs of chemicals into the reservoirs of the hydrosphere.

Canter, Larry W., Robert C. Knox, and Deborah M. Fairchild. *Ground Water Quality Protection.* Chelsea, Mich.: Lewis, 1987. A high-level text dealing with groundwater flow, water pollution, and water monitoring, analysis, and management.

D'Itri, F. M., and L. G. Wolfson. *Rural Ground Water Contamination.* Chelsea, Mich.: Lewis, 1987. While many of the classic problems in water pollution are related to the urban environment, this excellent book, dealing with nonindustrial sources of pollution, treats a wide variety of point and nonpoint sources of pollution having an impact on the quality of ground water in the United States.

Fetter, C. W. *Applied Hydrogeology.* 2d ed. Columbus, Ohio: Merrill, 1988. An introductory textbook that covers the whole range of groundwater studies. Contains an excellent section giving examples of groundwater pollution and groundwater restoration.

Moore, James W., and S. Ramanmoorthy. *Heavy Metals in Natural Waters: Applied Monitoring and Impact Assessment.* Berlin: Springer-Verlag, 1984. Discusses the chemistry of eight inorganic metals in the environment, along with monitoring approaches and political considerations of heavy metal contamination.

___________. *Organic Chemicals in Natural Waters: Applied Monitoring and Impact Assessment.* Berlin: Springer-Verlag, 1984. This review book discusses the chemistry of the different groups of organic pesticides in the environment, emphasizing their behavior in water, sediment, and organisms, and the strategies for prioritization of contamination risks.

Neely, W. B. *Chemicals in the Environment: Distribution, Transport, Fate, Analysis.* New York: Marcel Dekker, 1980. A general-to-specialty-level text that details much about geochemical cycling and the behavior of different organic and inorganic chemicals in the hydrosphere. Includes good discussions of mathematical models to aid in environmental studies.

Timothy P. Wilson
Donald F. Palmer

Cross-References

The Chemistry of Air Pollution, 61; Biological Compounds, 224; Concentration in Solutions, 526; Environmental Chemistry, 808; Hydrocarbons, 1071; Inorganic Compounds, 1125; Organic Compounds, 1689; Solutes and Precipitates, 2283.

WATER WAVES

Type of physical science: Classical physics
Field of study: Fluids

Waves are undulating forms that move along the surface of a body of water or at the interface between layers of different density within a body of water. While any kind of disturbance in water can generate waves, there are three principal causes: wind, gravitational pull of the Moon and the sun, and earthquakes.

Principal terms

DEEP-WATER WAVE: a wave traveling in water with a depth greater than one-half the wavelength so that it is little affected by the bottom

ORBIT: the path followed by a water particle affected by wave motion; deep-water orbits are nearly circular, and in shallow water they are flattened

REFRACTION: the bending process when a wave approaches the shore at an angle so that the part advancing in shallower water moves more slowly than that part in deeper water

SEICHE: a standing wave in an enclosed water body that does not progress forward but continues as a pendulum-like oscillation after the originating force has ceased

SHALLOW-WATER WAVE: a wave that is noticeably affected by bottom topography

SWELL: wind-generated waves that have traveled out of their generating area

TIDAL WAVE: the wave motion of the tides; periodic rising and falling of the sea that results from gravitational attraction of the Moon and the sun

TSUNAMI: a long-period sea wave produced by a submarine earthquake or volcanic eruption

WAVE HEIGHT: the vertical distance between a crest and the preceding trough

WAVE PERIOD: the time required for two adjacent wave crests or troughs to pass a stationary point

WAVELENGTH: the horizontal distance between two successive wave crests or troughs

Overview

Waves are disturbances of the surface of a fluid that typically occur on the surface between the atmosphere and a water body but may also occur on the surface that separates two water masses. Wave motion is periodic in that it is repetitive through fixed periods of time. At some stationary position, such as a pile off the seashore, a succession of wave crests (or troughs) pass at certain intervals of time, known as the wave period. The horizontal distance between successive crests is the wavelength. By dividing the wavelength by its period, the speed or velocity of wave propagation can be obtained easily. This relationship holds for all periodic waves. The other parameter used to describe simple periodic wave motion is height, the vertical distance from trough to crest. Wave height is independent of the other wave parameters, which accounts for its change as waves approach the shore.

If one carefully watches a cork floating on the water surface as waves pass, it will be observed that the cork rises and falls and at the same time appears to move back and forth, describing a circular motion whose diameter is the wave height and whose time to complete one cycle is the wave period. The cork makes no net advance in the direction of wave motion. Thus, waves may transfer energy and momentum across the water surface, often for thousands of kilometers, with negligible net drift of water itself. Water particles rotate in orbits, but they do not move along with the wave form.

Wind-generated waves are the most common type of waves on the oceans and large lakes. They are also the most variable. Because winds are, by nature, turbulent and gusty, there are local variations in the velocity and in air pressure on water surfaces. As a result, wavelets of all sizes are created simultaneously. The ultimate size and variety of the waves raised by the wind depend on three factors: the velocity of the wind, the open water distance over which the wind blows (known as fetch), and the duration of time the wind blows.

In deep water, as waves continue to grow, the surface confronting the wind becomes higher and steeper and the process of wave building becomes more efficient up to a certain point. There is a limit on how steep a wave can be. That limit is reached when the height of a wave is about one-seventh of the wavelength. Thus, a wave that is 7 meters between crests can be no more than 1 meter high. When small, steep waves exceed this limit, they break, forming whitecaps. When the surface of a lake or the sea is covered with such waves, it is said to be choppy. When the wind blows the top off a wave, causing it to break, most of the energy goes into longer, more stable waves. The result is a predominance of longer waves that can accept more energy and rise higher than shorter waves generated by the same wind. At the same time, new ripples and small waves are continually being formed on the slopes of the larger waves. Therefore, where the wind is moving faster than the waves, there is a wide spectrum of wavelengths.

As waves move away from the winds that generated them, their character changes. The original wind waves decay, in that the crests become lower, more rounded, and more symmetrical. They move in groups of similar period and height. Their form approaches that of a true sine curve; such waves are called swells. In this form, they can travel thousands of kilometers across deep water with little loss of energy.

Shallow-water waves, those that are traveling in water whose depth is less than one half the wavelength, interact with the bottom. As a wave moves into progressively shallower water, the surface orbits of the water particles begin to move faster than the wave propagation velocity, which has been slowed by bottom friction. Eventually, the wave steepens beyond the limit of stability and it breaks. In shallow water, waves can be diffracted, reflected, and refracted. Diffraction is the bending or spreading of waves around objects such that energy is transmitted behind a barrier. As waves pass a barrier—such as a protruding headlands along a rocky coast or an isolated island in deep water—some of the energy is transmitted sideways along the wave crest and extends into areas that might have been otherwise sheltered by the barrier. Waves that meet vertical walls, such as seawalls or jetties, are reflected back to sea with little energy loss. Waves that have their crests parallel to the reflecting surface may form standing waves, which move up and down but do not progress horizontally. Refraction of waves is the bending of wave fronts caused by the effects of shallow water. When one part of a wave reaches shallower water before another part, the part impeded by the shallow water is slowed relative to the other. When waves approach the coast at an angle, which is the usual case, longshore currents are generated to transport excess water away from the direction of the wave approach.

In the open ocean, the length of wind waves is normally several tens of meters, but storms regularly generate waves more than 100 meters in length. Swell waves tend to be longer, with lengths of 300 meters being common; in the Pacific Ocean, swell waves can exceed a kilometer. Near the shore, wavelengths are typically a few tens of meters. The periods of wind waves range from less than 0.1 second in capillary waves (the tiny ripples first formed by the wind) to the largest wind wave, called gravity waves, which have periods mostly in the 3- to 15-second range but can have periods up to five minutes. Wave heights can vary from minute capillary waves of less than a centimeter to giant storm waves of 35 meters.

Tides are long-period ocean waves, having a period of twelve hours and twenty-five minutes and a wavelength of half the circumference of the earth. The crest and trough of these waves are known as high tide and low tide. The wave height is called the range of tide, but because it is measured at the coasts, where it is influenced by the shape of the shore, it varies from place to place. The gravitational attraction of the Moon and the sun on the sea surface causes tides.

The Moon revolves around Earth once a month, during which there are various spatial relationships between the sun, Moon, and Earth that cause variations in the tides. During full moon and new moon, the sun, Earth, and Moon are aligned, thus producing the maximum distortion of the sea surface and therefore maximum tides. These are known as spring tides. First-quarter and third-quarter moons have the sun and Moon at right angles to the earth, thus working in opposition to each other producing minimal tidal ranges called neap tides.

Because of the difference in distance between the Moon on one side of Earth as compared to the other, there are unbalanced forces acting on the oceans. The gravitational attraction of the Moon exceeds the centripetal or rotational force on the side closest to the Moon, but the centripetal force exceeds the gravitational force on the far side of Earth. These forces produce the bulges on each side of Earth that are the oceanic tides. A lunar day is 24 hours and 50 minutes in duration, which means that the Moon passes a given point on Earth once in that period of time. Because of the two bulges in the ocean surface, two high tides and two low tides would be expected each day. Landmasses complicate the progression of tidal waves as Earth rotates, causing considerable deviation from this idealized plan at various positions along the world's coasts.

If the surface of an enclosed body of water such as a lake or bay is disturbed, long waves may be set up, which will rhythmically slosh back and forth. These waves, called wind tides, or seiches, have a period that depends on the size and depth of the basin. Tilting of a lake's surface, by wind or atmospheric pressure differences, is analogous to the up and down movements of a teeter-totter while the center remains stable. For example, on Lake Erie, northeast storms have produced a 5-meter difference in the water level at the eastern and western ends of the lake. This forced movement of the lake surface is known as wind tide, and the amount of rise produced is the wind setup. The resulting free oscillation of the lake surface, once the winds have abated, is the seiche. The major seiches on Lake Erie are parallel to the long axis of the lake and have a period of about 12 hours.

Disturbances along the boundary between two layers of different density (for example, cold and warm water masses) are known as internal waves. These waves often have an amplitude much greater than surface waves. Internal wave heights near 100 meters and wavelengths of several hundred meters, with periods of several minutes, are common in the sea. They can be generated by the same phenomena that cause tsunamis: earthquakes, submarine volcanic eruptions, and submarine slides. In the Great Lakes, they can be set in motion by storm surges or rapid barometric pressure changes that pile up water at one end of the lake and lower the level at the other. By way of example, a beaker containing half oil and half water, by virtue of their contrasting densities, will not mix well and will result in two distinct layers of liquid with a sharp boundary. A slight disturbance to the container will cause irregularities to the surface that are analogous to the formation of internal waves at sea.

Earthquakes, submarine slides, and other disturbances to the seafloor may generate large waves such as tsunamis or seismic sea waves. These waves have periods of several minutes and lengths of hundreds of kilometers. In the open ocean, they have wave heights of less than a meter. As they approach the shore, however, their long length causes the waves to steepen as the orbits interact with the seafloor and wave heights can rise to tens of meters with devastating effects as they strike the coast.

Applications

When waves eventually encounter a shore, all of their accumulated energy is released in a matter of minutes as the wave is changed into surf. If the surf is composed of swell that has traveled from distant storms, breakers will develop relatively near shore and the surf will be characterized by parallel lines of relatively uniform breakers. Yet, if the surf is composed of waves that have been generated by local storms, the waves may not have been sorted out into uniform groups, and the surf will be characterized by unstable, steep, high-energy waves. These waves will break shortly after "feeling" the bottom some distance from shore, and the surf will be rough and choppy with an irregular nature.

The breaking experienced in the surf results when the particle motion near the bottom of the wave has been interfered with by contact with the seafloor, and this tends to slow the waveform. Nevertheless, the particles that are orbiting near the ocean surface have not been slowed down as much. Thus, the top of the waveform begins to lean toward the shore, and the wave height increases. This motion can be seen particularly well in plunging breakers, which have a spectacular curling crest because the top has outrun the rest of the wave and there is nothing beneath to support their motion. Plunging breakers generally form on moderately steep beach slopes like those of Waimea Bay on the island of Oahu, Hawaii. The more commonly observed breaker is the spilling breaker that results from a relatively gentle slope of the ocean bottom, which more gradually extracts the energy from the wave, producing a turbulent mass rather than a curl. Spilling breakers are the type formed on the gentle shore of Waikiki, also on the island of Oahu.

Surfboards, small craft, and mammals, including porpoises and bodysurfers, can take energy out of the waves to propel themselves by sliding down the forward surface of an advancing wave. A surfboard is thrust forward by a downhill or slope force, which is balanced between gravity and buoyancy forces. The slope force is greater than the hydrodynamic drag, or water resistance, and the surfboard moves at wave-crest speed. The trick of surfing is to get the board moving and the weight properly balanced so that the slope force can take over the work of propulsion at the moment the wave passes beneath. Porpoises are neutrally buoyant and have learned to tilt themselves at the proper angle to take advantage of the slope force of an underwater constant-pressure surface. These mammals can ride beneath the bow wave of a ship indefinitely without appearing to exert any effort at all.

Any disturbance of the water surface, including the passage of ships, creates waves, which are sometimes called stern waves and bow waves. Much of the power expended in propelling a ship is converted to waves, and anything that can be done to reduce this power results in more efficiency. A ship moving through the water is accompanied by at least three pressure disturbances on each side, which produce several trains of waves. In order to improve ship designs and create hulls with a minimum loss of energy, model ships are tested by towing them through long tanks. Fortunately, the waves made by ship models are accurate predictions of those that will be created by the full-size ship they represent.

Extracting power from waves as they approach the shore is a possibility if significant problems can be overcome. Using the phenomena of wave diffraction and refraction, a large half-cone could be constructed to focus the energy of waves to a turbine at the apex of the structure. Such a device might extract up to 10 megawatts per kilometer of shore, but could produce significant power only when large storm waves existed at the structure. Thus, this system would operate best as a power supplement. Tides have proved to be a more effective use of wave energy. The world's first major tidal power system is located on the Rance River in France. Tides in this region have an amplitude of 13.5 meters. Yet, using tides for energy has some inherent difficulties, especially the fact that tides do not flow continuously but change directions several times a day and that they vary in strength from day to day on a two-week cycle. It has been possible to alleviate the problem to some degree by storing water in a reservoir at high tides, as is done at the Rance power plant, and letting it out slowly to turn turbines on a more continuous basis. Energy production at the Rance plant is 240 megawatts.

Context

Waves have attracted attention for untold centuries. Perhaps the first people to formulate ideas about wave mechanics were the ancient sailors of the Mediterranean Sea. They observed that a float tossed into the sea would rise and fall, move back and forth, as waves passed under it, but it did not move shoreward with the waves. With these simple observations, oscillatory waves were discovered. In the eighteenth and early nineteenth centuries, Daniel Bernoulli, often called the "father of modern fluid mechanics," and Franz Gerstner began the mathematical study of waves and produced the first wave theories. Later, Lord Kelvin, Sir George B. Airy, George G. Stokes, and William John Macquorn Rankine added to the theoretical understanding of waves. It is only within the later twentieth century, however, that the application of wave theory came to extensive, practical fruition.

The modern description of how the wind transfers its energy to the waves derives from the research of Harald Ulrik Sverdrup and Walter Munk of the Scripps Institution of Oceanography in La Jolla, California. During World War II, they worked on the problem of predicting the waves and surf that would exist on an enemy-held beach during amphibious landing operations. They were able to give the first reasonably quantitative description of how waves are generated, become swells, and move across the ocean to a distant shore.

In the past several decades, further studies of wave characteristics in the open sea and improved forecasting techniques have found numerous applications to the selection of shipping routes, as well as in the design of ships and coastal structures, such as piers, breakwaters, and jetties. Even today, however, it is recognized that substantial gaps exist between mathematical wave theories and actual waves observed at sea or in the laboratory, particularly under extreme conditions. The devastations that can be caused by such conditions illustrate the need for continuing research on water waves.

Storm waves are capable of great damage when they reach the shore. Concrete blocks weighing 65 tons have been torn loose from breakwaters and jetties by such waves. One of the most destructive waves is the tsunami. Most frequent in the Pacific Ocean, this type of wave is almost imperceptible in the open sea but can engulf boats, buildings, and people when they reach shore at heights of 30 meters or more. An added danger is that these seismic sea waves have a series of crests, which arrive about 20 minutes apart, with the third or fourth crest to arrive being the largest. Storm surges, usually accompanying hurricanes, are another dangerous wave type.

They periodically occur along the Atlantic and Gulf coasts of the United States and have been known to carry oceangoing vessels a kilometer or more inland. Six thousand people lost their lives in one such storm surge in Galveston, Texas, in 1900. Lakes are not immune from such disasters. In 1928, a hurricane-created seiche overflowed the dike surrounding Florida's Lake Okeechobee and drowned two thousand people.

Bibliography

Bascom, Willard. *Waves and Beaches: The Dynamics of the Ocean Surface*. 2d ed. Garden City, N.Y.: Doubleday, 1980. This is a thorough, but nontechnical treatment of the nature of waves and their interaction with the shore. An excellent source of information for nonscientists and a good review for professional oceanographers. Contains many helpful diagrams and tables, as well as numerous black-and-white photographs. Suitable for general audiences.

Davis, Richard A., Jr. *Oceanography: An Introduction to the Marine Environment*. Dubuque, Iowa: Wm. C. Brown, 1987. Chapter 6 provides a comprehensive overview of ocean waves, and chapter 7 describes tides. Contains many excellent diagrams and photographs of all aspects of deep-sea and shallow-water waves. A sidebar on measuring waves describes modern techniques for obtaining wave data. Suitable for college-level readers.

Fairbridge, Rhodes W., ed. *The Encyclopedia of Oceanography*. New York: Reinhold, 1966. This is an indispensable oceanography source book for the student and professional. Contains excellent sections on waves and the related topics of fetch, tsunamis, wave energy, wave refraction, and wave theory. A well-illustrated and carefully cross-referenced volume, which is aimed at readers with some technical background.

Gross, Grant M. *Oceanography: A View of the Earth*. 4th ed. Englewood Cliffs, N.J.: Prentice-Hall, 1987. A well-written and well-illustrated text that has been kept current with frequent revisions. Chapter 9 provides a comprehensive overview of all aspects of ocean waves, with good diagrams, tables, and photographs. The appendix includes a table of conversion factors and an excellent glossary. Suitable for general audiences.

Parker, Henry S. *Exploring the Oceans: An Introduction for the Traveler and Amateur Naturalist*. Englewood Cliffs, N.J.: Prentice-Hall, 1985. This book is designed to impart basic ocean science information in a clear and readable fashion without sacrificing the technical substance required for the serious student. Chapter 8 covers waves and tides with an exploratory approach. The text is well-supplemented, with good diagrams and photographs.

Trefil, James S. *A Scientist at the Seashore*. New York: Collier Books, 1984. Trefil invites the reader to walk along a favorite beach and deepen the appreciation of the experience by seeing it through his eyes. Chapters 6 through 9 contain fascinating accounts of waves and related features. The book is well illustrated with diagrams and photographs. Readers with some technical background will enjoy it most.

Tricker, R. A. R. *Bores, Breakers, Waves, and Wakes: An Introduction to the Study of Waves on Water.* New York: Elsevier, 1964. This is a comprehensive treatment of water-wave behavior, written with the aim of simplifying the mathematics so as to make the discussion understandable to the general reader. Chapter 17, "Ships' Wakes," provides an excellent coverage of the topic. There are many black-and-white and color plates illustrating various wave types. Suitable for college-level readers.

Charles E. Herdendorf

Cross-References

Centrifugal/Centripetal and Coriolis Accelerations, 332; Hydraulics, 1064; Seismics, 2170; Shock Waves, 2199; Sound Waves, 2311; Waves on Strings, 2720; The Physics of Weather, 2735.

WAVE-PARTICLE DUALITY

Type of physical science: Atomic physics
Field of study: Nonrelativistic quantum mechanics

Energy transport is accomplished via particles or waves. The realization, based upon symmetry considerations, that particle behavior can be manifested by a wave and wave behavior can be manifested by a particle represented a philosophical milestone that led to the development of quantum mechanical thought.

Principal terms

MATERIAL PARTICLE: a particle having nonzero rest mass

MATTER WAVE: wavelike characteristics of a material particle

MOMENTUM: a quantity of motion of a particle, usually expressed as the product of mass and velocity

PHOTON: a quantum of electromagnetic energy possessing no rest mass

QUANTIZATION: limitation of certain physical parameters to discrete values rather than a continuum of values

RELATIVISTIC MASS: the mass of an object when measured under conditions of motion relative to an inertial reference frame

REST MASS: the mass of an object when measured at rest with respect to an inertial reference frame

WAVELENGTH: the spatial repetition of a waveform; the distance covered by one complete wave oscillation

Overview

At the beginning of the twentieth century, physics began a revolution that would forever change the way in which scientists viewed the nature of space and time, energy and matter, and particles and waves. That revolution led to the development and acceptance of both quantum mechanics and relativity theory, both special relativity and general relativity. Prior to this revolution, particle behavior in energy and momentum transfer was considered separate from wave behavior in energy and momentum transfer. A number of phenomena studied in the latter part of the nineteenth century could not be explained in terms of classical electromagnetism as expressed concisely by James Clerk Maxwell. Both the Lord Rayleigh-Sir James Hopwood Jeans and the Wilhelm Wien description of black body radiation were inadequate to explain the experimentally observed data of spectral radiancy as a function of wavelength across the entire electromagnetic spectrum. A number of observed aspects of the photoelectric effect—the release of electrons from a metal surface when illuminated properly with electromagnetic radiation—were in total variance with predictions of classical electromagnetism. Classical theory did not predict the existence of a cutoff frequency, below which the photoelectric effect ceased regardless of illumination intensity, and the nearly instantaneous release of photoelectrons from a surface regardless of wave intensity.

One of the biggest philosophical steps that led to the explanation of both black body radiation and the photoelectric effect was the idea that an electromagnetic wave does not consist of a continuous flow of energy. The wave nature of electromagnetism was tied to the propagation of mutually perpendicular spatially and temporally dependent oscillating electric and magnetic fields, which did represent a continuous flow of energy. A wave has a well-defined wavelength, the spatial distance over which it completes one full cycle of oscillation, and a well-defined period, which is the time span during which a complete oscillation is made. A wave's frequency is the inverse of the period of oscillation and, as such, is the number of complete oscillations per second. To explain the noncontinuous flow of energy in an electromagnetic wave, a corpuscular (or particle) nature was attributed to it. A wave of a given frequency carries its energy in identical, massless packets of an energy that is proportional to the wave's frequency. The constant of proportionality between the energy and the frequency is called Planck's constant, h, which has a value of 6.63×10^{-34} joule-seconds. (In the limit that h approaches zero, quantum behavior reverts to classical behavior as espoused by Newtonian mechanics and classical electromagnetism.) The packet of energy is called a photon and is considered the quantum of the electromagnetic field. Energy transfer is accomplished via the exchange of photons.

This ambiguity begs the question as to whether electromagnetic radiation consists of particles or is a wave. The resolution of this ambiguity involves wave intensity and a probabilistic interpretation. The intensity of an electromagnetic wave at any location in space is not the wave's energy itself, but the probability of finding the wave's energy at that point in space. Where a wave intensity is high, large numbers of photons are likely to be detected, and where the wave intensity is low, fewer photons are likely to be observed.

Electromagnetic radiation has both a particle behavior, exhibited by the photon, and a wave behavior. Nevertheless, both behaviors cannot be exhibited at the same time. The equivalence of particle and wave behavior for electromagnetic radiation can be seen most easily in the energy equation. For a particle of zero rest mass, its energy is given by the product of its momentum and the speed of light c. For the wave nature of electromagnetic radiation, the energy is the product of Planck's constant and the wave's frequency. The equivalence of these two energy equations shows the ability of a wave to have momentum given by Planck's constant, multiplied by the wave frequency and divided by the speed of light. Since the dispersion relation for electromagnetic radiation gives the speed of light as the product of wavelength and frequency, the momentum of electromagnetic radiation can be expressed as Planck's constant divided by the wavelength.

One of the beauties of physics is its symmetries. If a wave could have a particle nature, Louis de Broglie postulated, then material (nonzero rest mass) particles (possessing energy and momentum) should have an associated matter wavelength given by Planck's constant divided by the momentum, the product of the particle's mass and velocity. The hypothesis of a wave nature for particles restored the symmetry broken by the acceptance of a particle nature of electromagnetic radiation. Whenever particles have a momentum **p**, associated with the motion of the particle is a wave of wavelength given as Planck's constant divided by the momentum **p**. Whenever there is a wave of given wavelength, the square of the wave's amplitude is proportional to the probability of observation of a particle with momentum **p** given by Planck's constant divided by the wavelength.

Therefore, waves are specified by wavelength, frequency, and a dispersion relation that gives the speed of propagation. Particles are specified by momentum, mass, and a particle speed. Applying this to electromagnetic radiation, Albert Einstein's relation between energy and mass ($E = mc^2$) allows one to calculate a relativistic mass for a photon in terms of the speed of light, Planck's constant, and the frequency. Applying this to material particles, a frequency of the associated matter wave can be calculated in terms of Planck's constant and the sum of the particle's kinetic and potential energies.

In everyday life, the wave nature of matter is not observed. Tennis balls get caught in a court net; they do not diffract through the way a light wave will diffract through a slit aperture or a grating. If a particle's matter-wave wavelength is comparable to an aperture, then diffraction effects would be seen for that particle.

Assignment of a speed **v** to a particle of given energy and momentum implies that E and **p** are localized in space and are transported through space at speed **v**. How does the particle speed compare to its associated matter-wave phase velocity? Phase velocity is the ratio of energy to momentum. If $E = mc^2$ and $p = mv$, then the phase velocity is greater than the speed of light, since $v < c$ for nonzero rest mass particles. The significance of this is simply that a monochromatic wave associated with a nonzero rest-mass particle is unobservable. Calculation of the phase velocity for electromagnetic radiation yields simply c. Here, the phase velocity and the wave speed are identical.

To describe a material particle, a group or packet of individual monochromatic waves differing in frequency and phase velocity must be used. Here, the group velocity—the rate of change of energy with momentum—will always be less than c and the group velocity is identical to the particle's speed **v**.

Wave and particle description for the same phenomenon are mutually incompatible. If a wave is to have its wavelength given with infinite precision, then it must have an infinite extent in space. Also, if it is confined to a limited region of space so its energy is confined to a localized position at any one time, then it resembles a particle by virtue of this localizability. Thus, one sees that wave-particle duality results from basic uncertainties inherent in nature itself. This principle is clearly demonstrated by the Heisenberg uncertainty principle.

Applications

Consider a 150-gram baseball thrown by a pitcher at 40 meters per second. What does the de Broglie equation predict for the baseball's associated matter wavelength?

The baseball's momentum would be 6 kilogram-meters per second. Dividing that into Planck's constant yields a wavelength of only 1.11×10^{-34} meters, a distance eighteen orders of magnitude below the smallest distance measurable by twentieth century physics. Thus, the wave character of this baseball is indiscernible.

If one considers a molecule of hydrogen at typical room temperature (300 Kelvins), its root-mean-square velocity is 1.84 kilometers per second. A calculation of the associated matter wavelength yields 1.08 angstroms, a distance easily detected. Indeed, this wavelength is typical of the wavelengths of X rays and the interatomic spacings in crystalline lattices. Wave effects can and have been observed for such molecules of hydrogen.

In 1927, Clinton Joseph Davisson and Lester Germer, working at Bell Telephone Laboratories, accidentally verified the existence of matter waves associated with particles. While working on secondary electron emission (a process in which a beam of electrons falls on a solid surface and liberates electrons originally bound to the solid) in nickel, they discovered what looked like a diffraction peak in the angular distribution of electrons emitted from the sample. This peak was observed after they had heated their nickel sample to high temperature in an attempt to clean the surface. The thermal energy of the heat treatment was sufficient to turn the original high-purity polycrystalline sample into single-crystal nickel, a nearly perfect face-centered cubic lattice structure throughout the sample. The interatomic spacing of the crystal was of the size of angstroms. (Crystallographers often use X rays of several angstroms wavelength to ascertain interatomic spacings in crystalline structures via diffraction techniques.) When Davisson and Germer illuminated the nickel sample with the electron beam (54 electronvolts of energy), they observed the superposition of a secondary emission spectrum and a diffraction peak (at an angle of 50 degrees).

A calculation of the matter wave wavelength for the electron beam indicated that, based upon the known interatomic spacing for face-centered cubic nickel, a first-order diffraction peak should be observed at the experimentally obtained angle for an X ray of equal wavelength. For the first time, diffraction with electrons, a material particle, had been experimentally demonstrated, and the results precisely followed the Bragg law for wave diffraction and the de Broglie relation for matter waves.

Since that accidental verification of particle diffraction, numerous other nonzero rest mass particles have shown diffraction effects. Consider a thermal neutron, a neutron at room temperature (300 Kelvins). The kinetic energy of this particle, by equipartition of energy (an energy of one-half the Boltzmann constant per unit Kelvins for each degree of freedom) is 6.2×10^{-21} joules. Equating this energy to $\frac{1}{2} mv^2$, the speed of such a neutron is found to be 2.7 kilometers per second. The matter wavelength would then be 1.4 angstroms. Therefore, diffraction peaks would be observed if thermal neutrons were incident on many varieties of naturally occurring crystalline solids.

Applications of wave-particle duality could be somewhat confusing. Both the wave and particle models were used to explain various phenomena for electromagnetic radiation and material particles. Nevertheless, the wave model and the particle model were never applied simultaneously. If one considers electromagnetic radiation, the wave model can be used to explain diffraction and interference. The photoelectric effect, Compton effect, pair production, and pair annihilation require use of the particle model. Thus, electromagnetic radiation exhibits both wave and particle aspects, but never at the same time.

This phenomenon is a result of the complementarity principle enunciated in 1928 by Niels Bohr. This principle states that wave and particle aspects are complementary rather than antagonistic. A full description of the nature of electromagnetic radiation, for example, requires a knowledge of the wave aspect and also the particle aspect. A knowledge of only one aspect would represent only partial understanding.

Context

An overriding concern in physics is symmetry. Symmetry principles are what provide physics with its intrinsic beauty of formulation. In the early development of quantum physics, particle behavior was attributed to what had previously been considered a wave phenomena. An explanation of black body radiation and the photoelectric effect was not within the capabilities of classical electromagnetic theory as summarized by Maxwell's equations. In 1900, Max Planck introduced the notion of quantization in order to interpret successfully experimental observations of black body radiation. In 1905, Albert Einstein introduced the concept of the photon—a particle quantum of electromagnetic energy with zero rest mass—to interpret successfully experimental observations of the photoelectric effect. (Actually, Einstein was not the first to attribute a particle nature to light. Sir Isaac Newton had espoused a corpuscular theory of light that failed to achieve universal acceptance.) Several other interactions between matter and energy were successfully explained, combining the notion of quantization with a particle nature attributed to electromagnetic radiation.

In 1924, a French graduate student in physics at the University of Paris proposed, principally on the grounds of symmetry considerations, that the movement of material particles is accomplished by a complementary matter wave whose wavelength was determined by the momentum of the material particle. Louis de Broglie restored symmetry in early quantum theory by attributing a wave nature to material particles, where the particle nature of electromagnetic waves had been established. The wave nature of the electron orbiting the proton in the nucleus of a hydrogen atom confirmed Bohr's postulates of atomic structure in a theoretical framework derivable from first principles. The electron could orbit the nucleus only in such a way that its matter wavelength, determined by its momentum, could be only integral multiples of the electron's orbital circumference. This condition mimicked a standing wave that has no net transfer of energy, thus explaining the stability of atomic structure. Only when the orbit was changed could the atom emit electromagnetic radiation, carrying away the energy and momentum. Two years after de Broglie's hypothesis of matter waves, Davisson and Germer experimentally verified diffraction of an electron beam in a single crystal of nickel.

Today, physicists believe in a wave-particle duality of existence. That which carries energy and has momentum has both a wave nature and a particle nature. When the wave nature is manifested, the particle nature is totally suppressed. When the particle nature is manifested, the wave nature is totally suppressed. Wave-particle duality has become the cornerstone upon which modern quantum mechanical theory is firmly constructed.

Bibliography

Beiser, Arthur. *Concepts of Modern Physics*. New York: McGraw-Hill, 1987. Excellent text on the development of nonclassical understanding of space and time, matter and energy. Easily accessible to the general reader. Illustrations and examples abound.

Burns, Marshall L. *Modern Physics for Science and Engineering*. San Diego: Harcourt Brace Jovanovich, 1988. Well-written account of modern physics. Although calculus-based and aimed at a college physics audience, qualitative descriptions of basic quantum phenomena are understandable to the layperson. Illustrations are few, but well chosen.

Halliday, David, and Robert Resnick. *Fundamentals of Physics*. 3d rev. ed. New York: John Wiley & Sons, 1988. This edition of the classic undergraduate physics text contains highly descriptive chapters on modern physics including a thorough elementary description of quantum physics concepts, such as wave-particle duality. Illustrations and sample problems abound. Excellent questions for further thought.

Jammer, Max. *The Philosophy of Quantum Mechanics*. New York: John Wiley & Sons, 1974. High-powered, but readable account of the philosophical implications of modern nonclassical thought about matter and energy.

Ohanian, Hans C. *Physics*. New York: W. W. Norton, 1985. Although calculus-based, the text is not mathematically rigorous. Accessible to those with modest mathematics skills. Excellent descriptions of difficult physical concepts. Well illustrated; provides practical examples.

Weidner, Richard T., and Robert L. Sells. *Elementary Modern Physics*. Boston: Allyn & Bacon, 1985. Beautifully written explanation of modern physics concepts such as the development of nonclassical thought. Wave-particle duality is demonstrated through discourse of matter and energy interactions. Well illustrated. Mathematics level is adequate for the layperson as well as the physics student.

Wilson, Jerry D., and John Kinard. *College Physics*. Boston: Allyn & Bacon, 1990. Basic text excellent for the reader unfamiliar with calculus. More qualitative than rigorous. Well-illustrated and thorough description of basic concepts without resorting to advanced mathematics. Excellent for high school physics instruction.

David G. Fisher

Cross-References

The Dirac Equation, 685; Electromagnetism, 750; Electrons and Atoms, 778; General Relativity: An Overview, 950; Mechanics, 1367; Nonrelativistic Quantum Mechanics, 1564; The Interpretation of Quantum Mechanics, 1986; Schrödinger's Equation, 2156; Special Relativity, 2333.

WAVES ON STRINGS

Type of physical science: Classical physics
Field of study: Mechanics

A study of the mechanical motion of pulses and standing waves in stretched strings is essential to understanding the physical behavior of the musical instruments of the string family, each of whose musical sounds derives from the bowing, plucking, or striking of strings under tension.

Principal terms

ELASTICITY: the tendency of solid objects to return to their original form after a stress is removed

FREQUENCY: the number of waves passing a fixed point in one unit of time; its common unit is the hertz, which is one per second

INERTIA: the tendency of a moving object to keep on moving

NEWTON: a common unit of force equal to the force required to accelerate a 1-kilogram mass 1 meter per second squared

NODE: a point that remains fixed in a standing wave; nodes separate the regions of maximum motion, the antinodes, in a standing wave

SUPERPOSITION: the addition of two or more waves that are present together in a medium

WAVELENGTH: the distance between adjacent points in a wave having exactly the same displacement and velocity at any given time

Overview

Waves and pulses are distinctive in that, once started in elastic substances, they propagate away from their points of origin entirely by themselves. A stone dropped into water, for example, creates a disturbance in which the water at the point of impact piles up slightly, and this local displacement is immediately transmitted to neighboring points on the surface, which in turn influence points still farther out, and so on. As a result, a localized group of waves propagate spontaneously outward. A cork floating on the surface notes the transit of the wave by bobbing up and down in the direction at right angles to that in which the wave is moving. This oscillatory behavior in the medium through which the wave is traveling is a characteristic of all wave motions. Locally, as a wave goes by a given point, the medium simply oscillates back and forth; it does not move bodily in the direction in which the wave disturbance is traveling. When the oscillatory motion is at right angles to the direction of the wave's travel, as in the case of waves on water, the wave is said to be transverse. Waves on strings are transverse. If the oscillatory motion is parallel to the direction of travel, as is the case for the motion of sound waves in air, the wave is said to be longitudinal.

The speed at which any wave moves forward depends on two properties of the medium. If the medium is very dense (a heavy rope, for example), its inertia will be high and it will oscillate relatively slowly, and so a traveling wave will also propagate slowly through it. The second factor has to do with the medium's elasticity. A material such as steel is highly elastic in the sense that very large forces come into play when it is deformed, while wood, on the other hand, is relatively inelastic. Hence, a sound wave traveling in steel will move more rapidly than in wood because the higher elasticity of steel more than outweighs the effect of its greater density. In a perfectly general way, the speed of transmission of any wave or pulse increases according to the ratio of the medium's elasticity to its density.

For waves on a stretched string, the speed of propagation can be shown to be given mathematically by the square root of the ratio of the tension in it (the relevant elastic factor, measured in newtons) to its mass density (the inertia factor in kilograms per meter of length). Imagine that there is a string under tension, tied at one end to a fixed support on a wall while the other end is held in the hand. One could initiate a pulse in the string by moving the free end rapidly up and down. Because a string can be thought of as a system of connected mass particles, the initial upward motion of the hand starts the particle next to it moving up, too. Since the first is attached to the number-two particle in line, it also experiences an upward pull and begins to move up, though its motion is slightly delayed in time with respect to that of the first. The upward displacement of successive particles in the string is identical, but delayed in time, and so the leading edge of the pulse is created in this way. Having reached its maximum displacement, the hand moves back toward its starting position. Again, this downward motion is transmitted successively to particles farther down the string, with a time delay, so that when the hand again arrives at its original position, a positive pulse will have been created that moves away at a speed determined by the tension and mass per length of the string. An identical, but downward, motion of the hand will then create the second half of what is recognized as a wavy configuration in the string: The whole shape would occupy what is termed one "wavelength" in the string. (If it appears exactly like the mathematical sine function, the wave is a sinusoid.) The maximum displacement of the hand would have fixed the maximum sideways displacement of the string, which is called the wave "amplitude." A wave transports energy, and it is not difficult to show that the energy it carries depends on the square of the amplitude.

Suppose that the hand is moved steadily up and down at a fixed rate—for example, three times each second: This is the source "frequency." If each full wavelength occupies a space of 2 meters, then the wave disturbance will have moved a distance $3 \times 2 = 6$ meters in that one second. That is, the speed of the wave will be 6 meters per second. In general, the speed at which a wave travels is given by the product of its frequency and wavelength. This speed is often referred to as the "phase velocity." Since the phase velocity is governed by the properties of the medium itself, increasing the frequency of the source (the hand, in the example above) will mean that the wavelength must decrease: It will do so in such a way that the speed, the product of frequency and wavelength, stays exactly constant. This means that all waves in the string, regardless of frequency, will travel at exactly the same speed. Such media are "nondispersive." There are media, however, in which the velocity of propagation does depend on frequency; in those media, two waves that differ in frequency get farther and farther apart as they move away from their source.

It can be shown mathematically that a well-localized, single pulse can be synthesized in a medium by simply adding together a large group of pure waves (sinusoids), each component of which differs slightly in frequency and wavelength from the next. If the medium is nondispersive, each component wave will travel at the same speed, and so the pulse will maintain its original shape, traveling at the same phase speed as each component. If, on the other hand, the medium were such that the component waves in the pulse had speeds that depended on their frequencies, then the faster components would get ahead of the slower ones, and the pulse shape itself would become more spread out, less well-localized. A medium of that kind is termed "dispersive." The speed at which the center of the pulse travels, which is a measure of energy flow, is not the same as that of any single component wave's phase velocity, or even of the average. It is termed the "group velocity" and is generally smaller than the average phase velocity of the components making up the group.

One might wonder what will happen in a medium if two independent waves arrive at the same place at the same time. Clearly, the medium can do only one thing. A person in a small boat, for example, might be subjected simultaneously to waves from two independent sources. At a given moment, one might have caused the boat to rise by three feet, while the other wave acting alone would have lowered it by an equal distance. It is found that the water simply sums algebraically the two individual displacements: +3 feet + (−3 feet) = 0. That is, at that particular instant, the boat suffers no displacement at all. This concept for the straightforward algebraic addition of individual displacements is called the "superposition principle."

Returning to the earlier example of a single, positive pulse traveling toward the fixed end of a string, one can see what happens when it arrives there. Until that moment, the various segments of string, being free to move, have simply responded to the forces imposed upon them during the passage of the pulse. Arriving at the fixed end, the leading edge attempts to pull up the wall at the point of attachment. While the rope is unable to cause any motion of the wall, its upward pull does generate a reaction force from the wall in the downward sense (Sir Isaac Newton's third law of motion—action/reaction), and so the leading edge of the pulse is pulled down, rather than up. The pulse itself is thus turned down and reflected back toward the hand. This mechanism explains why it is that, upon reflection from a fixed support, the crests of a continuous train of waves are returned as troughs, and the troughs as crests—or, one could say that a phase change had occurred upon reflection. Moreover, if there is no loss of wave energy, the incident and reflected waves will have identical amplitudes. Hence, when a wave travels down a string and is returned from a fixed end, the reflected wave has the same wavelength, frequency, amplitude, and speed.

The superposition principle allows one to determine the overall situation when two identical but oppositely moving waves exist simultaneously in the string. Exact analysis shows that, provided the tension is just right, the string divides into vibrating segments that are bounded by stationary points called nodal points, or nodes. Between the nodes are regions, antinodes, in which vigorous side-to-side motion is seen. (See figure; these motions can readily be studied by fixing a loose string at one end and shaking waves into it from the other.) A variety of patterns can be generated by making appropriate adjustments to the tension in the string, thus controlling the velocity—and therefore the wavelength at a given frequency—of the component waves. Since the pattern of nodes and antinodes remains stationary on the string, a wave of this type is called a "standing wave." In a standing wave, the distance between any two adjacent nodes is exactly one-half the wavelength of the component waves. The fixed ends of the string are, of necessity, themselves nodes.

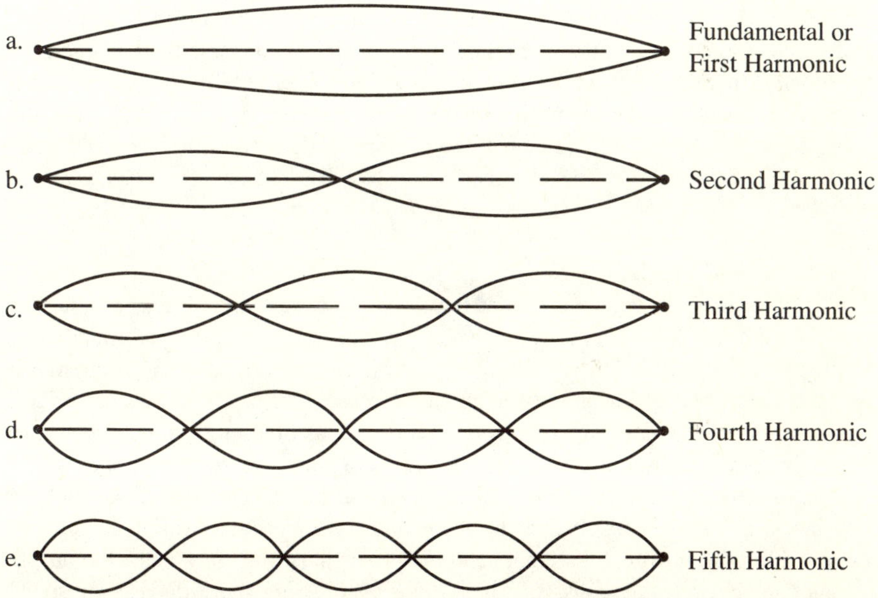

THE FIRST FIVE HARMONICS OF A STANDING WAVE IN A STRETCHED STRING

Since the product of wavelength and frequency gives the wave speed, an increase in frequency means a decrease in wavelength, and vice versa. Thus, the standing wave mode of longest wavelength is the one in which only a single antinode lies between the end nodes (see figure), and it is generated at the lowest source frequency. It is called the "fundamental." When the string is vibrating in its fundamental mode, one can easily find the wavelength of the component waves by multiplying the string length by two; then the speed of a traveling wave can be determined by finding the product of wavelength and frequency. The next higher mode (see figure) shows one wavelength standing in the string, the next one up three-halves, then four-halves, and so on. Since for a given medium (here, a string under fixed tension), the wave speed is fixed, and since each higher mode above the fundamental has progressively decreasing wavelength, the frequency of each mode increases correspondingly as one moves up in the series. For an ideal, perfectly flexible string, the modes form a harmonic series in which the frequency of each member is related as the ratios of the integers 1:2:3: etc. These are called the normal modes of oscillation.

Applications

There is no question that the most familiar practical application of the theory of traveling and standing waves in strings is to the various stringed musical instruments. The instruments of the string family—violin, viola, cello, and double bass—are among the most important members of the modern orchestra; the harp, piano, and guitar also depend on the motion of vibrating strings to create their characteristic sounds. In each of these instruments, a string under tension is either bowed, plucked, or struck. Each technique serves to start waves traveling away from the contact point in the string, and their subsequent reflections from the fixed ends lead to the establishment of complex standing waves. The string vibrations generate sound waves in the air as well as in the body of the instrument itself. The body's interior and outer surfaces serve as resonating members, whose functions are to amplify the sound of the strings and send it out to the listener.

In each of these instruments, the strings are held firmly under a tension that can be adjusted by means of a peg around which the free end of the string is wound, thus setting the frequency and pitch. The overall stress on the body of the instrument can be quite considerable: String tension amounts to some 50 or 60 pounds or so for a violin, while for a modern grand piano, it can reach as high as 20 tons. The strings themselves are usually made of catgut, nylon, or metal. The four basic open-string pitches of a violin (G, D, A, and E of fundamental frequencies 196, 294, 440, and 660 hertz) are clearly insufficient to account for the enormous variety of pitches found within the instrument's range, and the player must therefore make the proper changes in the string lengths, which is done by pressing them down against the fingerboard.

The pitch of a musical sound is directly related to the frequency of the vibrating string. The string impinges on the surrounding air and sets off a sympathetic vibration at the same frequency. High pitches are most easily generated by light, short strings held under high tension. Yet, for any given string, there is a whole series of standing waves that it can support. The lowest in frequency is the fundamental. The fundamental pitch is normally the most pronounced and has the greatest amplitude. The other standing wave modes (see figure) will also be excited to a greater or lesser degree: Surprisingly, an entire group of the string's harmonic series will normally be present simultaneously in the pattern of its motion. The quality, or timbre, of a note depends critically on the relative proportions of the string's harmonics in that overall mixture.

Bowing a string depends upon a phenomenon in which the hairs of the bow first grip the string, thus pushing it to one side, then slip, releasing it, then grip again, and so on. In this way, a steady vibration is set up. The spectrum of harmonics generated, and their relative strengths, depends on the point at which the bow contacts the string. For example, a string bowed near its midpoint will strongly excite the fundamental mode, since it has an antinode near its center (see figure). Conversely, the next harmonic will be relatively weakly stimulated, since it has a node at the midpoint (see figure). Other even harmonics will be strong for the same reason, while odd ones will tend to be weak. Thus, the position of the bowing or pluck point strongly affects the overall aural quality of the sound. The width of the pluck (its sharpness) also affects the nature of the sound produced by the instrument.

In a piano, there is normally a group of three closely spaced strings assigned to each key. It is found that a richer, more pleasant-sounding tone results when the piano tuner makes minute adjustments of tension in order to ensure that the pitch of each member of the set of three is slightly different. Another factor strongly influences the richness of the piano sound, and it comes about because, unlike the ideal, perfectly flexible string, the wires used in a piano—especially those of the bass notes—have a considerable intrinsic stiffness that tends to decrease their flexibility. For that reason, the ideal string's perfect harmonic series, whose frequencies increase in a strictly integral way, are somewhat compromised, and what are then termed the "partials" of the fundamental are not quite harmonic. This so-called inharmonicity is increasingly important for the higher partials of a given fundamental, and its presence adds a quality of richness to the piano's sound.

Context

The study of waves on strings is one of the oldest known to humankind. The Pythagoreans of the sixth century B.C., whose natural philosophy was founded on the integer numbers 1, 2, 3, and so on, are said to have discovered the pleasing qualities of the sounds emitted by stretched strings whose lengths were proportional to the integers. Nevertheless, a physicist's interest in studying wave motions generally, and standing waves in particular, is by no means restricted to simple, one-dimensional, mechanical systems, such as vibrating strings. They are naturally of greatest interest to musicians. Scientists in many fields have discovered an enormous variety of applications for the general theories of both traveling and standing waves.

Wave motion is, moreover, one of the great unifying concepts of physics. Traveling waves of all kinds are vehicles for the transmission of energy: Mechanical waves are observed as the precursors of often catastrophic seismic events on Earth; ocean waves are now regarded as reservoirs of energy that can be captured and utilized, and sound waves are responsible for the transmission and the perception of speech and music.

The wave behavior of light, confirmed experimentally during the early years of the nineteenth century, gradually led to the understanding of light as but one small part of the great electromagnetic spectrum that stretches from very short-wavelength X rays to the much longer-wavelength radio waves. All are fundamentally electric and magnetic in nature and all travel with the speed of light. Yet, as different as they would seem to be from simple waves on strings, they can be described in mathematical terms that are identical, and the concepts that were developed originally for mechanical waves are found to carry over in a very natural way. The startling discovery that matter itself can also exhibit a wavelike behavior led to a postulate by mathematical physicist Erwin Schrödinger (1887-1961) that material particles obey their own characteristic wave equation. This idea not only made possible the interpretation of certain puzzling experiments but also led to the notion that the electrons, which lie in the outer reaches of the atom surrounding the nucleus, could also be thought of as standing electron waves.

The basic idea of the one-dimensional standing wave—as exemplified by mechanical waves on strings—is readily expanded to three dimensions and to systems in which other than purely mechanical motions are involved. Numerous technological applications of these ideas have been exploited in virtually every subfield of physics, from radio transmitters to lasers.

Bibliography

Berg, Richard E., and David Stork. *The Physics of Sound*. Englewood Cliffs, N.J.: Prentice-Hall, 1982. This book contains an excellent discussion of the basic physics of traveling and standing waves and introduces the families of musical instruments, including the strings. Each chapter concludes with an annotated bibliography. Ancillary to this work is a set of videotapes made by the authors that is especially interesting when used in conjunction with the book.

Bolemon, Jay. *Physics: An Introduction*. Englewood Cliffs, N.J.: Prentice-Hall, 1989. This book is a general introductory college physics text written for students not majoring in a science. The discussion is often humorous, though full justice is done to the underlying principles, and the ideas are made accessible. Several chapters are devoted to vibrations, waves, and sound, while later chapters point up the notion of standing waves as applied to atoms.

French, A. P. *Vibrations and Waves*. New York: Norton, 1971. Another standard work that concentrates mainly on mechanical waves. It was written as one unit in the MIT introductory physics series and is both challenging and thorough in its approach. Contains many useful illustrations.

Pierce, John R. *Almost All About Waves*. Cambridge, Mass.: MIT Press, 1974. This is a classic introductory work on wave motion. Pierce attempts to present the material, some of it highly sophisticated, in reasonably nonmathematical language, although the knowledge of algebra is assumed. The topics extend beyond the normal discussion of simple mechanical waves to include seismic, electromagnetic, and ocean waves, and the discussion of phase and group velocities is unusual.

Rossing, Thomas D. *The Science of Sound*. 2d ed. Reading, Mass.: Addison-Wesley,

1989. This is a standard introductory text in the physics of sound and music. It is written in nontechnical language and is intended especially for the student without background in physics. Basic physics, which includes a discussion of traveling and standing waves, is covered in part 1, while the perception and measurement of sound, and the acoustics of instruments, are discussed in parts 2 and 3. The many illustrative photographs and drawings are exceptionally well done.

Van Bergeijk, Willem A., et al. *Waves and the Ear*. Garden City, N.Y.: Doubleday, 1960. This work, a part of Doubleday's Science Study Series, was conceived as a supplement to the standard introductory physics texts. The primary aim of the series was to capture and stimulate the interest of lay readers and beginning science students alike. Therefore, the discussion, while nonmathematical, is occasionally challenging. The authors, all researchers in the field, present the physical principles of wave motion in the early chapters, then move on to a thorough examination of the physiology and psychoacoustics of hearing.

David G. Fenton

Cross-References

Acoustics, 23; Electrons and Atoms, 778; Gravitational Waves, 1000; Lasers, 1227; Oscillating Systems, 1710; Phonons, 1785; Radar, 2020; Schrödinger's Equation, 2156; Producing and Detecting Sound, 2305; Sound Waves, 2311; Wave-Particle Duality, 2713.

WEAPONS MATERIALS REACTORS

Type of physical science: Nuclear physics
Field of study: Nuclear reactors

Nuclear reactors have been designed to use the neutron flux from a nuclear fission chain reaction in order to produce plutonium and tritium, which are suitable for the manufacture of nuclear weapons.

Principal terms

BETA DECAY: radioactive decay in which a nucleus emits either a positron or an electron (with the appropriate neutrino) and changes into a different chemical element with the same atomic mass

BURNUP: the percentage of fissionable nuclei in a sample that has undergone fission; measures the exposure of nuclear fuel elements to neutron flux

ISOTOPES: atoms of a particular chemical element that have different atomic mass numbers

NEUTRON CAPTURE: a nuclear reaction in which a nucleus absorbs a neutron and forms a heavier isotope of the same chemical element

NEUTRON FLUX: the number of neutrons crossing a unit area per unit time

PLUTONIUM: a man-made chemical element, an isotope of which (plutonium 239) is widely used in the manufacture of nuclear weapons

TRITIUM: an isotope of hydrogen which has one proton and two neutrons and which is used in the manufacture of boosted nuclear weapons

Overview

Nuclear reactors are based on nuclear fission. In nuclear fission, a nucleus absorbs a neutron and splits into two unequal parts and several extra neutrons. The fission fragments are frequently highly radioactive. To construct a nuclear reactor, fissile nuclei are arranged so that the neutrons from each fission trigger another fission. This process, called a chain reaction, releases both energy and neutrons at a steady rate. If each fission triggers more than one additional fission, then energy and neutron release rapidly accelerate, as is the case in a nuclear weapon. In a reactor, the chain reaction is controlled by inserting neutron-absorbing rods into the core of the reactor, and energy and neutrons are released at a constant rate.

The chemical behavior of an atom is determined by the number of electrons that surround its nucleus when it is not ionized. This number of electrons equals the number of protons in the nucleus of the atom. Thus, all carbon atoms have six protons in their nuclei, and all uranium atoms have ninety-two. The neutrons that also occupy the nucleus do not affect the chemistry of the atom, but they do play a major role in determining the stability of the nucleus. Nuclei with the same number of protons but different numbers of neutrons are said to be different isotopes of the same element. An isotope is designated by the total number of protons and neutrons in its nucleus. For example, the isotope carbon 12 has six protons and six neutrons.

The naturally occurring isotope uranium 235 undergoes fission, but it constitutes only 0.7 percent of the uranium found in nature. In order to construct a nuclear weapon, the concentration of uranium 235 must be increased to between 10 percent and 20 percent. Uranium that contains more than 0.7 percent uranium 235 is said to be enriched, while uranium that contains less than 0.7 percent is called depleted uranium. Most modern weapons-grade uranium is enriched until it is more than 90 percent uranium 235. The enrichment process is difficult and expensive, as the isotopes cannot be separated by chemical means. Nuclear reactors can operate with naturally occurring uranium that is 99.3 percent uranium 238 if they are cooled with graphite or heavy water. Reactors can operate with uranium enriched to 3 percent uranium 235 if they are cooled with normal water. Uranium 238, which constitutes 99.3 percent of naturally occurring uranium, does not fission unless it absorbs a high-energy neutron. In reactors, the neutrons are slowed down by collisions with the atoms of a substance called the moderator so that they do not have enough energy to cause fission in uranium 238.

A second fissionable isotope of element number 94, plutonium 239, does not occur in nature but can be produced in the cores of nuclear reactors. Plutonium 239 has a smaller critical mass than uranium 235 and produces slightly more neutrons in an average fission. Therefore, plutonium 239 offers some advantages over uranium 235 in the construction of nuclear weapons. Plutonium 239 is produced when a nucleus of uranium 238 captures a neutron to form uranium 239. Uranium 239 emits an electron and an antineutrino to form neptunium 239. This beta decay has a half-life of twenty-four minutes. The neptunium 239 in turn beta-decays with a half-life of 2.4 days to form plutonium 239. Thus plutonium is formed in the core of a nuclear reactor if targets of natural or depleted uranium are inserted in the reactor core. The longer the target remains in the core and the higher the neutron flux to which it is exposed, the more plutonium 239 will be produced.

Two processes compete with the formation of plutonium 239. First, plutonium 239 itself fissions so that, as plutonium 239 builds up in the reactor core, an increasing percentage of it absorbs neutrons and therefore fissions. Second, plutonium 239 can capture an extra neutron without fissioning to form the isotope plutonium 240. Plutonium 240, in turn, can capture neutrons to form still heavier isotopes of plutonium. The longer the target element remains in the core and the higher the neutron flux of the reactor core, the more heavy isotopes will be produced. Plutonium 240 fissions spontaneously without absorbing a neutron. Thus, large concentrations of it in nuclear weapons may cause a spontaneous chain reaction inside the weapon before it is triggered. The explosion would be a "fizzle" in relation to the designed yield of the weapon. Nevertheless, preignition from large concentrations of plutonium 240 would be disastrous. The U.S. Department of Energy defines weapons-grade plutonium as that which contains less than 7 percent plutonium 240. If it is necessary to separate plutonium 240 from plutonium 239, then the separation must be done by physical means and will involve the same difficulties and expense as enriching uranium.

The burnup of a fuel element measures the percentage of the fissionable nuclei in the element that has undergone fission. The burnup depends on the neutron flux in the core of the reactor and on the amount of time that the fuel element has been left in the core. The neutron flux depends on the rate of the chain reaction in the core and thus on the power that is being generated by the reactor. Reactor engineers typically express the burnup of a fuel element in terms of megawatt-days per metric ton of fuel. The amount of time that targets of uranium 238 are left in the core of a production reactor is measured in the desired burnup of the element. Exact times will depend on the detailed design of the reactor, which determines the neutron flux at the location of the target for different power levels of operation.

A second material critical to the production of modern nuclear weapons is the isotope hydrogen 3, or tritium. Tritium and hydrogen 2 (deuterium) are used to "boost" the yield of nuclear-fission weapons. The fission explosion triggers the fusion of a small amount of deuterium and tritium, which produces high-energy neutrons. These energetic neutrons trigger fission in a uranium-238 blanket surrounding the weapon, producing large amounts of energy in addition to the small amount of energy from the fusion. Deuterium is stable and found in naturally occurring water. Tritium has a half-life of 12.3 years, so supplies of this isotope must be constantly renewed. Like plutonium 239, tritium is produced in nuclear reactors. A target of the isotope lithium 6 is inserted into the reactor core. Lithium 6 captures a neutron and splits into a helium-4 nucleus and a tritium nucleus. Lithium 6 is seventeen times as likely as uranium 238 to capture a neutron, but tritium and plutonium-239 production rates are limited by the neutron flux in the reactor core. Therefore, the production of a mole of tritium requires approximately the same burnup as the production of a mole of plutonium.

Applications

Reactors that are designed for the production of materials for nuclear weapons contain the same basic elements as conventional reactors that are designed for the production of electricity. In fact, the first electricity generated from nuclear fission was produced as a by-product by a weapons materials reactor. The fissile fuel is inserted in fuel rods, which are lowered into the core. The neutrons from the fissions are slowed down by a coolant to thermal energies. The coolant or moderator also carries excess heat away from the core.

The first nuclear weapons reactors were literally piles of graphite blocks surrounding slugs of uranium metal or oxide fuel. Heat was carried away by air. The graphite served to moderate the energy of the fission neutrons, but it also absorbed many of the neutrons. Deuterium in the form of heavy water proved to be a much more efficient moderator than graphite: It not only absorbs fewer neutrons than the graphite but also slows the neutrons down faster. A deuterium nucleus has only twice the mass of a neutron, so neutrons lose more energy in each collision with a deuterium nucleus than in a collision with a more massive carbon nucleus. Weapons production reactors that are cooled with heavy water can be designed on a much smaller scale and sustain higher neutron fluxes than those cooled with graphite. Because they slow down faster in a reactor cooled with heavy water, neutrons are more likely to be captured by uranium-238 nuclei before they escape from the core. The reactor also uses control rods, which absorb neutrons and can be lowered into the core to retard the chain reaction.

Depending on the design of the individual reactor, target nuclei can be inserted as part of the fuel elements, used as part of the control rods (lithium 6 is particularly useful for this purpose, since it is an excellent neutron absorber), or placed in a blanket surrounding the reactor core. Nuclear weapons reactors have been run using a variety of fuels ranging from natural uranium to highly enriched uranium. Core designs and methods for loading fuel and target materials vary depending on the specific production campaign in progress. Nuclear weapons reactors can also be used to produce the fissionable isotope uranium 233, which does not occur naturally, or such exotic manmade nuclei as californium 252.

The target material must be easy to extract from the reactor core because operators wish to limit burnup in the targets in order to avoid the production of the heavy isotopes of plutonium. For example, a typical plutonium-239 production cycle at the Savannah River plant in Barnwell, South Carolina, calls for a sixty-day cycle at a power of 2,150 megawatts to produce plutonium that is 6 percent plutonium 240 and a thirty-day cycle for plutonium with 3 percent plutonium 240. Tritium production requires cycles on the order of two hundred days.

Adding or removing material from the core of the reactor tends to change the power level at which the chain reaction is stable. Thus, changing the target material in the reactor must be done with care in order to balance the supply of nuclear fuel and the target nuclei that are present. If there are not enough fissile nuclei in the core, then a chain reaction cannot be sustained. On the other hand, designers wish to maximize the number of target nuclei in the flux in order to breed as much plutonium as possible. Reactors are generally shut down during fuel and target changes because of the danger of instabilities developing during the process. In addition, the gradual accumulation of plutonium 239 inside the core increases the number of fissionable nuclei in the system. Fuel and target rods are usually closer together in the core of a production reactor than in a power reactor. This arrangement exposes the target nuclei to a higher neutron flux and helps to maintain a concentration of fissionable nuclei that is sufficiently high to sustain a chain reaction in the core. Engineers design cores known as charges for specific production tasks and burnup requirements.

After the target element has been irradiated and removed from the reactor, the plutonium must still be separated from the uranium 238 in the target. Because some of the plutonium in the target will have undergone fission, the material will be extremely radioactive from the fission fragments. The first step in extracting plutonium, uranium, or other isotopes for medical, military, or scientific uses is to place the target elements in cooling ponds for six months to allow short-lived isotopes to decay. Following the cooling period, the target elements must be chemically stripped of their protective coating or cladding and chopped into pieces for chemical processing.

There are several chemical processes that can be used to separate the elements in the still-radioactive rods. The most commonly used method in the United States is the "purex" process. The rods are dissolved in nitric acid, and the resulting solution is fed into a device in which the uranium and plutonium are concentrated in an organic solvent while the fission fragments are carried off in a weak solution of nitric acid. Uranium and plutonium are then separated from each other in a series of countercurrent processes using nitric acid and organic solvents. The purified plutonium is converted to an oxide or a metal for weapons manufacture.

The radioactive solutions are difficult to handle safely, and separation plants are expensive and difficult to design. The chemical process plants at Savannah River have two "canyons," each 9.0 meters wide at the top and 4.5 meters wide at the bottom and separated by four levels. The "hot" canyon is heavily shielded and contains equipment for the remote handling of radioactive material. The target material must be handled in this area until the highly radioactive fission products have been removed. After this removal, processing can be done in lightly shielded or unshielded areas. Tritium is recovered in a separate facility.

In addition to the danger that is intrinsic in handling highly radioactive materials, tight security must be maintained around nuclear weapons reactors to ensure that no one can steal the materials in order to manufacture a nuclear weapon.

Context

The first nuclear reactors developed during the Manhattan Project were nuclear weapons reactors. The second of two experimental graphite piles constructed in Chicago demonstrated plutonium production. An experimental air-cooled graphite pile at Oak Ridge, Tennessee, was the design prototype for the larger, water-cooled, graphite-moderated reactors constructed at Hanford, Washington, which produced the kilogram quantities of plutonium needed for the warheads detonated at the Alamogordo test site and over Nagasaki. The United States constructed other reactors at Hanford and Savannah River near Barnwell, South Carolina, during the 1950's. Each site also housed facilities for reprocessing fuel and target elements in order to separate out plutonium and tritium. Other nations, notably Great Britain, France, the Soviet Union, and the People's Republic of China, followed suit and constructed similar facilities in connection with the development of their own nuclear arsenals.

Many other nations have purchased or constructed nuclear reactors for electric-power production or for peaceful nuclear research. Experts have long been concerned that these reactors could be used to produce weapons-grade plutonium. Spent fuel from reactors that are designed to produce electric power contains a high concentration of plutonium 240, making it a poor candidate for a source of weapons material. The operation of electric-power reactors could be modified, however, so that fuel elements could be extracted after a much briefer processing period. Similarly, natural or depleted uranium targets could be irradiated in the cores of research reactors. Using a research reactor, India produced enough weapons-grade material for a nuclear test in 1974. It is believed that India has reprocessed enough fuel from civilian power reactors to construct a nuclear arsenal. Other nations may follow a similar model.

Concern over the spread of nuclear weapons prompted the signing of the Treaty on the Non-Proliferation of Nuclear Weapons in 1968. Although this treaty instituted international safeguards for the nuclear facilities of participating nations under the direction of the International Atomic Energy Agency, many of the nations whose potential for developing a nuclear arsenal is considered greatest have refused to sign the treaty. As of 1990, these nations included France, China, Israel, South Africa, India, and Pakistan. India had accepted international safeguards on the reactor fuel from which the material for its nuclear test device was extracted but seems to have developed nuclear weapons anyway. World reaction to India's nuclear test was not strong enough to deter other nations from doing likewise.

In the United States, nuclear weapons reactors were constructed following World War II. The materials from which they were constructed have undergone years of high-level irradiation. Many of these reactors have little or no external containment, and both Hanford and Savannah River are known to have severe problems with radioactive contamination of the environment. The aging reactors are becoming dangerous rapidly, forcing the United States either to rely on existing stocks of plutonium and tritium or to deploy a new generation of reactors for producing these materials quickly.

Bibliography

Cochran, Thomas B., William M. Arkin, Robert S. Norris, and Milton M. Hoenig. *U.S. Nuclear Warhead Production*. Vol. 2 in *Nuclear Weapons Data Book*. Cambridge, Mass.: Ballinger, 1987. Section 5 of this useful compendium of information contains a clear, simple summary of the operation of nuclear weapons reactors, including the design of the reactors and the principles governing the separation technologies. Written for nonspecialists, it provides a succinct, comprehensive picture of the operation of the reactors.

____________. *U.S. Nuclear Warhead Facility Profiles*. Vol. 3 in *Nuclear Weapons Data Book*. Cambridge, Mass.: Ballinger, 1987. This volume provides detailed portraits of the nuclear weapons reactors in the United States, including very specific data on the design of the cores used for various types of production projects and routine operating procedures for the reactors themselves. The data are probably more numerous than the casual reader will need, but they present a useful picture of U.S. production reactors.

De Volpi, Alexander. *Proliferation, Plutonium, and Policy: Institutional and Technological Impediments to Nuclear Weapons Propagation*. New York: Pergamon

Press, 1979. A more technical introduction to the problems of proliferation. Presents some of the practical steps that can be taken to ensure that plutonium intended for civilian use cannot be diverted to weapons production. This volume provides a very good introduction to the problems in weapons construction that are associated with heavy isotopes of plutonium.

Nero, Anthony V., Jr. "The Weapons Connection." In *A Guidebook to Nuclear Reactors*. Berkeley: University of California Press, 1979. This volume presents the production of materials for weapons from the point of view of a reactor specialist. Describes the origin of the problem of nuclear proliferation in connection with reactors designed for civilian use.

Patterson, Walter C. *The Plutonium Business and the Spread of the Bomb*. San Francisco: Sierra Club Books, 1984. This description of the production of plutonium for civilian use and the possibility of its diversion for nuclear weapons is very readable. It is strongly biased, however, against the widespread use of civilian nuclear power.

Rhodes, Richard. *The Making of the Atomic Bomb*. New York: Simon & Schuster, 1986. This excellent history not only of the Manhattan Project but also of the physics that underlies nuclear weapons provides an excellent introduction to the technology of the production of plutonium for use in weapons. The discussion in part 2 includes the history of the first plutonium production reactors and presents their basic principles in a clear and readable manner.

Spector, Leonard S. *The Undeclared Bomb*. Cambridge, Mass.: Ballinger, 1988. This volume traces the production of nuclear weapons and the materials that contribute to their manufacture throughout the world. A study in nuclear proliferation, it clarifies the relationship between civilian and military reactors that exists in practice. Includes an appendix on weapons material production.

Ruth H. Howes

Cross-References

Breeder Reactors, 259; Fission and Thermonuclear Weapons, 892; Nuclear Reactions and Scattering, 1605; Nuclear Reactors: Design and Operation, 1613; Power Reactors, 1928; Radioactive Nuclear Decay and Nuclear Excited States, 2057; Reactor Fuels and Waste Products, 2079.

THE PHYSICS OF WEATHER

Type of physical science: Classical physics
Field of study: Thermodynamics

The thermal structure of the earth's atmosphere determines the development of large-scale weather systems, the dispersion of atmospheric pollutants, and the potential for severe weather. A clear understanding of the thermodynamic processes of the atmosphere is important in many applications, including weather prediction, the completion of environmental impact studies, and the modification of select weather phenomena.

Principal terms

AVAILABLE POTENTIAL ENERGY: that part of the total potential energy of the atmosphere that is available for conversion to the kinetic energy of atmospheric winds

DRY ADIABATIC LAPSE RATE: the rate at which the temperature of a dry air parcel decreases with altitude as it rises through the atmosphere

EQUIVALENT POTENTIAL TEMPERATURE: the potential temperature an air parcel would have if its water vapor were condensed and the latent heat added to the parcel

HYDROSTATIC BALANCE: the balance between the downward force of gravity and the upward force resulting from the decline in air pressure with altitude

LATENT HEAT: heat released during the condensation of water vapor to form liquid water or the vapor deposition of water vapor to form ice

POTENTIAL TEMPERATURE: the temperature an air parcel would have if it were adiabatically expanded or compressed to a standard pressure of 100 kilopascals

TOTAL POTENTIAL ENERGY: the sum of the internal energy and the gravitational potential energy of an air column

WET ADIABATIC LAPSE RATE: the rate of temperature decrease with altitude of a rising air parcel in which condensation occurs

Overview

Earth's atmosphere is a giant heat engine that converts radiant energy from the sun into the kinetic energy of atmospheric winds. Rising air motions associated with atmospheric circulations are responsible for the formation of important weather phenomena such as clouds, precipitation, and severe storms. Depending upon the vertical variation of air temperature, rising air motions may be either enhanced or suppressed by buoyant forces, so the thermal structure of the atmosphere is a key factor in the development of weather.

For the larger-scale motions of the atmosphere, the acceleration of air in the vertical direction is always small. Therefore, an approximate balance exists between the downward force of gravity and the upward force because of the decline in air pressure with height. Motions that satisfy this approximate balance are said to be hydrostatic. The rate at which pressure decreases with height depends on the air density, and, since air behaves as an ideal gas, the density at any given pressure level is determined by the air temperature. Assuming a hydrostatic balance, the pressure at any level of the atmosphere may then be determined from a knowledge of the surface pressure and the vertical distribution of temperature.

Vertical air motions and air temperature are strongly interdependent. This may be understood by considering the buoyant force that acts on a rising air parcel—a small, isolated mass of air that neither mixes nor exchanges heat with the surrounding air. As the parcel rises through the atmosphere, it expands in response to the decrease in atmospheric pressure. During expansion, the parcel works to push back the air in the surrounding atmosphere. If the expansion can be considered adiabatic—that is, if the parcel does not lose or gain heat—the work is done at the expense of the energy associated with the random motions of the parcel's air molecules, the internal energy. Therefore, the temperature of the parcel will decrease as it rises through the atmosphere. In a hydrostatic atmosphere, the first law of thermodynamics may be used to show that the temperature of the parcel will decrease with increasing altitude at a rate of 9.8 Kelvins per kilometer. This rate of temperature decrease is called the dry adiabatic lapse rate.

The buoyant force experienced by the air parcel will depend on how the density of the parcel compares with that of its environment. If the temperature of the air parcel is less than that of the surrounding air, it will experience a downward buoyant force, which acts to return it to the level at which it originated. Similarly, if the air parcel has a temperature that is greater than its surroundings, it will experience a buoyant force, which acts to accelerate its upward motion.

Since the temperature of a parcel changes with its motion, it is useful to define a related thermodynamic variable that is conserved during adiabatic air movements. Potential temperature is defined as the temperature an air parcel would have if it were adiabatically expanded or compressed to a standard pressure of 100 kilopascals. Air temperature and pressure uniquely determine the potential temperature. While the potential temperature of a rising air parcel is constant, that of the surrounding atmosphere may differ from one level to another. The variation of the potential temperature of the atmosphere with height determines whether vertical air motions will be enhanced or suppressed by buoyant forces. If the potential temperature increases with height, the atmosphere is said to be stable, since rising or sinking air parcels will experience buoyant forces that act to return them to their original level. Similarly, an unstable atmosphere is one in which the potential temperature decreases with height.

The presence of water vapor has only a slight effect on the thermodynamic properties of air unless condensation takes place. If the temperature of an air parcel becomes cool enough, water vapor in the parcel will condense to form droplets of liquid water. The potential energy lost by the condensing water molecules is transferred to the surrounding air molecules as latent heat, so the temperature of a rising air parcel in which condensation takes place will decrease less rapidly than the temperature of a parcel in which it does not. The so-called wet adiabatic lapse rate is not constant but depends on temperature and pressure. The altitude at which a rising parcel of surface air begins to experience condensation is called the lifting condensation level. This level is often a good estimate of the altitude at which the bases of convective clouds are found.

If the lapse rate of the atmospheric environment—the rate at which air temperature decreases with increasing altitude—is intermediate between the wet adiabatic lapse rate and the dry adiabatic lapse rate, the atmosphere is said to be conditionally unstable. In such an environment, a rising parcel of surface air will experience a downward buoyant force unless it is lifted high enough that condensation occurs and the release of latent heat causes its temperature to exceed that of its surroundings. The altitude above which the parcel becomes positively buoyant and continues to rise without an external source of lifting is called the level of free convection.

During condensation and cloud formation, the potential temperature of an air parcel is not conserved. Yet, it is possible to define an equivalent potential temperature that is conserved. Equivalent potential temperature is the potential temperature an air parcel would have if all of its water vapor were condensed and the latent heat added to the parcel. When the equivalent potential temperature of the atmospheric environment decreases with height, the atmosphere is said to be convectively unstable. Under these conditions, if a layer of air is lifted through a large enough distance, the bottom of the layer will experience a greater heating because of the release of latent heat than will the top. Thus, the layer will become unstable even if the original lapse rate within the layer was stable. Most outbreaks of severe weather occur when the atmosphere is convectively unstable.

Once clouds form, some liquid water droplets and/or ice crystals within the cloud may grow large enough to fall to the earth as precipitation. At middle latitudes, most clouds contain some ice, and the growth of ice particles dominates the production of precipitation. This results from the fact that the amount of water vapor that is needed within a cloud for the growth of an ice crystal is significantly less than that needed for the growth of a liquid cloud droplet of the same size. Ice crystals, therefore, will grow at the expense of liquid droplets within the same cloud. The droplets evaporate and supply water vapor to the growing ice particles. Once the ice crystal has grown to a large enough size by vapor deposition, it will fall through the cloud and grow further by the collection of other cloud particles until it reaches precipitation size.

Atmospheric conditions that control local weather phenomena, such as clouds and precipitation, are determined by the large-scale motions of the earth's atmosphere, which are, in turn, driven by radiant energy from the sun. Near Earth's equator, the total solar energy incident at the top of the atmosphere exceeds the energy emitted by the atmosphere to space in the form of infrared radiation. By contrast, the polar regions of Earth's atmosphere lose more energy to space than they receive from the sun. The net effect is to increase the internal energy of the atmosphere near the equator and to deplete that energy near the poles. The winds of Earth's atmosphere play a dominant role in transporting excess energy poleward to balance the energy deficit at high latitudes.

In a hydrostatic atmosphere, the gravitational potential energy of an air column bears a constant ratio, about 0.4, to the internal energy. If internal energy is either added by heating or converted to kinetic energy, the gravitational potential energy changes to maintain the same ratio. Therefore, it is convenient to treat gravitational potential energy and internal energy together as the total potential energy. The amount of potential energy available to drive atmospheric circulations depends on the distribution of potential temperature. In the case of a stable and horizontally uniform atmosphere, no potential energy would be available for conversion to kinetic energy. Thus, the north-south temperature gradient created by the latitudinal variation in solar heating makes energy available to atmospheric circulation systems.

The portion of the potential energy that is available for conversion to kinetic energy during an adiabatic redistribution of air to create a horizontally uniform and stable atmosphere is called the available potential energy. Such a redistribution of air requires the sinking of cold air and the rising of warm air so that gravitational potential energy and internal energy are released simultaneously. At middle latitudes, these rising and sinking motions occur in association with large-scale circulations around low-pressure systems that develop along fronts. As warm air collides with and is lifted over cold air along the frontal boundary, potential energy is converted into kinetic energy from the developing circulation. Most significant weather events at middle latitudes are associated with this type of circulation system.

Applications

The concepts of atmospheric thermodynamics are of considerable practical importance in many areas of atmospheric science, including weather analysis and forecasting, the assessment of environmental impact, and weather modification.

Much of Earth's weather results from the interaction of adjacent air masses having different characteristics of temperature and moisture content. It is therefore important to be able to follow the motions of air masses by identifying characteristic properties that are conserved during their motion. Since most air motions are approximately adiabatic, air parcels will tend to move along surfaces of constant potential temperature. If the mixing ratio—the ratio of the mass of water vapor to the mass of dry air in an air parcel—is plotted on a map representing a surface of constant potential temperature, the motions of air masses become apparent, since mixing ratio is a conserved quantity outside regions where condensation occurs. Such an analysis is called an isentropic analysis.

Severe weather in the United States is most commonly associated with the southeasterly flow of warm moist air from the Gulf of Mexico near the surface of the earth and the westerly flow of cool dry air aloft. Under these conditions, the atmosphere is convectively unstable, and the change in wind direction with height favors the development of long-lived thunderstorms. If the afternoon heating of

Earth's surface is sufficient, the resulting convective motions may trigger the formation of severe weather. One measure of atmospheric stability that is used in severe weather forecasting is the lifted index. To calculate this index, consider an imaginary parcel of air with a temperature equal to the forecast high temperature for the day and a moisture content equal to that averaged over the lower atmosphere. If one imagines such a parcel to rise to the middle of the atmosphere, about 50 kilopascals, its temperature there can be calculated on theoretical grounds. As it is lifted, the parcel's temperature decreases at the dry adiabatic lapse rate until it reaches its lifting condensation level. From that point upward, its temperature decreases more slowly at the wet adiabatic lapse rate. At its destination, the calculated temperature of the parcel is subtracted from the measured temperature of its environment to obtain the lifted index. A large negative lifted index indicates a high probability of severe weather. Lifted indexes more negative than about -2 are correlated with tornado-producing storms.

Atmospheric stability also has a profound influence on the rate at which atmospheric pollutants are diluted by mixing with air. Turbulent eddies, which are responsible for most of the mixing, may be enhanced by an unstable environment or suppressed by a stable environment. At night, the radiational cooling of Earth's surface and adjacent air typically produces a stable atmosphere in which the dilution of atmospheric pollutants is very slow. Pollutants may be transported large distances from their source by the wind before their concentrations are appreciably reduced. During the daylight hours, the lower atmosphere is generally unstable, and convective motions driven by solar heating mix pollutants upward efficiently.

Often, thc layer mixed by convection is capped by a temperature inversion, a stable layer of air in which temperature increases with height. The depth of the atmospheric layer through which pollutants are mixed is an important factor in determining their concentrations. Therefore, accurate modeling of air quality requires a knowledge of the level at which the upper-level inversion occurs. Normally, the height of the inversion is determined from temperature measurements taken at several altitudes by an instrumented balloon called a radiosonde.

Strong upper-air inversions are usually formed by sinking air motions, subsidence, associated with systems of high pressure. The convective mixing in the lower atmosphere prevents these sinking motions from extending all the way to Earth's surface, so only air parcels above the mixed layer move downward, and the adiabatic temperature increase the sinking motions experience results in the formation of a so-called subsidence inversion. Most severe air pollution episodes are caused by subsidence inversions created by slow-moving high-pressure systems.

Since weather is such an important factor in many economic endeavors, including agriculture, the prospect of weather modification has been the focus of much attention. Most attempts at weather modification are based on the fact that ice crystals, if they exist within a cloud at the correct concentration, will grow at the expense of water droplets in the same cloud and may eventually produce precipitation. If a cloud whose upper parts are at a temperature below 0 degrees Celsius has too little ice to produce precipitation efficiently, it may be possible to enhance the precipitation by introducing ice nuclei into the cloud particles, which promote the growth of ice crystals. The addition of ice nuclei into a cloud is referred to as cloud seeding. Unfortunately, most studies on the enhancement of precipitation by cloud seeding have not been carried out with enough care to allow proper statistical evaluation of the results. With further study, however, this technique may eventually prove to be a useful tool for weather modification.

Context

By the early eighteenth century, it was already widely accepted that atmospheric winds must somehow be the result of differences in solar heating between the equator and the poles. Nevertheless, the exact relationship between surface winds and the latitudinal variation in air temperature was not generally understood. In 1735, George Hadley proposed a model of the general circulation of the earth's atmosphere, in which atmospheric winds were driven by a single thermal circulation cell between the equator and the poles. The trade winds—the easterly winds in the tropics—were explained in terms of the deflection of winds in the circulation cell by the rotation of the earth. This model more firmly established the connection between solar heating and large-scale air motions. William Ferrel modified Hadley's model in 1856 to include three main circulation cells in order to account for observed winds at higher latitudes.

In 1903, when Max Margules first introduced the idea of total potential energy, an understanding of the energetics of middle-latitude cyclones or low-pressure systems began to emerge. Nevertheless, it was not until 1918 that Vilhelm and Jacob Bjerknes developed the polar front theory of cyclones and convincingly showed that those storms receive their energy from the interaction of air masses of different temperatures. In 1926, Sir Harold Jeffreys showed that middle-latitude cyclones play an important role in maintaining the general circulation. The concept of available potential temperature and its usefulness in the study of the general circulation were presented by Edward Lorenz in 1955.

The forecasting of weather by solving the fundamental equations describing the physics of the atmosphere has been realized only in recent years. The difficulty lies in part in the fact that the equations are very complicated and general solutions have not been forthcoming. In 1904, Vilhelm Bjerknes first suggested that forecasts might be made by solving the equations numerically. Lewis Fry Richardson showed, in 1922, that equations describing the state of the atmosphere could be reduced to many approximate algebraic equations using appropriate techniques. Since digital computers were not yet available, however, the computations needed to obtain solutions were too lengthy to allow the technique to be useful for weather forecasting. In 1937, Carl-Gustaf Arvid Rossby developed a simple equation for forecasting the movement of large-scale troughs and ridges of pressure in the middle atmosphere by using the concept of vorticity—the spin of air around a vertical axis. Rossby's equation assumed that the atmosphere was barotropic, that is, that the temperature of the atmosphere was uniform in the horizontal. Thus, the equation was unable to predict changes in the strength of atmospheric disturbances that result from the release of available potential energy. Nevertheless, the first successful numerical forecasts performed in 1950 were based upon Rossby's work. By 1953, forecast models capable of predicting the development of middle-latitude cyclones were being integrated numerically. As more powerful generations of digital computers become available, numerical forecast models continue to include more detailed descriptions of the physics of weather.

Bibliography

Battan, Louis J. *Weather*. 2d ed. Englewood Cliffs, N.J.: Prentice-Hall, 1985. This book provides a concise qualitative treatment of the physics of weather. Battan has written several books on the subject of meteorology that are accessible to the general reader.

Friedman, Robert Marc. *Appropriating the Weather: Vilhelm Bjerknes and the Construction of a Modern Meteorology*. Ithaca, N.Y.: Cornell University Press, 1989. The author traces the development of meteorological thought that led to a modern understanding of middle-latitude cyclones. The relationship between the development of the commercial applications of meteorology and that of the science itself is explored. The career of the Norwegian meteorologist Vilhelm Bjerknes is central to the discussion.

Houghton, David D., ed. *Handbook of Applied Meteorology*. New York: Wiley-Interscience, 1985. This handbook is designed as a comprehensive reference for the nonmeteorologist. In addition to a thorough treatment of the fundamentals of the theory, extensive discussions of most modern applications of atmospheric science are included. Mathematics is employed frequently; however, much of the material requires little mathematical sophistication.

Sutton, O. G. *The Challenge of the Atmosphere*. Reprint. New York: Greenwood Press, 1969. This highly readable account of weather and climate provides a good overview of atmospheric physics at an elementary level. Sutton presents a well-integrated view of atmospheric circulation systems and their associated weather.

Wallace, John M., and Peter V. Hobbs. *Atmospheric Science: An Introductory Survey*. New York: Academic Press, 1977. This text gives a quantitative treatment of a wide range of topics in atmospheric science at an introductory level. The chapters on atmospheric thermodynamics, clouds and storms, and the general circulation are most relevant. Numerous biographical footnotes provide a historical background for the discussions.

R. D. Russell

Cross-References

The Behavior of Gases, 935; Phase Changes, 1777; Planetary Atmospheres, 1836; Stability, 2347; Laws of Thermodynamics, 2493; Thermodynamics: An Overview, 2486; The Von Neumann Machine, 2690.

WHITE DWARF STARS

Type of physical science: Astronomy and astrophysics
Field of study: Stars

Stars in their late stage of evolution of about a solar mass, but with diameters comparable to that of Earth and thus having ultrahigh densities, are known as white dwarfs. Typically, such stars have high surface temperatures and very low luminosities.

Principal terms

ASTRONOMICAL UNIT: average distance between the sun and Earth, which is equal to 149.59 billion meters

BINARIES: a system of two stars revolving around the common center of mass

DEGENERATE ELECTRON GAS: an assembly of electrons occupying all the quantum mechanically allowed space, found in ultrahigh densities within white dwarfs, where the electron pressure does not depend on temperature

DENSITY: mass per unit volume measured in grams per cubic centimeter or equivalently, kilograms per cubic meter

HERTZSPRUNG-RUSSELL DIAGRAM: a plot of luminosity against surface temperature of a collection of stars, named after Ejnar Hertzsprung and Henry Norris Russell

MAGNITUDE: a numerical value of the intensity of an object; in this scale of assigned values, the larger the number, the smaller the intensity

PROPER MOTION: the motion of a star on the surface of the celestial sphere

STELLAR LUMINOSITY: energy emitted by a star per second, measured in ergs per second or, equivalently, joules per second (watts)

Overview

White dwarfs are a unique class of stars. With approximately 1 solar mass and planet-sized diameters, they have densities ranging from 10 million to 1 billion kilograms per cubic meter. Compared to the main sequence stars of the same mass, their luminosities are very low, which makes them difficult to observe. Also, white dwarfs, in comparison with the main sequence stars of the same luminosity, would be characterized by much higher surface temperature. Since these faint stars, of necessity, must be relatively nearby to render them visible, they are observed to have large proper motions.

In 1862, Alvan Graham Clark discovered that the bright star Sirius had a faint companion, Sirius B. Soon, astronomers had determined the separation distance between the two stars to be 19 astronomical units, and Sirius B had about the sun's mass and 1/400 of solar luminosity. Walter Sydney Adams found in 1915 that the companion to Sirius had a high-surface temperature of 8,000 Kelvins. A normal star at this temperature would have three times the solar luminosity. It follows that the surface of this star is twelve hundred times smaller than that of the sun with 1/40,000 of the solar volume and a mean density of 50 million kilograms per cubic meter. About this time, Harvard astronomers discovered O Eridani B that had a surface temperature of 10,000 Kelvins, 0.44 solar mass, and 1/400 the luminosity of the sun, with the same density as Sirius B. Shortly thereafter, Adriaan Van Maanen found a lone star (not a part of the binary system), similar to the two earlier ones, which was subsequently named after him. Because of their small size, low luminosity, and high surface temperature, this class of stars is called white dwarfs. The unusual properties of white dwarfs have spurred numerous continuing theoretical studies. Sir Arthur Stanley Eddington, Ralph Howard Fowler, and Subrahmanyan Chandrasekhar were among the pioneers who studied and theoretically explained the numerous anomalous features of the white dwarfs.

White dwarfs are intrinsically faint (their visual magnitude ranges from 8 to 23, and absolute magnitude from 9 to 19), which places severe restrictions on their discovery and observation. White dwarfs occur among binary star systems as well as singly. One usually looks for faint blue stars with large proper motion in searching for them. Willem Jacob Luyten, who has spent his career studying these stars, estimates that 2 percent of the stars within 20 parsecs are white dwarfs. Even with improved techniques, their visibility is limited to 100 parsecs, and only incomplete information on a vast majority of white dwarfs is obtainable. It is highly likely that taking into account the increased density of stars among clusters and proximity to the galactic center, the incidence of white dwarfs may exceed 2 percent. Varying estimates indicate that 15 to 20 percent of galactic mass constitutes white dwarfs.

White dwarfs have complex spectral features. High surface gravity leads to broadening of lines, and in many cases one sees them as continuous spectra, rather than discrete lines, as in Wolf 489. They also show a very large Stark effect (named for Johannes Stark) caused by the presence of a strong electrostatic field that results from high densities and accompanying degeneracy. The majority of white dwarfs possess only hydrogen spectral lines, while others show only helium lines, and in rare cases, as in Van Maanen's star, hydrogen, potassium, calcium, and iron lines are observed.

Spectroscopic analysis of white dwarfs in principle yields surface temperature, surface gravity (in special cases, line-of-sight velocity), and visual magnitude. For stars of known distance and surface temperature, the radius and the mass can be computed. If the distance to the star is unknown, the mass can be calculated from the known surface gravity and the mass-radius relation. Spectroscopically knowing the gravitational redshift, using mass-radius relation from Albert Einstein's general theory of relativity, one can obtain the mass of the white dwarf. Jesse Leonard Greenstein has extensively discussed the problem of determining masses of white dwarfs belonging to binaries, such as Sirius B, O Eridani B, and a host of others. It is now well established that the progenitor of the white dwarf is the central star of a planetary nebula. From the known evolutionary tracks of the objects, masses of the central stars, and hence those of the white dwarfs, can be deduced. Relatively low-mass stars are likely to evolve into white dwarfs without becoming or being seen as planetary nebulas. Once the mass and radius of the star are known, its density is readily computable.

The theory of white dwarfs began with Eddington, who in 1924 suggested that the ultrahigh density of white dwarfs may be caused by the total ionization of matter within, resulting from extreme temperatures, so that more nuclei would be packed into the available volume of space. This theory leads to a famous paradox, which implies that when such a star cools, it must expand, requiring a source of energy that does not exist (the problem was later solved by Fowler).

In 1927, Fowler, applying the newly developed Fermi-Dirac statistics (named for Enrico Fermi and Paul Adrien Maurice Dirac), showed that at extreme densities within white dwarf stars, electrons are completely degenerate; that is, having occupied every available energy state, they behave collectively like incompressible fluid. At this juncture, Chandrasekhar developed what may be described as the "classical" theory of white dwarfs on the basis of the assumptions that electrons, which outnumber the nuclei and behave like incompressible fluid, are highly degenerate throughout the volume of the star; and the partial pressure exerted by the nuclei (or ions) is negligible in comparison with that of the electron gas. Degenerate electrons at high densities are more energetic than the nuclei. For example, at 10 million Kelvins, electron kinetic energy is 150 times that of the nucleon and thus exerts far greater pressure. Furthermore, the totally ionized state prevailing inside a white dwarf implies a far greater number of electrons in comparison with the number of nucleons. Hence, as opposed to the conditions in a normal star, it is the outward-directed degenerate electron gas pressure rather than the nucleon pressure (or atomic pressure in cooler stars) that balances the inward-directed gravitational pressure. Finally, the chemical composition is uniform within the star, so that the molecular weight per electron is constant throughout. Chandrasekhar's theory of white dwarfs has generally explained most of the observed features, in addition to providing important predictions. The theory points to 1.44 solar masses as the upper limit for the mass of a white dwarf. This results from an upper limit to the core temperature, which prevents further nuclear burning processes from continuing. The range of observed mass lies between 0.30 to 1.20 solar masses, with a peak at 0.60. At these low masses, the thermonuclear furnace in the center of a white dwarf would have ceased to exist, because the core temperature would not have been high enough to go beyond the helium burning stage. The theory yields several astrophysically important relationships between mass-radius, pressure-mass-radius, density-radius, temperature-mass, and the like. The radius of a white dwarf is inversely proportional to the cube root of its mass, an often-used functional relationship.

The central temperature of white dwarfs may range from 10 to 100 million Kelvins, and it is this value, which is dependent on the mass, that determines the cooling-off period, or the life of the star. Several competing physical processes and limiting factors play a crucial role in the life of a white dwarf. In addition to the usual emission of light, the star can lose energy by neutrino emission. It is suspected that as the white dwarf approaches a certain temperature, nuclei form body-centered cubic crystals that, although they behave more like a liquid, delay the cooling process through release of internal heat. The star traces a straight line from left to right, parallel to the temperature axis in the Hertzsprung-Russell diagram, through the period of cooling from white dwarf to black dwarf, taking several billion years, since its radius remains constant.

In some white dwarfs, the spectral analysis indicates the presence of a magnetic field 10 million times stronger than that of Earth. It is also suspected that the intense magnetic field of these stars, combined with rapid rotation, could be a prime source of cosmic rays. Given their large number and the interstellar magnetic fields in the galaxy, white dwarfs may be the most important high-energy source of cosmic rays. It is suspected that certain binaries are symbiotically coupled to dense, hot white dwarfs, forming accretion discs (rapidly spinning discs) around them, resulting in an intense source of ultraviolet radiation and jets of hot plasma. The study of white dwarfs, because of their faintness, presents a great challenge to astronomers.

Applications

Eddington first suspected that the answer to the puzzle of low luminosity, coupled with the high density of white dwarfs, may be related to one of the observable effects of Einstein's general theory of relativity. Light emitted in the vicinity of an inordinately dense object, exerting a powerful gravitational pull on the atoms (such as in the case of Sirius B), would be redshifted because of the elongation of wave lengths. At Eddington's request, Adams at Mount Wilson Observatory showed that it was as expected, confirming Eddington's suspicion. The confirmation of a prediction of the general theory of relativity through observation of gravitational redshifts of white dwarfs was a major achievement. The impetus that this successful result brought to the study of the new state of matter can be measured only by the progress made during the following decades.

Bose-Einstein statistics (named for Satyendra N. Bose and Albert Einstein) were discovered in the context of quantum systems and microsystems. Fowler, in applying the concept of degeneracy of electron gas—namely, that electrons must obey the Wolfgang Pauli exclusion principle (no two electrons can have the same energy, an esoteric notion at that time)—successfully explained the internal contents and pressure dynamics of white dwarfs. Thereafter, Chandrasekhar developed a complete and elaborate theory of the degenerate electron gas in white dwarfs, accounting for most of their observed features. Thus, the degeneracy of matter in macroscopic systems, such as in a special class of stars, was accepted as the norm rather than the exception. The idea was later extended to neutron star matter, where one comes across densities a billion times those of white dwarfs with analogous properties, such as the lack of an internal source of energy, incompressibility, and fixed radius.

The modern theories of condensed matter, solid states, superfluidity, and superconductivity have benefitted greatly from the theory of white dwarfs. In recognition of his enormous success, generality, and importance, Chandrasekhar was awarded the Nobel Prize in Physics in 1983, which he shared with William Alfred Fowler. The discovery and spectacular success in the theoretical study, followed by a series of very difficult observations of white dwarfs, mark the beginning of an exciting period in modern astrophysics and stellar astronomy. There appears to be no doubt that white dwarfs are the final stage of most stars with less than 2 to 3 solar masses. Since smaller-mass stars predominate, the Milky Way and other galaxies may contain many more brown and black dwarfs that cannot be observed. It is suspected that white dwarfs can be found both among the older stars (globular clusters) and relatively newer groups of stars. Faintness of these stars has led astronomers to adopt ingenious and novel techniques of observation with supersensitive instruments.

Astronomers have found that contrary to expectations, white dwarfs of lower luminosity cool more rapidly, and the observed radii are always somewhat larger than theoretically predicted. The effects of magnetic fields, rotation, and the possible existence of partially degenerate, unstable outer shells pose a few of the many aspects of the problem. Recent studies of neutron star binaries, as well as neutron and massive white dwarfs, present the application of the knowledge of white dwarfs in binaries to a new type of binary system. It is well known that mass accretion on white dwarfs from a normal binary companion is one of the possible mechanisms taking place in a recurring nova.

Theoretical studies of the possible further evolution of certain types of massive white dwarfs (containing mainly carbon and oxygen) are being conducted. It is not known, for example, whether such stars would collapse into neutron stars or whether oxygen and carbon burning would commence in their core, leading to explosion. Theoretical explorations are therefore essential to the understanding of the evolutionary tracks of stars at their limiting masses.

Context

The idea that the stars are born, live out their lives, and die is relatively new in astronomy. Until the discovery of neutron stars, which astronomers firmly believe is another evolutionary track to the final phase of more massive stars, white dwarfs were considered the only objects that had reached their end point. Herein lies the importance of the study of these stars, and in an attempt to understand and explain their profoundly unique features, one had to appeal to the fundamental aspect of nature—the properties of one of the basic building blocks of nature, the electron. The first step in the collapse of matter subject to the weakest of natural forces—the gravitational force—is the white dwarf formation. Astronomers learned later that the same force inexorably crushes matter into denser neutron stars and finally into superdense black holes.

New techniques, larger telescopes, and more sensitive instruments enable astronomers to penetrate new areas rapidly. The next generation of observers will no doubt be able to obtain images of far fainter objects and will probably discover many more white dwarfs among them. It is more than likely that greater focus will be placed on features that are already known or suspected, in addition to new ones.

For the first time, astronomers have come across stellar density and size never thought possible, and this alone has contributed to a widening of the horizons, so essential for progress in any field. Faintness of white dwarfs will always remain a challenge to the astronomer. While it is now known that only massive stars can explode as supernovas, having created heavier elements within, the observed presence of only the lighter elements in white dwarfs reassures the astrophysicist that the notion of the formation of heavy elements from hydrogen is essentially correct.

There are compelling reasons for the scanning of fainter objects in the infrared, ultraviolet, and X-ray regions of the spectrum, making use of orbiting astronomical instruments and targeting the coldest and the hottest of dwarfs. When correlated with the data from visible spectra, unknown aspects of these stars are likely to be revealed, and solutions to many of the puzzles should become evident. For example, information gathered at ultraviolet regions might enhance an understanding of the progenitors of the white dwarf, while data gathered at the infrared regions will no doubt add to the knowledge of their end points.

A complete and thorough knowledge of all stars, including the white dwarfs, is a goal astronomers cannot and should not hope to achieve soon. White dwarfs still remain a puzzle with many missing pieces. For one thing, the understanding of high-density matter, matter at limiting temperatures, the ultimate building blocks of matter, and the like, is intricately related to superdense stars. Astronomers are entering an era of great excitement and expectations resulting from new types of telescopes that operate above Earth's atmosphere; complementing their findings are physicists who are unraveling the mysteries of matter and the forces holding it together. Possibly, the answer to the origin of matter is intimately related to its end point, the present quest of astronomers and physicists alike.

Bibliography

Chandrasekhar, Subrahmanyan. "On Stars, Their Evolution and Their Stability." *Reviews of Modern Physics* 56 (1984): 137-147. Although this is a technical article, it is included for historical reasons. The general reader may easily skip over the mathematical equations and concentrate on the text, which is highly informative, it being the Nobel lecture delivered by Chandrasekhar in 1983. The article also includes references to several important works on the subject.

Gursky, Herbert. "Neutron Stars, Black Holes, and Supernovae." In *Frontiers of Astrophysics*, edited by E. H. Averett. Cambridge, Mass.: Harvard University Press, 1976. Even though this is a technical article, it is easy to read and nearly non-mathematical. The book also contains other articles, which adds to its diversity.

Kafatos, Minas C., and Andrew G. Michalitsianos. "Symbiotic Stars." *Scientific American* 251 (July, 1984): 89-94. An informative article that describes some of the intriguing aspects of binary systems in which a red giant loses its mass and then accretes around a hot white dwarf.

Luyten, Willem J. "White Dwarfs." In *Advances in Astronomy and Astrophysics*. Vol. 2. New York: Academic Press, 1963. Luyten succinctly introduces all the observed properties of white dwarfs and major problems facing astronomers. Having spent his career identifying and studying the strange and unusual features of these stars, Luyten presents a very readable introduction with a list of some thirty-seven white dwarfs and relevant references.

Van Horn, Hugh M. "Physics of White Dwarfs." *Physics Today* 32, no. 1 (1979): 23. An elementary exposé of the mathematically involved subject matter should be useful in the understanding of what occurs inside a white dwarf. Van Horn gives a list of references that should prove quite useful.

Weidemann, V. "Masses and Evolutionary Status of White Dwarfs and Their Progenitors." *Annual Reviews of Astronomy and Astrophysics* 28 (1990): 71-103. This is a lengthy article with numerous references describing the development in observational aspects of the masses and evolution of white dwarfs. The same issue also contains an article on the cooling of white dwarfs by Francesca D'Antona and I. Mazzitelli. The articles, although technical in nature, are very readable.

V. L. Madhyastha

Cross-References

Binary Stars, 218; Black Holes, 238; Neutron Stars, 1503; Novas, Burstors, and X-Ray Sources, 1572; Red Giant Stars, 2092; Supernovas, 2434.

X-RAY AND ELECTRON DIFFRACTION

Type of physical science: Condensed matter physics
Field of study: Surfaces

Diffraction occurs when a wave encounters an aperture with a characteristic dimension that is roughly equivalent to its wavelength. Separations between crystalline planes can be determined using beams of electromagnetic radiation (in the form of X rays) or electrons of appropriate energies.

Principal terms

CRYSTAL PLANE: a two-dimensional distribution of atoms having a particular symmetry

CRYSTAL STRUCTURE: atoms arranged in a periodic fashion at lattice sites; only fourteen crystal types occur in nature

DIFFRACTION: a wave phenomenon that occurs when a wave encounters an aperture with a characteristic dimension similar to the wavelength of the incident wave

ELECTRON: a fundamental particle of matter upon which resides the fundamental charge of nature (*e*)

ION: an atom that has either lost or gained electrons

LATTICE: a regular array of points in space with periodic locations within a three-dimensional volume

LATTICE SPACING: the distance between similar crystal planes in a lattice

X RAYS: electromagnetic radiation with wavelengths ranging from 0.01 nanometer to 10 nanometers

Overview

Diffraction is a wave phenomenon that occurs when a wave encounters an obstacle or aperture in its path of propagation that is of the order of the wave's periodic spatial variation (or wavelength). If a trough of uniform width is filled to a reasonable depth with water and the end of a wooden board is dipped into the water repetitively at a constant frequency, plane waves will be set up in the water propagating forward from the source of the disturbance. Wave crests will be represented by parallel lines. If two pieces of wood are inserted into the trough but do not traverse the full width of the trough, an opening is created that is smaller than the trough width, allowing the water waves to pass through and propagate beyond the barrier. If the opening at this barrier is large relative to the wave's wavelength, then the plane waves created by the disturbance can pass through the opening relatively unaltered. If the aperture is roughly the same size as the wave's wavelength, then the plane waves must diffract through the opening and the waves that pass through the opening will cease to be plane waves. Wave fronts propagating through the opening are no longer represented by parallel lines. Rather, they have a cylindrical symmetry and are represented by circular arcs of ever-increasing radius as the waves propagate away from the barrier.

Diffraction explains why a teacher lecturing in a classroom can be heard but not seen by people down the hall from the classroom if the door is open. The teacher generates sound waves when speaking. These sound waves have wavelengths that are roughly of the order of the width of the door opening and thus the sound can diffract around the corner. Light waves, on the other hand, have wavelengths that are many orders of magnitude smaller than the width of the door opening and thus travel straight through the door without alteration.

Diffraction of light can be explained in exactly the same way as the previous discussion of water waves; however, the width of the slit that light must pass through in order to diffract must be the same size as the wavelength of the incident light. A diffraction grating is a specially made slit consisting of many closely spaced grooves. Diffraction of incident light occurs as the light passes through the grating.

Materials that form compounds in regular arrays such as crystalline structures provide natural three-dimensional diffraction gratings for the portion of the electromagnetic spectrum with wavelengths roughly equal to the separation between crystal planes. Most crystalline structures have intercrystalline plane distances ranging between several individual to several tens of angstroms (10^{-10} meters). Electromagnetic radiation having wavelengths of this size range are called X rays. Visible light passing through crystalline structures would not exhibit diffraction, but X rays incident on crystalline structures would.

Bragg diffraction is the diffraction of X rays or an equivalent energetic beam of particles from a crystal. A single crystal with a simple lattice structure of cubic symmetry has a unit cell composed of atoms placed at the corners of a cube of side d. In a plane, all atoms are a linear distance d away from their nearest neighbors along mutually perpendicular directions in the plane. One such plane is a distance d away from an identical plane above and below it.

The atoms in the crystal lattice act as scattering centers for incident radiation. If an incident beam of X rays strikes a crystal sample at an angle θ with respect to the first Bragg plane, the remainder of the incident ray is transmitted to the second Bragg plane. At this second plane, it suffers partial reflection at angle θ with respect to that plane. The remainder of the ray is transmitted to the third Bragg plane, and so on. All reflected rays are parallel and are scattered by an angle 2θ from the original direction of the incident X ray. To reach a detector, the reflected ray from the second Bragg plane must travel a longer path length than the reflected ray from the first Bragg plane. The reflected ray must travel a distance $d \sin \theta$ to reach the second Bragg plane and another distance $d \sin \theta$ to return to the first Bragg plane. From there to the detector, reflected rays from the first and second Bragg planes travel equal distances.

For a diffraction pattern to be observed at the detector, the path length difference for two rays must be an integral number of wavelengths. Thus, the Bragg law of diffraction requires $n\lambda = 2d \sin \theta$, where λ is the X-ray wavelength and $n = 1, 2, 3,$

and so on and is called the order of diffraction. To use the Laue method for crystal plane spacing determination, a crystal sample is illuminated with X rays of known wavelength. The resulting diffraction pattern is recorded and analyzed for symmetry patterns and intensity peaks. The diffraction pattern is often recorded on a flat X-ray sensitive film. This pattern will reveal the symmetry of the crystal; a crystal with six-fold symmetry will display a diffraction pattern with six-fold symmetry on the exposed X-ray film. One-, two-, three-, four-, and six-fold symmetry occur naturally in crystals. Some quasicrystals have been produced recently that exhibit a type of five-fold symmetry. Diffraction techniques would work on these exotic materials as well. The simple Laue method has been replaced by more sophisticated analytic techniques for the determination of crystallographic planes.

Applications

In 1912, Max von Laue proposed that naturally occurring crystalline materials could be used to demonstrate diffraction effects with X rays. In a presentation to the Bavarian Academy of Sciences in Munich, Germany, von Laue elucidated a theory to explain how a periodic atomic structure could diffract beams of X rays. Experimental results accompanied the theoretical discussion. Armed with a tool to study crystalline structures, researchers firmly developed the discipline called solid-state physics.

In another area of applied physics, analysis of a resulting diffraction pattern could be used to determine experimentally the wavelength of the incident X rays if the interplanar crystal spacing was known independently. That knowledge can be obtained easily for the simpler crystalline structures. All that is required is the molecular weight and density of the substance forming the simple crystal structure, the number of atoms per unit cell, and Avogadro's number (the number of molecules in a mole of any substance).

Diffraction of electron beams by crystalline materials was discovered accidentally in 1927 by a pair of Bell Laboratory researchers, Clinton Joseph Davisson and Lester Halbert Germer, who were investigating secondary emission of electrons from a nickel sample's surface. Secondary emission is one of several processes whereby electrons can be liberated from a solid surface. Some of the other methods are the photoelectric effect (successfully explained in 1905 by Albert Einstein), field emission, and thermionic emission. In the photoelectric effect, incident photons with sufficient energy to overcome the binding energy of an electron free up electrons with kinetic energies ranging from zero to a maximum value given by the difference between the incident photon energy and the binding energy of the atom. Field emission liberates electrons from a surface by exposing the surface to intense electric fields. Thermionic emission liberates electrons from a surface using thermal energy. In secondary emission, an energetic beam (a few tens or hundreds of electronvolts) is incident upon a solid surface and the incident electrons collide with bound atomic electrons, transferring enough kinetic energy to liberate electrons from the solid surface. This process is important for electronic applications. For certain solids and over certain incident electron energy ranges, the number of liberated electrons exceeds the number of incident electrons. This effect is used to amplify electrical signals in a photomultiplier tube.

Davisson and Germer were interested in detecting secondary emission from a polycrystalline nickel sample. The number of secondary electrons as a function of angle (defined with respect to the normal to the solid surface) was measured experimentally. The angular distribution was expected to be proportional to the cosine of that angle. A large fraction of liberated electrons was not expected at any particular angle of observation. Initial experiments verified the anticipated angular distribution of secondary electron emission. During the course of the investigations, a nickel sample was accidentally contaminated. To correct the situation, Davisson and Germer heated the nickel sample at high temperature for a prolonged time. After cooling the cleaned nickel sample, they continued the experiments; however, the experimental results now had an added feature.

In addition to the expected angular distribution of secondary electron emission, Davisson and Germer observed higher numbers of electrons at specific angles. This effect was most prevalent when the incident electron beam energy was 54 electronvolts. A peak on top of the secondary emission was observed at 50 degrees. It was clear that something in addition to secondary emission was occurring. Originally, the nickel sample was polycrystalline, meaning that the sample contained myriad microcrystals randomly oriented throughout the bulk. The heat treatment provided sufficient energy to transform the sample into a single crystal configuration, a nearly perfect (apart from impurities and point defects), face-centered cubic lattice structure. Excess electron emission at a specific angle suggested the type of diffraction that had been demonstrated using X-ray beams. The atoms in the crystal lattice appeared to behave as scattering centers for the incident electrons. From the first-order results, an electron wavelength was determined using Bragg's law. The experimental wavelength precisely matched the de Broglie wavelength predicted by the electron beam's energy.

Diffraction of other material particles has been demonstrated in solids with crystalline structures. A thermal neutron has a kinetic temperature of approximately 300 Kelvins. If the neutron's thermal kinetic energy, $\frac{3}{2} kT$, is equated to $\frac{1}{2}mv^2$ or $\mathbf{p}^2/2m$, the neutron's momentum can be calculated. The neutron's wavelength, from the de Broglie relation, is Planck's constant divided by the momentum (h/mv). This operation yields a wavelength of 1.4×10^{-10} meters, or 1.4 angstroms. This wavelength is useful for diffraction studies in many solids. Specifically, 1.4 angstroms falls within the X-ray portion of the electromagnetic spectrum if the particles are photons.

Context

Wilhelm Conrad Röntgen is credited with the discovery of X-ray radiation. His work was considered so significant that he was awarded the first Nobel Prize in Physics in 1901. Von Laue first used X rays to demonstrate diffraction in crystals. Beginning in 1914 and continuing for the next three years, the Nobel Prize in Physics was awarded to researchers who investigated X-ray behavior in solids. Von Laue was granted the 1914 Nobel Prize for demonstrating X-ray diffraction. In 1915, William Henry Bragg and Lawrence Bragg shared the Nobel Prize for developing a technique to study crystal structure using X-ray radiation. Charles Glover Barkla won the 1917 Nobel Prize for his discovery of characteristic X rays of elements.

The use of X rays to ascertain crystalline structure represented the logical application of a thorough understanding of the properties of electromagnetic radiation to a specific problem. The accidental discovery of electron diffraction in a single crystal of nickel by Davisson and Germer in 1927 represented a watershed experiment that verified the existence of wave-particle duality.

In 1924, Louis de Broglie used symmetry arguments to postulate that material particles could be assigned a wavelength. This assumption removed the asymmetry in the concept of particles and waves that was caused by the realization that certain aspects of electromagnetic radiation could be properly explained only by attributing a particle nature (photons) to electromagnetic radiation. Experimental verification of electron diffraction proved the correctness of de Broglie's hypothesis. De Broglie was awarded the Nobel Prize in Physics in 1929 for his theoretical work on the wave nature of particles. Davisson shared the 1937 Nobel Prize in Physics with George Paget Thomson for the discovery of electron diffraction in crystals.

X-ray and electron diffraction have become powerful tools in determining crystalline structures and structures of solids in general. The diffraction process also can be used with amorphous and pseudocrystalline materials to provide clues into both topological short-range ordering and quasicrystalline behavior in such materials. Whenever a solid exhibits some degree of structural symmetry, the Laue diffraction method can be used to reveal that symmetry.

Bibliography

Burns, Marshall L. *Modern Physics for Science and Engineering*. San Diego: Harcourt Brace Jovanovich, 1988. A well-written account of modern physics. Although calculus-based and aimed at a college physics audience, qualitative descriptions of basic electromagnetism and quantum theory are understandable to the layperson.

Halliday, David, and Robert Resnick. *Fundamentals of Physics*. 3d. rev. ed. New York: John Wiley & Sons, 1988. Considered a classic textbook for investigating elementary calculus-based physics. The section on diffraction (as applied to both X-ray and electron beams) is heavily illustrated and thoroughly explained.

Myers, H. P. *Introductory Solid State Physics*. London: Taylor & Francis, 1990. A very thorough text on the physics of the solid state. Two chapters are particularly useful in understanding crystallographic structures and the diffraction of X-ray or electron beams in solids.

Ohanian, Hans C. *Physics*. New York: W. W. Norton, 1985. Although calculus-based, the text is not mathematically rigorous. Accessible to those with modest mathematical skills. Highly descriptive and well illustrated.

Sells, Robert L., and Richard T. Weidner. *Elementary Modern Physics*. Boston: Allyn & Bacon, 1980. A thorough treatment of modern physics using algebraic discourse. Covers wave-particle duality from both a theoretical and experimental viewpoint. Discusses the Davisson-Germer experiment and uses of X-ray and electron diffraction in crystallographic analyses.

David G. Fisher

Cross-References

Determining Crystal Structures, 605; Crystal Symmetries, 612; Diffraction, 671; Interference, 1161; Optics, 1673; Wave-Particle Duality, 2713.

X-RAY AND GAMMA-RAY ASTRONOMY

Type of physical science: Astronomy and astrophysics
Field of study: Observational techniques

X-ray and gamma-ray astronomy involves the observation of events in the universe that occur at energies far greater than what is normally shown by visible light or other forms of astronomy. The state of X-ray and gamma-ray astronomy is less advanced than that of other forms (partly because it is younger), but it promises great advances in understanding the formation and fate of matter in the universe.

Principal terms

ELECTROMAGNETIC RADIATION: energy encompassing an electric element and a magnetic element that travels in the form of "rays," as distinct from radiation composing high-speed atomic nuclei or particles; light and radio waves are the best known forms

GRAZING INCIDENCE: reflection where the incoming radiation strikes at an extremely shallow angle (typically less than 1 degree)

SCINTILLATION: light emitted when high-energy radiation is absorbed by matter then re-emitted at lower energies

SPECTROSCOPY: measurement of the intensity of light at specific wavelengths (energy levels) in the spectrum

Overview

X-ray and gamma-ray astronomy involves the observation of events in the universe that occur at energies far greater than what is normally shown by visible light or other forms of astronomy. Both X rays and gamma rays are electromagnetic radiation, as light is, but at much shorter wavelengths (and correspondingly higher frequencies). Because of this extreme difference, it is much more difficult to collect large quantities of X rays and gamma rays and to detect a source. In general, the field is called high-energy astrophysics, a name that also includes cosmic rays (atomic and subatomic particles spewed by various nuclear reactions).

Because of the extremely short wavelengths involved, X rays and gamma rays are both measured in terms of their energy, from thousands to billions of electronvolts. Although there is no firm demarcation, X rays span the range of 1,000 to 100,000 electronvolts and gamma rays upward from 100,000 electronvolts (visible light is around 1-2 electronvolts). X rays are emitted by the innermost electrons of an atom releasing energy. Gamma rays, by definition, are emitted by reactions in the nuclei of atoms or by subatomic reactions. In truth, there is some overlap between the two spectral bands, and it may be impossible to tell if an emission was from an energetic electron or a nuclear reaction.

Since X rays and gamma rays both pass through solid matter, it is difficult to focus them onto a detector and difficult to make them interact with the matter in the detector itself. A common method of "focusing" high-energy radiation is to use collimators to exclude all radiation except that from a particular direction. The basic technique can be imitated by looking through a cluster of straws held at arm's length: The view is narrow and restricted. Essentially, all light that is nearly "on axis" (that is, parallel with the centerline of the straws) passes through, while light that is "off axis" strikes the sides of the straw. The longer the straws, the narrower the field of view. A variety of complex collimation techniques have been developed to allow only radiation from a desired source to fall directly on a detector. While simple, this method generally yields images of the sky that lack the finer resolution of optical telescopes and thus leaves much uncertainty about the location of energy sources.

The most direct solution to this problem is the use of grazing-incidence mirrors to focus "soft" (low energy) X rays. Grazing incidence occurs when radiation strikes a surface at an extremely shallow angle and is reflected rather than being absorbed or scattered. The effect is readily seen when visible light strikes a windshield or pond at less than the critical angle and is reflected to cause glare. Because of the energy of X rays, however, the angle of incidence must be extremely shallow, typically less than 1 degree, and the surface must be exceptionally smooth to allow an image to form. X-ray telescopes generally use two reflectors—a parabolic primary and a hyperbolic secondary—to focus the radiation into an image without aberration. To provide the shallow angle of incidence, the mirrors resemble tubes (the segment of the paraboloid or hyperboloid surface is more distant from the focus than with conventional telescopes). This also requires that the secondary mirror be mounted directly behind and precisely aligned with the primary. Such an arrangement is known as a Wolter Type I telescope. A number of variations are available. Because only a small region of the radiation is intercepted by the mirrors, modern X-ray telescopes typically will use a nesting scheme in which up to six complete telescopes are built within each other.

Another type of X-ray telescope is the Kirkpatrick-Baez, which uses curved plates of glass in banks one behind the other. This arrangement will focus the light in one axis, then the other. While lacking the fine resolution of Wolter telescopes, the Kirkpatrick-Baez arrangement is well suited to all-sky surveys.

Grazing incidence telescopes were developed in the 1960's and 1970's; the 1980's saw the development of a radically new approach: normal-incidence X-ray mirrors. In Bragg crystals, the internal structure of a crystal can refract X rays almost as effectively as glass refracts light. Normal-incidence mirrors are built up in layers of microscopic crystals that intercept the X rays and reverse their direction. When laid upon parabolic or hyperbolic surfaces, one can build an X-ray telescope that resembles a conventional reflector telescope and provide comparable resolution for low- to medium-energy X rays.

At higher energies, though, the efficiency of reflectors decreases until virtually no image is being focused. At very high energies, a variation on the pinhole camera technique may be used. In this approach, the front end of the telescope is a plate punctured by holes in a pseudorandom pattern. Each pinhole allows an X-ray image to be projected onto the detectors behind the plate. Yet, because so many images are projected at once, the scene must be mathematically deconvolved (decoded) to produce a true picture of the scene.

At present, there is no way to focus gamma rays. Their energies are simply so high that they would pass unhindered through the mirrors. Instead, gamma-ray "telescopes" constitute instruments that are designed to detect interactions of gamma rays with matter.

X rays and gamma rays, because of the energies involved, are detected less directly than light; specifically, the energy yielded by interaction with some intermediate object is detected. The first and simplest X-ray detector is photographic film. Exposure to X rays will cause the film to darken (that is, produce a negative image). Film is used only in X-ray telescopes that can be recovered, such as manned satellites and sounding rockets; various electronic detectors are used in unmanned satellites. The two are complementary in use. Film provides the highest spatial resolution and reveals fine details of the source and its position relative to other objects but over a broad energy range. Electronic detectors can measure energy levels with a fair degree of accuracy but at the expense of spatial resolution. In both cases, the energy resolution can be enhanced by placing filters in front of the detector, but this generally establishes a low-energy cutoff and higher-energy radiation will still pass through.

Most development work has gone into electronic detectors that measure the energies yielded when an X ray or gamma-ray photon strikes matter and releases enough electrons to generate a current, or stimulates electrons to release lower-energy photons as visible light, which can then be detected by conventional devices.

Proportional counters are similar to Geiger-tube counters. A high-energy photon enters a gas-filled chamber, intercepts a gas atom or molecule, and generates an ion and an electron. These are attracted to the electrodes (anode and cathode) in the tube and cause an increase in the electrical current that is measured by the instrument's electronics. In its simplest form, the counter simply registers the arrival of a photon regardless of direction. Most proportional counters, though, are actually gangs of many small counters mounted behind collimators to narrow the field of view. They also use shielding to reduce the chance of photons entering from behind the detector and giving a false signal. In some cases, the collimators move so as to make sources wink on and off and thus improve the precision of its location and even to deduce the structure of extended sources.

To increase the chances of detecting an X ray, detectors sometimes are electrically charged so that a single X ray will produce a shower of electrons that is readily detected (this is comparable to electric eyes that generate an electric current when struck by light). The electrically charged device is a microchannel plate, which is like a microscopic collimator with its channels as a slant. This ensures that an X ray will strike the wall of the channel, which has an electric charge not quite high enough to cause an electric arc. The incoming X ray provides the needed extra jolt and releases a cascade of electrons that exits the back of the plate to be measured by the wires of a grid behind the detector.

In the late 1980's, progress was made on developing X-ray charge-coupled devices (CCDs) that could provide a more efficient means of creating X-ray images. CCDs are electronic analogs of the retina and comprise thousands of small photosensors that are read individually by computer. Unlike conventional CCDs, an X-ray CCD will be illuminated from behind, again so that the incident X rays will generate light or an electrical charge that is detected by the CCD.

Spectral measurements in X rays use proportional counters, too, but require a grating or crystal to spread the X rays. The familiar visible-light analog is a wedge prism that breaks white light into the color spectrum. With X rays, a similar effect is desired since the intensity of the X rays at different energies is a measure of the activity at the source. Gratings are surfaces that have a microscopically ribbed surface. The degree to which radiation scatters depends on the energy of the photon. A proportional counter can then be moved along an arc and radiation can be measured. Spectral and imaging instruments can be combined by placing a transmission grating or crystal in the telescope's line of sight. The focused radiation then is spread across the imager and its intensity measured (normally this works only for strong sources).

In scintillation detectors, the incoming radiation is trapped in a crystal and causes a flash of light, which is measured by a photomultiplier tube. The most sensitive of these instruments require that the crystal be supercooled by liquefied or solid gas so that the body temperature of the crystals do not cause false readings. Such crystals can be used in spectrometers. A variation of the effect is Compton scattering, in which a high-energy X ray strikes matter and is scattered at lower energy and with release of an electron whose energy corresponds to the original energy. At higher energies, however, the efficiency of this effect declines. Gamma-ray interactions can also produce electron-positron (antielectron) pairs that can be detected in spark chambers, which cause sparks between electrified plates or wires. These are then registered by photomultiplier tubes.

Because gamma rays cannot be focused, different techniques are used to determined the shape of a source. The instrument can be tilted back and forth and the flux change measured to determine its origin. Or, the instrument can be built as a pair, one behind the other, and the two sets of signals collated to determine the origin of each gamma ray.

X-ray and gamma-ray detectors, in general, have anticoincidence detectors that are designed to detect cosmic rays and thus allow the signals they generate to be subtracted from the instrument signal, somewhat like filtering the noise in a radio.

Applications

X-ray and gamma-ray astronomy is conducted above the atmosphere of the earth since it absorbs all X rays; even the thin layer above 36,000 meters absorbs soft X rays. While both fields are best conducted from satellites, both started on (and still use) suborbital platforms—X-ray astronomy from suborbital rockets and gamma-ray astronomy from balloons because the lower fluxes required larger, heavier detectors. Suborbital rockets (also called sounding rockets) can expose a payload to the space environment for several minutes depending on the weight of the instrument and the power of the rocket. Suborbital rockets were a major tool during much of early X-ray astronomy and continue to serve the same purpose as new telescopes are developed and tested. They also have filled the gap between flights of major X-ray satellites. Gamma-ray instruments, though, have to be larger and spend more time at altitude to intercept a sufficient flux from gamma-ray sources, so unmanned balloons have been their preferred suborbital platform. In either case, though, satellites are the best means of operation since they provide essentially indefinite observing time for medium to heavy instruments.

Several X-ray astronomy satellites have been launched, the most notable being the High Energy Astrophysical Observatory (HEAO) series, EXOSAT, and ROSAT. Wolter Type 1 X-ray telescopes were carried by three of these satellites (HEAO-2, EXOSAT, ROSAT) to produce images and spectra of the heavens.

The principal X-ray observatory for the twenty-first century is expected to be the Advanced X-ray Astrophysics Facility (AXAF), conceived as an X-ray complement to the optical Hubble Space Telescope. AXAF will have a set of six nested Wolter Type I mirror pairs (each pair equivalent to a single telescope) to produce X-ray images with resolutions of 0.1 arc-second or better (comparable to what the Hubble should ultimately deliver). It will carry advanced instrumentation derived from the HEAO-2 suite of instruments to produce images and spectra of X-ray objects. Also included will be a unique X-ray CCD camera taking advantage of the latest in solid-state detectors.

Fewer gamma-ray satellites have been flown, thus holding that field at an earlier stage of development. The most notable X-ray satellites were COS-B (Europe, 1975) and HEAO-3, which included a gamma-ray spectrometer.

The Gamma Ray Observatory (GRO) scheduled for launch in 1991 will carry four different but complementary gamma-ray instruments. Because gamma rays cannot be focused, they are not arranged in a common focal plane, like those of the Hubble Space Telescope but stand alone on the satellite bus. GRO's instruments will span a broad energy range, from 10,000 to 10 billion electronvolts. The Burst and Transient Source Experiment will record gamma-ray "flashes" that have puzzled scientists since 1967. These flashes may be caused by matter falling into neutron stars or black holes, but their unpredictability has hampered efforts to locate them. The burst experiment involves eight detectors that will stare at the entire sky so that any burst will be recorded and located so that optical telescopes and other types of telescopes can try to identify candidates for the burster. The Oriented Scintillation Spectrometer Experiment will measure the spectra of objects as gamma rays cause scintillations in crystals. The four scintillation units can be tilted so the "empty" sky next to a source can be observed and its background noise subtracted to obtain the signal of the object being observed. The Imaging Compton Telescope is actually two detector arrays using the Compton effect. Light is scattered and recorded as the gamma ray encounters a scintillation liquid in the first detector, and the gamma ray is re-emitted at lower energy to be detected in the same manner at the next level. The energy of the two emitted photons will reveal the energy of the original gamma ray, and the locations of the two "hits" will point back toward the source.

The Energetic Gamma-Ray Experiment Telescope is a large instrument designed to detect even low fluxes of the highest-energy gamma rays. A gamma ray entering the telescope strikes a tantalum sheet and creates an electron-positron pair that then travels through two spark chambers and finally into a crystal scintillator. A complex anticoincidence system discounts sparks caused by cosmic rays. The energy of the gamma ray is recorded and its direction is revealed by the paths of the two particles it creates.

Context

X-ray and gamma-ray astrophysics has become one of the most revealing disciplines in astrophysics since its development following World War II. Extraterrestrial cosmic rays and X rays were detected in the 1800's by balloon crews who carried electrostatic instruments and cloud chambers aloft. Even when their true meaning became known, it was not appreciated that the sources might be caused by stars. As laboratory physics developed an understanding of nuclear fusion and how matter decays, it became obvious that X rays and gamma rays were generated by the stars. Nevertheless, it was doubted that the flux (or, total energy flow) would be great enough to be measured, at least for any star other than Earth's sun.

With the availability of captured V-2 rockets to carry instruments aloft in tests after World War II, however, some rudimentary instruments were flown, followed by larger instruments aboard unmanned balloons starting in the 1950's.

It was discovered that as sensitivity increased, what could be seen became richer and more detailed. In the early 1960's, it appeared that the sky was suffused with a strong background glow of X rays. The Uhuru (meaning "freedom" in Swahili) satellite carried an all-sky survey detector that had sufficient resolution to detect more than three hundred discrete sources among the background glow. This led to the development and launch of a series of three High Energy Astrophysical Observatories. HEAO-1 and 3 carried detectors that mapped the entire sky in X rays through gamma rays (HEAO-3 also carried cosmic-ray detectors). HEAO-2 carried the first stellar X-ray telescope and discovered that the X-ray background was composed largely of point sources that could not be distinguished at lower resolutions. This led to the discovery of the X-ray components of known visible objects and of previously unknown objects. In many cases, what was seen in X rays matched very nicely with the visible and radio components. In other cases, it seemed as though two different objects were being viewed.

Comparable work has been done in solar physics. The Orbiting Solar Observatories (OSO) carried a number of X-ray and gamma-ray instruments in the 1960's and 1970's, and the Skylab space station in 1973-1974 included X-ray spectrometers and imaging telescopes in its array of eight solar telescopes. The Solar Maximum Mission (SMM) satellite (1980-1988) carried X-ray burst detectors and gamma-ray spectrometers to measure the output of the sun.

In general, astronomers have found that the universe is far more energetic than previously believed. The Crab nebula—the remnant of a star that exploded in 1054—was found to pulse in X rays at the same rate as its visible pulsar, thus suggesting a very compact object. X rays have been found coming from the cores of quasars and most "normal" galaxies. The heart of the Milky Way seems to be the site of matter-antimatter annihilation. Instruments have measured X rays being emitted at an energy of 511 kiloelectronvolts, which corresponds to the conversion of electrons and positrons (antielectrons) into energy.

Gamma-ray astronomy provides a different view of the universe. Specifically, it reveals the creation and destruction of matter (properly, its conversion from energy to matter and back) in supernovas, neutron stars, pulsars, black holes, quasars, active galaxies, and other objects. Gamma-ray astronomy has confirmed that the heavier elements are created in the blast furnace of a supernova when a star self-destructs. The famous Supernova 1987A produced gamma-ray lines indicating that cobalt 56 was decaying into iron. Cobalt 56 is unstable and had to have been created shortly before the observation—that is, when the star exploded.

The most important result from high-energy astronomy—as well as from radio and infrared—is the increasing awareness that objects must be studied in terms of their total output rather than as emitters in different spectral bands. What puzzles scientists in one band may be solved in another, or at least a new line of investigation can be illuminated.

Bibliography

Hirsch, Richard F. *Glimpsing the Emerging Universe: The Emergence of X-Ray Astronomy.* Cambridge, England: Cambridge University Press, 1983. History of X-ray astronomy through the High Energy Astrophysical Observatories. Written for the well-informed reader.

Kaufmann, William J., III. *Universe.* New York: W. H. Freeman, 1985. College-level introductory text covering the field of astronomy. Contains descriptions of astrophysical questions and their relationships. Informative.

McLean, Ian S. *Electronic and Computer-Aided Astronomy: From Eyes to Electronic Sensors.* Chichester, England: Horwood, 1989. A survey of the history and applications of electronics and astronomy. While much of the book is a technical survey, several chapters provide a general history and explanation of CCDs.

Tucker, Wallace. *The Cosmic Inquirers: Modern Telescopes and Their Makers.* Cambridge, Mass.: Harvard University Press, 1986. Literate, well-written descriptions of modern observatories and the scientific questions that led to their construction.

__________. *The Star Splitters: The High Energy Astronomy Observatories.* NASA SP-466. Washington, D.C.: Government Printing Office, 1984. History of the HEAO satellite program and an overview of its results. Includes descriptions of X-ray optics and detectors.

Dave Dooling

Cross-References

Black Holes, 238; Electromagnetism, 750; Optical Detectors, 1657; Pulsars, 1958; Quasi-Stellar Objects, 2014.

X-RAY DETERMINATION OF MOLECULAR STRUCTURE

Type of physical science: Chemistry
Field of study: Chemistry of molecules: structure

X-ray crystallography is the science that describes and interprets the diffraction of X rays by crystalline substances. Such diffraction patterns can be used to identify the structure of a molecule, and in fact have led to the development of many important structural concepts in a wide variety of sciences.

Principal terms

CRYSTAL: a substance that is crystalline in three dimensions and bound by plane faces

CRYSTALLOGRAPHY: a branch of mineralogy involved mostly with the recognition, description, and classification of naturally occurring crystal species

DIFFRACTION: the spreading of waves as they move past an obstacle

INTERFERENCE: the superposition of two or more waves, resulting in a net displacement of their amplitudes; the resultant displacement is the sum of the original amplitudes

UNIT CELL: an imaginary solid figure drawn around a molecule in a crystal to indicate a repeating unit

X RAY: radiation produced when streams of electrons are allowed to bombard a metal, falling between 6×10^9 hertz and 1×10^{17} hertz on the electromagnetic spectrum

Overview

X rays are defined as short-wavelength electromagnetic radiation produced by transitions of electrons in the inner orbitals of atoms. The wavelength range is from approximately 10^{-5} angstrom to 100 angstroms. For analytical purposes, the X ray can be obtained in one of three ways: by bombarding a metal target with a stream of electrons; by subjecting a target to a primary beam of electrons in order to produce a secondary beam of fluorescent X rays; or by using a radioactive source that, upon decay, yields an X-ray beam.

Primarily, the interest in X rays as a means of determining molecular structure arises from the fact that a crystalline material will produce a unique and regular diffraction pattern when subjected to X-radiation. A crystalline substance is defined as a homogeneous solid having an ordered internal atomic arrangement and a definite composition. Since no two chemical materials form structures in which the spacing of both the atoms and the planes in which they sit is identical, X-ray diffraction patterns can be used to determine the structure of the original diffracting molecule.

The term "crystal," Greek for "ice," refers to a regular three-dimensional arrangement of atoms, ions, or molecules. This regularity gives rise to what is called the unit cell, which essentially outlines the simplest repeating unit found in a crystal. A simple analogy for the unit cell would be to consider the regular, repeating brick units in the wall of a house. The wall would be considered the crystal structure, with each brick representing a unit cell. Each wall in the house could therefore be described by the repeating dimensions of the individual bricks that make it up.

A unit cell is defined by six parameters. These parameters express the characteristics of the size and shape of the cell. Three parameters (a, b, and c) describe the sides of the cell, while the remaining three (alpha, beta, and gamma) describe the angles between those sides. The possible combinations of these parameters result in a variety of shapes that a unit cell or crystal might assume. Analysis has shown that there are seven possible geometric shapes in which a unit cell may exist. These possibilities are pictured in Figure 1. Notice that for some, the unit contains the same atom at the corners of the cell, but not in the center of the face (the side of the cell formed by imaginary lines connecting the atoms) or the center of the body of the cell. These cells are referred to as primitive cells (P). Those in which the same type of atom occupies the corners of the cell and the intersection of the diagonals through the body of the cell are given the name body-centered cells (I). Others, called faced-centered cells (F), have the same type of atom occupying the corners and the intersection of all diagonals through a face of the cell. Finally, the end-centered cell (C) is that in which the corners of the cell and the intersection of the face diagonals for parallel faces are occupied by the same atom. In all, fourteen varieties of a unit cell exist and are called the Bravais lattices (named for August Bravais).

A second concept fundamental to the use of X rays in structural determination is that of diffraction. Diffraction is the bending of waves into the shadow regions of obstacles. This might at first glance appear to be a rather obscure concept, but, as an example, consider hearing sound around the corner of a building from where a person is standing. The sound waves could not have followed a straight line path. In fact, they have spread out into the geometric shadow regions of the building's edge. The same effect describes the experience of viewing a bright light through a common wire mesh similar to those used in sorting powders. When the light hits the mesh, it is scattered and spreads around the mesh into its shadow region, creating a fuzzy pattern of light of varying intensity that can be seen on a screen behind the mesh. This blurring of the light results from the interference of the rays with one another as they are scattered by the mesh.

When X rays fall on a body, the atoms scatter the incident radiation in much the same manner. In a crystalline structure, where the atoms are arranged in a regular pattern, there exists a definite phase relationship between the scattered rays from the neighboring atoms. If a monochromatic X-ray beam falls on a crystal, the rays are reflected by each face of the crystal. Each reflected ray interacts with the other reflected rays. If the rays are out of phase, they tend to destroy one another, with the net result being no emerging ray or a dark spot. Constructive interference of the rays, in which reflected rays reinforce one another, is known to occur when the

System	Parameters	
Cubic	a = b = c	α = β = γ = 90 degrees
Tetragonal	a = b = c	α = β = γ = 90 degrees
Hexagonal	a = b ≠ c	α = β = 90°γ = 120 degrees
Rhombohedral	a = b = c	α = β = γ ≠ 90 degrees
Orthorhombic	a ≠ b ≠ c	α = β = γ = 90 degrees
Monoclinic	a ≠ b ≠ c	α = γ = 90 degrees ≠ β
Triclinic	a ≠ b ≠ c	α ≠ β ≠ γ ≠ 90 degrees

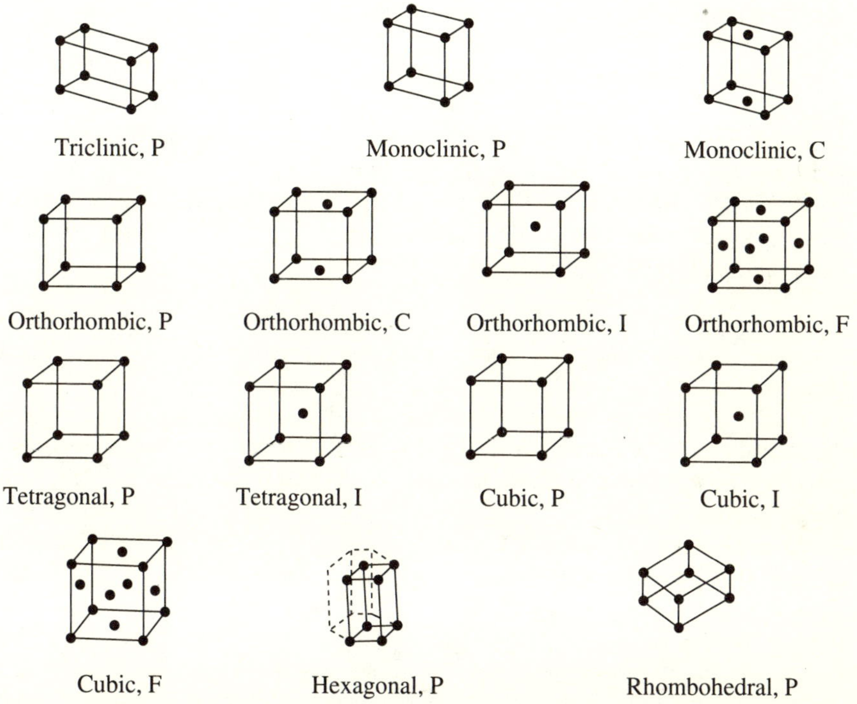

Figure 1. CRYSTAL LATTICE SYSTEMS AND BRAVAIS LATTICES

conditions of Bragg's equation ($n\lambda = 2d \sin \theta$) are met. Where λ is the wavelength of the incident X-ray beam, *d* is the distance between crystal planes, and θ is the angle of the incident ray.

Consequently, a crystal will generate a series of diffraction lines from each plane for a series of wavelengths. The sum of these lines is a diffraction pattern of light and dark spots, from which it is possible to determine the different distances between the crystal planes as well as the angles between the planes. Based on such information, the physical dimensions of the crystal can be deduced.

Applications

X-ray diffraction methods are one means of providing the details of molecular structure. They play a major role in numerous aspects of modern science, from biochemistry to engineering. Empirical drug design, molecular modeling, protein-substrate interactions, and numerous other applications all depend on the identification of a molecule's structure. Materials scientists, whose research deals with the relationship between a material's structure and its properties, are especially dependent on X-ray methods. Material engineering is concerned with the modification of properties and the performance of the modified material. The ability to determine a material's structure—not only the crystal structure but also the electronic structure and the structure of boundaries and interfaces—is fundamental to such an operation.

Typically, a diffraction pattern is obtained by mounting a crystal so that it may be subjected to X-radiation. The crystal is oriented, usually by computer-controlled instrumentation, about different axes successively, so that a complete distribution of reflected X-ray beams is obtained from each plane of the crystal. This process will give a complete diffraction pattern. Figure 2 shows a typical diffraction pattern for an inorganic salt crystal. By careful measurement of the distances between the spots and by consideration of the intensities of these spots, the crystal can be identified.

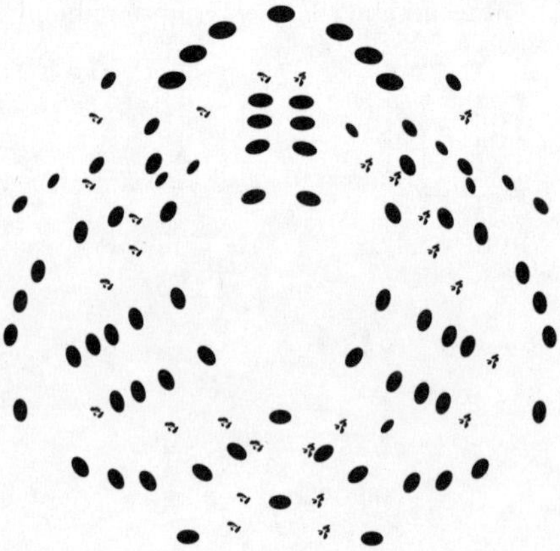

Figure 2. Diffraction Pattern of an Inorganic Salt Crystal

One method is to construct from a diffraction pattern a series of rays drawn perpendicular from a common point to each plane of the crystal. Such a drawing is called the reciprocal lattice, because the distance of each point from the origin is reciprocal to the interplanar spacing of the plane it represents. Connection of these rays leads, essentially, to the construction of a sphere, which in turn yields, in two dimensions, a pattern of circles. The pattern for a particular species is always the same. An alternate method is to replace the single crystal with a large collection of very small crystals that are randomly oriented. Such a method is called a powder method, and the resulting pattern is called a powder diffraction diagram.

The powder diffraction diagram is sufficient data if the identification of a sample is the only information required. Qualitative identification of the substance can be made by matching the diffraction pattern to known patterns. Computer matching of diffraction patterns has significantly improved such structural identifications. It is common to run the unknown sample and then compare the diffraction pattern to those of standards. The method simply requires that a library of standard patterns is available. A second means of qualitative identification entails the measurement of the distance between the planes of the crystal, the *d*-value. These values are calculated from the diffraction diagram of the unknown substance and again compared with those in a data file. The file is arranged so that it can be searched first by the distance between the crystal planes, and then according to the intensity of the reflections. A correct match requires agreement in both of these areas.

If the unknown substance contains a mixture of components, each must be identified separately. One way of doing this is to treat a list of *d*-values and intensities as a single component, and then match as many of the reflections as possible to a known substance. After a suitable match is made for this component, these lines are omitted and the remaining lines are matched. This requires that the intensities of the remaining reflections are adjusted to indicate their brightness, as in the pure component. The same matching procedure is followed for the second component.

A more complete structural analysis of a molecule or crystal would include a precise location of all atoms involved. X-ray diffraction methods are capable of determining the placement of an atom to within one-hundredth of an angstrom. This allows for the actual reconstruction of the crystal or molecular lattice. It also provides a means of calculating such parameters as the angle between bonds and the distance between the atoms involved in a bond.

X-ray diffraction is adaptable to quantitative work as well as qualitative identification; however, this process is difficult, and complications are abundant. The intensities of a diffraction pattern are proportional to the fraction of the material present in a mixture. Theoretically, this should allow for the same calibration-type processes common to other absorption methods; however, the intensity measurement and the direct comparison of those measurements is most difficult. Interferences are frequently encountered. Internal standards are sometimes used, but they cannot solve all the problems. At the present time, X-ray diffraction remains largely a qualitative technique for the structural identification of compounds.

Context

The first X-ray photograph was taken in 1921. Since that time, X-ray techniques have developed into one of the most informative experimental methods available to the scientific community. Given their capability to describe not only the macroscopic picture of molecular structure but also the microscopic properties, such as the geometric realities within that structure, X-ray methods have aided and improved many science and engineering techniques.

The initial application of X-ray techniques to chemical structural identification is credited to Max von Laue, who in 1912 realized that the crystal's internal regularity was a potential source of reflectance. Following this, William H. Bragg and Lawrence Bragg extended Laue's work to link the intensities of the diffraction pattern with the atomic arrangement of the molecule.

Most of the early work in X-ray crystallography focused on the structures of known compounds. X-ray methods were applied to research of bond angles, bond forces, and bond lengths. The Nobel Prizes in Chemistry in 1936 (Peter Debye) and 1954 (Linus Pauling) are examples of the importance of X-ray diffraction and crystallography to these concepts. Debye was able to establish an understanding of the molecular dipole, while Pauling completed his masterful work on the chemical bond and chemical characterization. The field of physics has also benefited from X-ray methods, as witnessed in three Nobel Prizes in Physics being awarded to Laue (1914), the Braggs (1915), and Clinton Joseph Davisson and George Paget Thompson (1937, for his theories of electron diffraction).

One of the most famous examples of the importance of X-ray crystallography to the science world is credited to James D. Watson, Francis Crick, and Maurice Wilkins, for their discovery of the molecular structure of nucleic acid and its significance to the genetic code. It was the diffraction pattern of DNA (deoxyribonucleic acid) that allowed Watson to conclude that the regularity of the molecule was a result of a helical nature. Their ability to interpret the diffraction pattern produced is nothing short of brilliant.

X-ray crystallography has continued to grow and enjoys a rather prominent position in many science fields. Many see crystallographic methods as a bridge between biological, physical, chemical, and materials sciences. Its importance as a means of structural identification cannot be stressed enough. In fact, when the 1985 Nobel Prize in Chemistry was awarded to Herbert A. Hauptman and Jerome Karle for their work on a direct-method approach to solving crystal structures, the award speech stated, "In order to understand the nature of chemical bonds, the functions of molecules in a biological context, and the mechanism and dynamics of reactions, knowledge of the exact molecular structure is absolutely essential." It is likely, therefore, that X-ray diffraction will continue to be a fundamental and leading technique for structural concepts in the future.

Bibliography

Barrett, Jack. *Atomic and Molecular Structure*. New York: John Wiley & Sons, 1970. A simple and easily read account of the principles of theoretical chemistry. Very little math is needed to benefit from this text.

Denny, R. C. *A Dictionary of Spectroscopy*. New York: John Wiley & Sons, 1982. A wonderful reference tool for any analytical technique. The author defines in a simple manner those terms most frequently encountered in spectroscopy. The dictionary is directed to the nonspecialist.

Holmes, K. C., and D. M. Blow. *The Use of X-Ray Diffraction in the Study of Protein and Nucleic Acid Structure*. New York: John Wiley & Sons, 1966. A clearly written text that introduces X-ray diffraction and applies it to the field of biochemistry.

Ladd, M. F. C., and R. A. Palmer. *Structure Determination by X-Ray Crystallography*. New York: Plenum Press, 1977. A basic introductory text on the subject of X-ray crystallography. Provides a thorough introduction to structure determination. The text is generously illustrated and contains problem sets for each topic.

Russell, P. A. *Electron Microscopy and X-Ray Applications*. Vol. 2. Ann Arbor, Mich.: Ann Arbor Science, 1981. The second volume of a series of books dedicated to spectroscopic methods. Somewhat detailed, but provides several practical examples of X-ray methods.

Watson, J. D. *The Double Helix*. New York: Atheneum, 1968. A superbly written personal account of the investigations of DNA. Easily read and quite enjoyable, the book puts the technique of X-ray diffraction in simple terms for the nonscientist.

Whiston, Clive. *X-Ray Methods*. New York: John Wiley & Sons, 1987. One in a series of open learning books, the text presents X-ray crystallography at a beginner's level. The goal of the text is to help the reader develop a working knowledge of the basic theory involved in X-ray diffraction. A problem-centered approach is used.

Mary Beth McGranaghan

Cross-References

Chemical Bond Angles and Lengths, 367; Determining Crystal Structures, 605; Diffraction, 671; Molecular Configurations, 1409; Molecular Spectra, 1437; Calculations of Molecular Structure, 1443; X-Ray and Electron Diffraction, 2749.

GLOSSARY

Absolute zero: The lowest possible temperature (assigned a value of zero on Kelvin scale), at which atomic motion and randomness of atomic arrangement are at a minimum.

Absorption spectrum: A continuous spectrum, such as formed by white light, but containing dark lines caused by the absorption of certain frequencies by a substance through which the radiation has passed.

Acceleration: The rate at which an object's velocity changes (in speed and/or direction).

Accretion: A gravitational accumulation of matter that can result eventually in the formation of a planet or smaller-sized body.

Accuracy: The intrinsic, objective truth of a measurement or calculation, or the degree to which calculations or measurements approach true values.

Acid: A substance that loses a proton (hydrogen cation) during a chemical reaction, usually donating the proton to a chemical base.

Acoustic modes: In a polyatomic crystal, those normal modes in which atoms in the same unit cell vibrate in phase with one another.

Actinides: A series of fourteen transitional elements in period 7 of the periodic table, corresponding to the lanthanides and listed right below them.

Activation energy: The energy, in excess of the average energy of reactant molecules, that is required in order for molecules to react.

Adhesion: The chemical attraction of unlike substances for each other, which causes resistance or sticking between the molecules composing each substance.

Adiabatic process: A process undergone by a system without addition or extraction of heat.

Adsorption: A process by which a molecule or an atom is trapped on a surface.

Affinity: The degree of attractiveness or adhesion between unlike substances; high affinity indicates considerable attraction, whereas low affinity indicates very little attraction.

Albedo: The portion of light reflected by a body.

Alcohol: Compounds with a hydroxy (−OH) functional group attached to a carbon atom.

Aldehyde: Any organic compound characterized by the group CHO.

Algebraic equation: An equation in positive integral powers of a variable *x*, with coefficients as integers.

Algorithm: A fixed and describable method for achieving some result, usually a list of steps to be carried out in sequence.

Alkali metal: The elements lithium, sodium, potassium, rubidium, cesium, and francium (but not hydrogen), which appear in the first column of the periodic table.

Alkaline: A substance that is basic (rather than acidic).

Allotropy: The occurrence of multiple forms (allotropic forms) of a chemical element, usually as solids with different crystal structures.

Alloy: A metal consisting of two or more metals or a metal and a nonmetal.

Alpha particle: A helium nucleus, consisting of two protons and two neutrons, given off by nuclei in radioactive decay.

Alternating current: An electric current that reverses its direction in a circuit at regular intervals.

Amino acid: Any one of twenty types of organic nitrogen-containing compounds that can be bonded together to form macromolecular chains called proteins.

Amplitude: The magnitude of displacement of an oscillating system from its equilibrium point; the amplitude of an electromagnetic signal is related to its intensity.

Amplitude modulation (AM): The process of altering the amplitude of a radio frequency in accordance with the amplitude of the modulating signal.

Analog circuits: Electronic circuits that are made to amplify, generate, or operate on an input signal, consisting of sinusoidal waves.

Analytical geometry: The study of geometrical and metric properties of the figures using algebraic operations on angles and coordinates of positions.

Angular momentum: The product of an object's rate of rotation and its moment of inertia around the axis of rotation. A measurement of the total rotational motion of a system; measured in units of mass multiplied by length squared per unit time and assigned a direction along the axis of rotational motion.

Angular velocity: The rate of change of position of a body moving in a circle, commonly measured in rotations per minute.

Anion: A negatively charged atom, compound, or molecule that has gained an electron to stabilize an outer atomic energy level.

Anisotropic: A substance whose properties, such as light velocity, vary with direction.

Annealing: The process whereby manufactured glass is cooled in such a way that no region does so at a faster or slower rate than any other.

Anode: The electrode at which oxidation takes place in an electrochemical cell.

Antenna: A device used for radiating and receiving radio waves.

Anthropic principle: The philosophical viewpoint that the universe is structured so that galactic and stellar evolution inevitably will lead to the evolution of life, including intelligent life.

Antimatter: A form of matter which has the same mass as normal matter (electrons and protons) but reversed charges.

Antiparallel: The orientation of two vectors, which are parallel to each other but point in opposite directions.

Antiparticle: An elementary particle with the same mass and spin (intrinsic rotation) as another particle, but with opposite charge and magnetic moment; a particle and its antiparticle annihilate into radiation when they come together.

Aphelion: The farthest position of a planet from the sun.

Apogee: The point of an orbit at which the object is the farthest from the earth.

Apparent magnitude: The brightness of a star as measured from Earth's surface.

Approximate (inverse) solutions: Solutions providing only limited precision and accuracy, as a result of simplifying assumptions concerning the physical process or mathematical technique representing the process.

Aqueous mixture: A mixture of one or more compounds with water.

Armature: In an electric motor or generator, the part that revolves; a series of coils of wire wound around an iron core.

Assembly language: A set of instructions using memorable symbols and data addresses that translate into one equivalent computer machine instruction.

Asteroid belt: The region between the orbits of Mars and Jupiter, which is filled with stony or metallic bodies called asteroids.

Astronomical unit (AU): A measure of astronomical distance equal to the distance from the Sun to Earth: 150 million kilometers.

Astrophysics: The study of the physics and chemistry of celestial objects and forces.

Asymptotic freedom: A feature of the quark theory in which interactions become weaker at shorter distances or at higher momenta.

Atmosphere: The layer of air around the earth's surface that supports life; a unit of pressure equal to the force per unit area that the air exerts at the surface of the earth at sea level.

Atom: The smallest unit of a chemical element, composed of a tiny but massive central nucleus surrounded by a characteristic number of electrons.

Atomic clock: An extremely precise clock that operates on the basis of natural vibrations within the atomic structure.

Atomic mass unit: The unit employed in the scale of relative atomic masses of the elements; defined as one-twelfth the mass of an atom of carbon 12, equal to 1.6606×10^{-24} grams.

Atomic number: The number of protons in an atom of an element and the "order number" by which elements are arranged on the periodic table of elements.

Atomic orbitals: The electrons associated with an atom found to exist in discrete orbitals, or states, having quantized energies; the electron charge density for the lowest-energy, "*s*-type" orbitals is spherical; the higher-energy, "*p*-type" orbitals are two-lobed (dumbbell-shaped); and the "*d*-type" orbitals have four lobes (cloverleaf-shaped).

Atomic radius: The radius of an atom; the distance from the center of an atomic nucleus to the outermost orbiting electrons of the atom.

Atomic spectrum: Light of characteristic wavelengths emitted by a heated chemical element that serves as a means of detecting the element and distinguishing it from other elements.

Attenuation: The loss of power suffered by radiation as it passes through matter.

Aurora: The bright glow in Earth's atmosphere near the magnetic pole, which is produced by charged particles from the sun striking and exciting atoms in the atmosphere.

Avogadro's number: That number of particles (or units) that is the same number as there are atoms in one gram-atomic mass of any element; the numerical value is 6.02×10^{23}.

Axioms: Statements regarded as basic or "given."

Azimuth: An arc of the horizon measured between a fixed point (as true north) and the vertical circle passing through the center of an object clockwise from the north point through 360 degrees.

Band gap: The minimum energy required to release an electron from a particular atom in a crystal and allow it to move about the crystal.

Baryons: Particles that are composed of three quarks, such as neutrons and protons.

Base (chemistry): A substance, also called an alkali, that gains a proton (hydrogen cation) during a chemical reaction, usually receiving the proton from an acid.

Base (electronics): The thinnest of three parts of a transistor that is doped opposite to the other two parts, acting as the input portion of the transistor.

Beta decay: A radioactive decay process in which an electron and a neutrino are emitted, transforming a neutron into a proton.

Beta particles: Energetic electrons, a product of radioactive decay.

Big bang theory: The predominant cosmological theory of the creation of the universe, which maintains that the universe emerged from a singularity (that is, black hole) approximately 10 to 20 billion years ago.

Binary stars: A system of two stars that orbit a common center of gravity; occasionally one star will eclipse the other relative to Earth viewers.

Binding energy: A measure of the energy of attraction for constituents of a bound system; the sum of the rest mass energies ($E = mc^2$) of the components of the system minus the mass energy of the system is positive for bound systems.

Biochemistry: The study of molecules and chemistry found in living matter; chemistry applied to biological problems and vice versa.

Biological membrane: A natural barrier consisting of a double layer of lipid molecules.

Bit: The amount of memory required to store one binary digit.

Black body radiation: Radiation created by the random thermal motions of atoms; it rapidly strengthens as the temperature increases.

Blueshift: When the spectral lines have shifted toward shorter wavelengths than the laboratory reference set of lines; this is caused by the fact that the object is moving toward the observer.

Body waves: Seismic waves that travel through materials; they are of two types, compressional or P waves, and transverse or S waves.

Boiling point: The temperature at which the first gas bubbles form in a liquid.

Boltzmann constant: The fundamental constant of statistical mechanics, designated k and equal to 1.38×10^{-34} joule/Kelvin.

Boltzmann distribution: The way that atoms and molecules are distributed among their allowed energy levels in a macroscopic system at equilibrium.

Bond: One or more pairs of electrons that are responsible for holding atoms together; in carbon compounds, these are shared pairs or covalent bonds.

Bond angle: In a three-atom group, A–B–C, B may be considered as the vertex of an angle formed by the bonds to A and C; this is the bond angle around B.

Bond length: The center-to-center distance between the nuclei held together by a covalent bond.

Born-Oppenheimer approximation: A principle stating that the nuclei in a molecule are fixed so that the Schrödinger equation can be solved for the motion of the electrons about the nuclei; used to calculate potential energy curves.

Bose-Einstein statistics: A form of quantum mechanical statistics used to describe an assembly of "bosons," which are particles whose spin is an even multiple of Planck's constant, h, in a noninteracting configuration at low temperatures.

Bosons: The group of elementary particles that have integral spins and transmit the basic forces of nature, including the photon, gluon, and intermediate vector bosons.

Bremsstrahlung: The energy loss of charged particles (usually electrons) resulting from the electromagnetic radiation emitted in the violent accelerations that occur during collisions with atomic electrons.

Buffer: A substance that resists changes in acidity, or pH.

CAD/CAM: Acronyms for computer-aided design and computer-aided manufacture; the use of computer graphics to view and/or alter industrial and other working designs and models.

Calorie: The unit of heat defined as the amount of heat needed to increase the temperature of 1 gram of water by 1 degree Celsius.

Capacitance: The property of an electrical circuit element that allows it to store an electrical charge when a voltage is placed across it.

Capacitor: An electrical device that stores charge when voltage is applied to it.

Catalyst: A substance that enhances the rate of a reaction without being either a reactant or a product, and without affecting the final equilibrium point of the reaction.

Cathode: The primary source of electrons in an electron vacuum tube. The electrode at which reduction takes place in an electrochemical cell.

Cathode-ray tube: A vacuum tube for television or computer image production in which a hot cathode emits an electron beam accelerated through a high-voltage anode, which is focused and made to fall on a fluorescent screen.

Cation: A positively charged atom, compound, or molecule that has lost an electron from an unstable outer atomic energy level, thereby eliminating that energy level.

Center-of-mass frame: The (usually moving) coordinate system in which the interaction is at rest.

Centrifugal force: The outward force exerted on an object moving in a circular path.

Centripetal force: The inward force required to keep an object moving in a circular path.

Cepheid variable star: A variable star whose period of variation is related to its absolute brightness (luminosity); used to measure distances to star clusters and galaxies.

Chain reaction: A reaction sequence in which one event causes the next event, which causes the next event, and so forth.

Chaos: The tendency for ordered natural processes to become disordered, and vice versa, often because of the influence of minute external factors.

Charge: A basic property of matter which occurs as either positive (+) or negative (−); opposites attract and likes repel according to Coulomb's law.

Charm: A quantum mechanical property of baryons and mesons which is conserved in strong nuclear interactions but not in weak interactions.

Chemical bond: An attractive force between atoms and/or molecules, usually involving the sharing or transfer of electrons and variable electromagnetic charges.

Chemical reaction: Any process in which one or several new substances (reaction products) are produced as a result of chemical changes in chemical starting materials (reactants).

Cherenkov radiation: The radiation, or light, emitted by a charged particle moving through a medium faster than the speed of light.

Chondrite: A type of meteorite that has nearly the same abundances of heavy elements as the sun and may represent original solar system material.

Chromatography: A general term that describes a range of methods used to separate mixtures of chemicals into their components, either spatially or in time.

Classical mechanics: The branch of physics that deals with the action of force on objects of mass; classical mechanics accounts for this action in terms of Sir Isaac Newton's three laws of motion.

Coercivity or coercive force: The strength of a magnetic field at opposite polarity that is necessary to demagnetize a magnetized material.

Coherent radiation: Radiation in which components of the signal are in phase with one another.

Collector: One of three parts of a transistor, doped opposite to the base part and usually connected to the chassis for thermal energy dissipation.

Colloids: Particles with a high ratio of surface area to mass (diameter of about 10^{-7} to 10^{-5} centimeters) that are electrically charged, repel one another, and resist coagulation.

Combustion: A chemical reaction in which a fuel reacts with the surrounding atmosphere, releasing substantial heat energy.

Compiler: A computer program that translates the entire program written in source code into a form that can then be directly executed by a computer.

Complex conjugate: A number $x - iy$ that is closely related to the complex number $z = x + iy$ (obtained by replacing i with $-i$).

Complex number: A number of the form a $+bi$, where i represents $\sqrt{-1}$.

Compound: Matter composed of atoms of two or more different elements chemically combined in fixed proportions.

Compressibility: The degree to which the volume occupied by a substance decreases as the pressure acting on the substance increases.

Compton effect: The reduction of a photon's energy that is caused by its interaction with a free electron; part of the photon's energy is transferred to the electron.

Computer program: A sequence of instructions to a computer.

Concentration: The number of molecules of a substance per unit volume.

Condensation: A phase change from the gas (vapor) phase to the liquid phase, in which particles of matter lose energy and move closer together.

Conduction band: A series of very closely spaced energy levels; electrons of these energies can move freely throughout a conductor.

Conductivity: The measure of a material's ability to carry electrical current.

Conductor: A substance through which heat or an electric current can flow at a rapid rate.

Conservation of angular momentum: The total amount of angular momentum of a group of objects that remains the same.

Conservation of energy: The principle, applicable in classical and quantum physics, which states that the total energy of a system must remain constant.

Continuity: The property of having no breaks or gaps.

Continuous function: A function f with the property that as the number u gets close to x, the functional value $f(u)$ gets close to $f(x)$.

Continuum mechanics: That branch of mechanics dealing with macroscopic phenomena whereby the medium is thought of not as consisting of individual particles but as a continuous medium.

Convection: Flow of material caused by temperature differences that causes warm, light material to rise and cool dense material to sink.

Convergence: The condition, in a progressive or iterative process, when successive calculated values approach closer to observed data values; mathematically, the condition whereby computed values approach definite limits as the number of computational terms used increases.

Coordinate system: A grouping of magnitudes that are used to define the positions of points in space.

Copenhagen interpretation: The standard interpretation of quantum mechanics, which claims that it makes no sense to ask what a particle is doing or what properties it has when it is not being measured or observed.

Coriolis effect: A force that deflects moving objects to the right (in the Northern Hemisphere) or to the left (in the Southern Hemisphere) because of the earth's rotation.

Correlation function: A statistical measure of the tendency of two random signals to vary together in a correlated fashion; auto- and cross-correlation functions are, respectively, measures of the similarity of a signal with itself and with some other signal.

Cosmic background radiation: Microwave radiation from the glowing of the hot early universe, now cooled by the expansion of the universe to a temperature of about 3 Kelvins.

Cosmic rays: Ionizing radiation, or high-energy atomic particles, from space.

Cosmology: The study of the large-scale structure of the universe and its movements, origin, evolution, and ultimate fate.

Coulomb force: The force of interaction involving charged particles separated by distance.

Coupling constant: The strength of an interaction between two particles is given by the coupling constant, while the range over which the interaction takes place is given by the mass of the virtual boson that is exchanged between them.

Covalent bond: A force of attraction that exists between two atoms as a result of sharing a pair of electrons.

Critical temperature: The temperature above which a gas cannot be liquefied by compression and the liquid phase cannot be distinguished from the vapor phase.

Cross-correlation: A mathematical measure of the similarity, or correspondence, between matching members of two or more data series.

Cross section: The effective area that has to be attributed to a particular atom or nucleus to account for its interaction with an incident beam of radiation.

Cryogenic: Having to do with, or characterized by, very low temperatures.

Crystal: An orderly arrangement of atoms or molecules to produce a regular solid having definite dimensions and physical properties.

Crystal lattice: The regular array of points marking the equilibrium positions of the atoms in a crystalline solid.

CT scanner: A sophisticated X-ray imager capable of taking detailed X-ray pictures of thin slices of opaque objects, usually body parts, via a rotating X-ray source-detector assembly and a computer that deciphers the obtained data.

Current: The rate of electrical charge movement through a circuit element; usually measured in units of coulombs per second or amperes.

Curved space: A space which does not obey the rules of Euclidean geometry; may be positively curved or negatively curved space.

Cyclotron: A device that accelerates ions in a spiral path by the influence of electrical and magnetic fields until they emerge at great speed, at which the ions can then bombard other elements for the creation of other atoms.

Damping force: Any force that causes a loss of energy (slowing or stopping) in a mechanical or electrical system; generally friction and electrical resistance, respectively.

Dark matter: A hypothesized mass of invisible interstellar hydrogen gas that may account for up to 90 percent of the universe's total mass.

Degrees of freedom: The number of modes or types of possible motion for a molecule, normally partitioned into translational, vibrational, and rotational contributions; reflective of the type of bonds holding the molecule together.

Density: The mass of a substance per unit volume, measured in kilograms per cubic meter.

Dependent variable: A variable quantity whose value depends on (is a function of) another quantity of other quantities, the independent variable or variables.

Derivative: The slope of a curve.

Deuterium: A heavy isotope of hydrogen, having one proton and one neutron in its nucleus.

Dielectric: An electrical insulator, which is a substance that will not allow electric charge to flow.

Dielectric constant: The factor by which replacing a vacuum between the plates of a capacitor with a particular dielectric will increase the capacitance.

Differential equation: An equation in which the unknown is a function, or dependent variable quantity, and in which appears the derivative, or some partial derivatives, or subsequent derivatives of the derivative(s) of the unknown function.

Diffraction: A wave phenomenon that occurs when a wave encounters an aperture with a characteristic dimension similar to the wavelength of the incident wave.

Diffusion: A net movement of molecules from one region to another in response to a concentration gradient.

Digital signals: A signal format in which the information is transmitted by a series of binary-coded pulses rather than by analog fashion.

Diode: A two-electrode electron tube in which electron flow is from cathode to anode, in that direction only; flow in the opposite direction, as happens with alternating current, is highly resisted.

Dipole: A pair of equal magnetic poles of opposite sign that are separated by a small distance; an arrangement of two equal but opposite charges separated by a fixed distance, such as a bar magnet.

Direct current: A current that does not change its magnitude or direction in time; measured in amperes.

Dispersion: An expression for the variability of the index of refraction with the wavelength of light.

Displacement: The distance of an atom from its equilibrium position.

Dissociation energy: The energy required to break a molecule into smaller molecular fragments or atoms.

Doppler effect: A change in the observed frequency of electromagnetic or sound waves, caused by relative motion between the source and the observer.

Dynamics: The study of motions, taking their causes into account.

Efficiency: The useful work divided by the heat input for an engine.

Eigenvalue: A characteristic value for a system, such as a characteristic frequency.

Elasticity: The tendency of solid objects to return to their original form after a stress is removed.

Electric field: One component of electromagnetic radiation; a static electric field is a change in space, caused by an electric charge, such that another electric charge experiences a force.

Electrolysis: The separation of a substance into its constituent chemical elements on passage through it of an electric current.

Electromagnet: A magnet created by an electric current moving through loops of wire wound around a core of highly susceptible material.

Electromagnetic field: The combined name of the electric and the magnetic field.

Electromagnetic radiation: Waves of electrical and magnetic energy that travel through space at the speed of light; the entire range of energy states from radio waves through infrared, visible light, ultraviolet, X-ray and gamma-ray radiation is known as the electromagnetic spectrum.

Electromotive force (emf): The potential difference between two points (or terminals) of a device that is used as a source of electrical energy.

Electron: An elementary particle with a negative charge of magnitude 1.602×10^{-19} coulomb and a mass of 9.109×10^{-31} kilogram. This stable elementary particle (lepton) is that negatively charged constituent of ordinary matter which orbits the nucleus of the atom; the antiparticle of the positron.

Electronegativity: The tendency of certain substances to attract electrons more strongly than other substances; elements located to the upper right of the periodic table of elements have high electronegativities.

Electrons and holes: Charges in a solid can be carried either by negatively charged particles or by regions of empty space in the solid from which a negative charge has been removed; the latter are called (positive) holes.

Electronvolt: The amount of energy gained by an electron when it is accelerated by only 1 volt.

Electrostatic: Characterized by or having electric charges at rest.

Element: A pure substance that contains only one kind of atom.

Ellipse: A closed oval figure defined on the basis of two evenly spaced points in the interior called foci; the sum of the distances from each of the two foci to any point on the ellipse is always the same.

Emitter: One of three parts of a transistor, doped like a collector.

Energy: A measure of the total physical work that can be obtained from a system; most frequently measured in electronvolts when describing atoms and subatomic particles.

Energy levels: The allowable energies that an electron or particle can occupy in quantum theory.

Entropy: A measure of the energy of a system that is unavailable for work; a measure of the randomness in the microscopic state of a macroscopic system; or a mathematical measure that quantifies information content of a probabilistic experiment.

Equation of state: The relationship between thermodynamic variables (usually pressure, volume, and temperature) for a specific substance.

Equations of motion: The set of mathematical equations (from Newton's laws of motion) that can be solved to determine the position and velocity of a particle at any point in time.

Equilibrium: A condition of stability whereby the state of a system does not change.

Equipartition theorem: One of the fundamental results of classical statistical mechanics requiring that, at thermal equilibrium, all independent modes of vibration in an oscillating system have the same average energy.

Erg: The amount of kinetic energy of a mass of 1 gram moving at 1 centimeter per second; a mosquito in flight possesses about an erg of energy.

Event: A fundamental "point" of space-time, specified not only by a place but also by a time of occurrence.

Event horizon: A zone around a black hole within which time stops, and from which nothing, not even light, can escape.

Excitation: The process by which a molecule gains energy and undergoes a transition to an energy level that is higher than its ground state.

Fermi-Dirac statistics: A form of quantum mechanical statics used to describe an assembly of "fermions," which are particles whose spin is a half-integral multiple of Planck's constant.

Fermions: Particles with half integral values of spin and that obey the exclusion principle; examples include protons, neutrons, and electrons.

Field: A physical parameter defined at all points in space and time in a particular region, such as an electromagnetic field.

First law of thermodynamics: Energy cannot be created or destroyed, only transformed from one form to another.

Fission: The splitting of a heavy atomic nucleus into lighter atomic nuclei.

Flat space: A space of any number of dimensions which obeys the rules of Euclidean geometry.

Fluid: A substance that flows, including liquids, gases, and free charge clouds in plasmas.

Flux: The value of the electric (or magnetic) field multiplied by the surface area that it crosses.

Focal length: The distance from the center of a convex lens to the focal point, or the distance from the surface of a mirror to its focal point.

Force: A physical phenomenon capable of changing the momentum of an object; the four fundamental forces in the universe are gravity, electromagnetism, the strong nuclear force, and the weak nuclear force.

Fourier analysis: A transformation of functions that permits expression of any sectionally continuous function in terms of orthogonal functions such as the trigonometric sequence $\{\sin nx\}$ and $\{\cos nx\}$.

Fractal: A geometric object that can be described as having a dimension not equal to a whole number. Occurs when the value obtained from a measurement depends on the size of the measuring device, that is, "the length of the coastline of England."

Frame of reference: A system of space-time coordinates used by an observer to measure the physical world.

Frequency: The number of periodic wave pulses passing by in a unit of time; the usual unit of frequency is the hertz; one hertz is one wave pulse per second.

Friction: The resistance that occurs between the surfaces of two materials that are moving against each other.

Function: A rule that assigns some definite number to every value in some given set of numbers.

Fusion: The process of combining light nuclei to form heavier ones, with a resulting release of energy.

Galaxy: An aggregate of stars, dust, and gas with a more or less definite structure.

Gamma radiation: A high-energy form of light that is emitted by unstable nuclei in the process of decaying toward stability.

Gas: A phase of matter in which particles have their greatest energy and motion such that they occupy no definite shape or volume.

Gauge theories: A class of theories for the fundamental forces whose equations remain unchanged in form by certain symmetry transformations at any point in space and time.

Gauss: A unit of magnetic field strength.

General theory of relativity: Albert Einstein's theory of gravity, proposed in 1916, which states that the curvature of space-time causes gravitational fields.

Geometry: A set of rules which describes the structure of a region of space; traditional geometries describe space as flat, while relativistic geometries describe it as curved.

Gluon: The massive, colored carrier of the strong nuclear interaction; the eight different gluons all have one unit of spin.

Gradient: A gradation with distance in the rate of change of a physical quantity such as temperature, concentration, or pressure.

Grand unified theories: Theories of modern physics that attempt to unite three of the four fundamental forces of nature (the electromagnetic, weak, and strong nuclear forces) into a single unified force, such that these three forces would have been essentially identical during the early moments after the big bang.

Gravitational constant, G: The constant in Newton's law of gravity that determines gravitational attraction of two masses separated from each other.

Gravitational field: The strength of the gravitational force that a given mass experiences per location.

Graviton: The massless, chargeless carrier of the gravitational interaction that is believed to have two units of spin and has not yet been observed experimentally.

Greenhouse effect: A phenomenon where an environment is heated by the trappings of infrared radiation; responsible for the high surface temperatures of Venus.

Ground state: The lowest energy state of a molecule, atom, nucleus, or ion.

Group (chemistry): A vertical column of elements in the periodic table; the members have similar chemical and physical properties.

Group (mathematics): A set S together with an operation $*$ on S subject to $a*(b*c) = (a*b)*c$, existence of an identity element in S, and existence of an inverse element for each element in S.

Hadron: Any particle that participates in the strong interaction; a hadron is divided into baryons, which obey the Pauli exclusion principle, and mesons, which do not.

Half-life: The time that is needed for one-half of any amount of a radioisotope to undergo decay; a constant characteristic of the particular radioisotope.

Halogen: The elements fluorine, chlorine, bromine, iodine, and astatine, which appear in the seventh column of the periodic table.

Heat: The energy that flows between substances because of their difference in temperature.

Heat capacity: The amount of heat an object can absorb divided by the temperature change in that absorption.

Helicity: The handedness (left-handedness or right-handedness) of a particle, which is determined by how it spins with respect to the direction of its momentum.

Hertzsprung-Russell diagram: A plot of luminosity against surface temperature of a collection of stars, named after Ejnar Hertzsprung and Henry Norris Russell.

Hole: The absence of an electron in the valence band, which behaves effectively as a positively charged particle.

Horizontal branch: A stage of helium core burning; in evolving star clusters, this appears as a nearly horizontal grouping in the Hertzsprung-Russell diagram.

Hubble's law: The spectral redshift of light from a distant galaxy is a measure of its velocity, which is dependent on the galaxy's distance.

Hugoniot elastic limit: The greatest stress that can be developed in a material without permanent deformation remaining when the stress is released.

Hydrocarbons: Chemical compounds composed of hydrogen and carbon only; the major chemical components of petroleum and coal.

Hydrogen bond: An attraction, less powerful than a covalent bond, between a hydrogen atom attached to an atom like nitrogen, oxygen, or sulfur, and an unshared pair of electrons on some other atom.

Hydrolysis: A chemical reaction in which a molecule is cleaved by water.

Hydrophilic: Materials that are attracted to and that attract water or that are "water loving."

Hydrophobic: Materials that have the property of being excluded from water or that are "water hating."

Hysteresis: The lag in the response of a magnetic material to a changing applied magnetic field.

Ideal gas law: An equation of state obeyed by all gases in the limit of low density.

Imaginary part: The real number y in $z = x + iy$ is called the imaginary part of the complex number z; if the imaginary part is zero, then the number z is a purely real number.

Index of refraction: The ratio of the velocity of light in a vacuum to light velocity in a particular medium.

Inductance: The property of an electrical circuit element that produces a voltage across that element when the current through the element is changing.

Inertia: The tendency of an object at rest to remain at rest, or, if moving, to continue moving in the same direction.

Inertial frame of reference: A frame of reference which moves at a constant velocity and for which Isaac Newton's laws of motion are valid.

Infrared: Invisible rays of light (electromagnetic radiation) with wavelengths just beyond red on the visible spectrum.

Insulator: A substance that is a poor electrical conductor.

Integer: Any whole number that is either positive, negative, or zero; similarly, any number computed from zero by a finite number of additions of $+1$ or -1.

Interference: The superposition of two or more waves, resulting in a net displacement of their amplitudes; the resultant displacement is the sum of the original amplitudes.

Interferometry: A method used for measuring very small dimensions in which a beam of light is split in two; the two resulting beams are channeled through different paths and then compared.

Intermediate vector boson: The massive, charged carrier of the weak nuclear interaction; the three different vector bosons (the W^+, W^-, and Z^0, with charge of $+1$, -1, and 0, respectively); all have one unit of spin and have been directly observed in high-energy accelerator experiments.

Interpreter: A computer program that translates and immediately executes one line of source code at a time.

Interstellar medium: Gas and dust between the stars that eventually are incorporated into new generations of star formation.

Intrinsic angular momentum: Also known as the spin angular momentum; in the terms of the natural unit given by Planck's constant divided by 2π, the intrinsic angular momentum has a magnitude of either an integer or an integer plus ½.

Invariant: Unchanged by a transformation of coordinates, such as the interval between two events in space-time.

Ion: An atom or a molecule that has either lost or gained electrons.

Ionic bond: A force of attraction between oppositely charged atoms or groups of atoms (ions); responsible for the stability of solids like sodium chloride.

Ionization energy: The change in energy when an electron is removed from a gaseous atom or molecule; "ionization potential" is an older synonym.

Ionosphere: An upper layer of the earth's atmosphere above 60 kilometers in altitude, containing ionized particles and electrons.

Isomers: Chemical compounds with the same chemical formula (identical atomic composition) but with atoms bonded together in different ways; they often exhibit distinct chemical and physical properties.

Isotopes: Two or more nuclei of the same element having different masses, or numbers of neutrons.

Joule heat: Irreversible heat dissipated within an electrical conductor by virtue of current flow.

Joule-Thomson effect: The change in temperature observed when a gas is expanded under conditions where there is no net flow of heat into or out of the gas.

Kelvin: An absolute temperature scale where 0 is set at absolute zero; in this system, a unit of temperature, or Kelvin, is equal to a degree Celsius.

Kinetic energy: The energy of a system stemming solely from its motion.

Kinetic theory: The attempt to relate the physical properties of matter in bulk to the motion of the constituent atoms or molecules.

Kirchhoff's laws: Two laws governing the behavior of current and voltage in electrical circuits; equivalent to the physical principles of conservation of energy and charge.

Lagrange points: Stable points in the orbit of an intermediate body around a larger body where small particles may accumulate; these are sometimes referred to as L4 and L5.

Lanthanides: The group of fourteen 4*f* transition elements in period 6 of the periodic table, corresponding to the actinides and listed right above them, and with highly similar chemistry.

Laser: A device which produces coherent *l*ight *a*mplification by *s*timulated *e*mission of *r*adiation.

Latent heat: Heat released during the condensation of water vapor to form liquid water or the vapor deposition of water vapor to form ice.

Lattice: A regular array of points in space with periodic locations within a three-dimensional volume.

Lepton family: A grouping of the lightest subatomic particles, which includes the electron, the muon, the tau particle, and the associated neutrinos and their antiparticles; leptons do not participate in the strong nuclear interaction.

Ligand: A molecule or ion bonded to the central atom or ion in a coordination compound.

Light: The generic term for any type of electromagnetic wave or radiation; common usage refers to the portion of the electromagnetic spectrum called visible light, which ranges between infrared and ultraviolet wavelengths.

Light-year: A unit of distance equal to the distance light travels in a vacuum in one year; at a speed of approximately 300,000 kilometers per second, a light-year is about 10 trillion kilometers.

Linear polarization: Light in which the electric field oscillates in one plane.

Liquid: A phase of matter in which particles have less energy than in a gas and more than in a solid, so that they are freely flowing but occupy a specific volume.

Liquid crystals: Liquids which exhibit many physical properties characteristic of solid crystals, such as nonisotropic (directional) molecular orientation and optical birefringence.

Lithosphere: The rigid outer layer of a planet, as contrasted with less rigid layers beneath; may include the crust and part of the mantle.

Lorentz force: The net force resulting from a combined electric and magnetic field acting on a charged particle.

Luminosity: The intrinsic brightness or light output of a star as distinct from its apparent brightness; the amount of energy a star radiates in one second.

Mach number: The ratio of the velocity of an object to the speed of sound in the surrounding medium.

Machine language: A set of computer instruction codes in the form of sequences of binary digits (1's and 0's); these codes correspond to specific actions within the computer.

Macroscopic: Pertaining to matter in bulk.

Magic number: A number associated with the quantity of nuclear particles required to form a stable shell or layer of protons or neutrons in an atomic nucleus.

Magnetic domain: A small volume within a ferromagnetic or antiferromagnetic solid where there is a spontaneous alignment of electron spin direction.

Magnetic field: A condition produced when two or more moving charges each produce kinetic energy and momentum as the result of an electric current or magnet. A vector field describing the force generated by currents of either macroscopic or microscopic origin.

Magnetic field intensity: A vector quantity that measures the strength of a magnetic field; this value is often expressed in terms of the density of lines of force representing the field.

Magnetic monopole: A particle predicted to exist in nature but not yet observed; it would have a magnetic charge of only one kind (either north or south) and would be a source charge for the magnetic field.

Magnitude: A scale for measuring brightness, for which each magnitude is approximately two and a half times brighter; the lower the magnitude, the brighter the star.

Main sequence star: Typically, a stable star in steady state, consuming hydrogen by thermonuclear fusion; "main sequence" describes the position of a star on the Hertzsprung-Russell plot.

Mantle: The intermediate layer of a planet, between the core and the crust.

Mass: The total amount of material in any object, determined by its gravity or inertia.

Mass spectrometry: An experimental technique that allows for the determination of molecular mass by separating a stream of charged particles according to the mass-to-charge ratio.

Matrix: A rectangular array of numbers arranged in rows and columns; the numbers within a matrix are called elements or entries of the matrix.

Maxwell's equations: Vector equations that relate the partial derivatives of electromagnetic fields to electric charge and current density.

Mean free path: The average distance a conduction electron travels before it collides with a lattice ion.

Melting: A phase change from the solid phase to the liquid phase as the result of an increase in energy, thus breaking the chemical bonds holding the particles together.

Mesons: A subset of a class of subatomic particles intermediate in mass between electrons and protons.

Metals: A class of elements that are usually lustrous solids and are good conductors of heat and electricity.

Meteorites: Generally small, solid masses which have fallen to the ground after colliding with the earth's atmosphere, consisting of varying ratios of metal to silicate minerals; classified accordingly as stones, stony-irons, and irons.

Microscopic: Pertaining to matter on the atomic or molecular level.

Microwave radiation: Electromagnetic radiation of wavelengths ranging from 0.03 to 30 centimeters (frequency of 1 to 100 gigahertz).

Milky Way: The bright, dense band of stars seen overhead on a summer's night; the name for the galaxy in which our solar system is located.

Model: A conceptual representation, in mathematical form, by which one or more behaviors or properties of a given data set can be organized, simulated, and predicted.

Moderator: A chemical substance which slows down neutrons for better absorption into a fissile nucleus.

Mole: A unit of quantity equal to the number of atoms in a 12-gram sample of isotopically pure carbon, approximately 6×10^{23}.

Mole fraction: A measure of the concentration of components in a mixture based upon the relative number of molecules of each constituent.

Molecular weight: The relative weight of a molecule, based on a scale in which the weight of an oxygen atom is expressed as 16.

Molecule: A group of atoms bonded together to form a single chemical species.

Moment of inertia: A property of an object's mass distribution around a given axis of rotation; the angular momentum of the object is the product of the moment of inertia and the angular acceleration of the object.

Momentum: The vector physical quantity, conserved under the condition of no net force, defined by the product of the mass and velocity of a particle or system (mv).

Muon: A charged lepton with a mass about 207 times the electron mass, which decays into an electron and two neutrinos.

Natural frequency: A preferred frequency of oscillation of an object of a medium; depends on the properties of the system.

Neutrino: A stable, uncharged lepton that has a zero or near-zero rest mass, interacts by the weak force, and has a left-handed spin relative to its motion at or near the speed of light.

Neutron: An elementary particle found in the nucleus of an atom possessing unit mass and no charge.

Neutron star: A collapsed remnant of a supernova whose high density crushes all matter, including protons and electrons, into neutrons.

Newton's first law of motion: An object at rest remains at rest and an object in straight-line unaccelerated motion remains that way unless acted upon by unbalanced forces.

Newton's second law of motion: Simply stated, force equals mass times acceleration ($F = ma$).

Newton's third law of motion: To every action, there is an equal and opposite reaction.

Noble gases: Helium, neon, argon, xenon, krypton, and radon.

Normal distribution: A well-known continuous probability distribution, popularly described as a "bell-shaped curve."

Normal mode: A simple type of motion into which a complex system (a system composed of several joined masses or circuits) may be separated; there may be many normal modes in a complex system, each behaving like an independent oscillating system with its own natural frequency.

Nuclear electric quadrupole moment: The charge distribution characteristic of the excited nuclei of many atoms with protons that are nonspherically symmetric.

Nuclear magnetic resonance (NMR): The phenomenon exhibited by the magnetic spin systems formed by nuclei of certain atoms, whereby nuclei absorb energy at specific natural (resonant) frequencies when subjected to alternating magnetic fields.

Nucleic acids: Deoxyribonucleic acid (DNA) and ribonucleic acid (RNA), the biomolecules that control all cellular functions, including growth, metabolism, and reproduction.

Nucleon: Those atomic particles that reside in an atom's nucleus, namely, protons and neutrons.

Nucleosynthesis: The process by which heavier elements are produced from hydrogen and helium in stars.

Nucleus: The central structure of an atom, composed of tightly bound, positively charged protons and neutral neutrons; nuclei are much heavier and slower than electrons and attract electrons to their positive charge.

Ohm: A unit measure of resistance equal to the resistance of a circuit, where 1 volt of electromotive force maintains 1 ampere of electric current.

Oort cloud: A cloud of comets existing as a spherical cloud with an approximate diameter of 100,000 astronomical units from the sun.

Organic compound: A compound that contains carbon.

Orthogonal: Geometrically, means a pair of vectors are perpendicular to one another; can be extended beyond three dimensions and to vector functions that can represent physical quantities.

Oscilloscope: A device used to observe voltage waveforms in electrical circuits; includes a cathode-ray tube, sweep generator, horizontal deflection amplifier, and vertical deflection amplifier.

Oxidation: The chemical process of removing electrons from an atom or molecule, often involving the introduction of oxygen.

Oxidation state: The charge an atom in a compound would have if all the bonding electron pairs belonged to the more electronegative atom involved in the bond.

Oxides: Binary chemical compounds that include oxygen.

Ozone: A gas that is a major constituent of the upper atmosphere and consists of three oxygen atoms bonded together.

P-n junction: The contact area between two regions of a semiconducting material, a positively charged region (*p*) and a negatively charged region (*n*).

Pair production: The creation of an electron and a positron that is caused by the interaction between a photon and an atom.

Parsec: The distance at which the stellar parallax is 1 second of arc, or approximately 3.26 light-years.

Partial derivative: The derivative of a function of several variables with respect to only one of the variables, the remainder being kept constant.

Partial differential equation: An equation involving derivatives of a function of more than one independent variable; a solution must satisfy the partial differential equation and its boundary conditions.

Particle accelerator: A device for imparting energy to a charged particle, bringing it to high energies and velocities that are close to the speed of light.

Pauli exclusion principle: The principle that no two particles of the same type can occupy precisely the same quantum state is obeyed by baryons, leptons, and quarks, but not by photons, mesons, or gluons.

Perigee: The point of an orbit at which the object is the nearest to the earth.

Perihelion: The closest position of a planet to the sun.

Period: The time interval between successive recurrences of the same event, such as the eclipsing of binary stars or the brightening of Cepheid variables.

Periodic table of the elements: An arrangement of the known elements in order of increasing atomic number, in which certain chemical and physical properties recur periodically and allow the elements to be organized in vertical groups of elements with similar properties.

Permeability: A measure of the change in the force of attraction or repulsion between two magnetic poles; this depends on the properties of the specific material being affected by the magnetic field.

Permittivity: A key electromagnetic parameter characterizing the ease with which a material conducts electricity.

Perpetual motion machine of the first kind: A machine that violates the first law of thermodynamics by obtaining energy from nowhere.

Perpetual motion machine of the second kind: A machine that violates the second law of thermodynamics by running at 100 percent efficiency.

pH: A measure, on a scale of 1 to 14, of the degree of acidity; 1 is most acid, 7 is neutral, and 14 is most basic.

Phase: A region that has the same chemical and physical properties throughout; for example, an oil-water mixture separates into two phases. Referring to wave motion, the position of a wave with respect to a reference wave.

Phase velocity: The speed at which the crest of a wave appears to travel, referenced to a specific direction in space.

Phonon: A quantum of energy associated with the disturbance, or vibration, of a crystalline lattice.

Photoelectric effect: An electron emission caused by complete energy transfer from a photon to an atomic electron.

Photon: A quantum of electromagnetic wave whose energy and momentum are proportional to the frequency and the wave number of the electromagnetic wave.

Photosphere: The dense visible surface of the sun, which has a temperature of about 5,000 Kelvins.

Pi-meson: The lightest strongly interacting particle; the carrier of the nuclear force of longest range; has three charge states, positively charged π^+, negatively charged π^-, and neutral π^0; often called pion or π.

Piezoelectric effect: When certain crystals or ceramic materials are compressed or distorted, they acquire an electric voltage across the material that is proportional to the amount of compression or deflection.

Pixel: A picture-cell or smallest part of an image represented on a computer.

Planck's constant: The fundamental unit of action in quantum mechanics, designated h and equal to 6.616×10^{-34} joules per second.

Plasma: A gaseous fourth state of matter that is a cloud of ionized particles, usually electrons and ions.

Plasmon: A quantum of energy moving longitudinally through a crystal lattice as a wave.

Point group: A combination of symmetry operations that acts around a single point.

Polarity: The unsymmetrical distribution of electrons in molecules that results when one atom attracts electrons more strongly than another.

Polarization: The direction of oscillation of the electric (or magnetic) field in reference to the propagation direction of the light beam.

Polaroid plates: Plates, made of transparent materials, that allow passage of only those light waves vibrating in one direction; two Polaroid plates placed at right angles to each other (crossed) do not allow any light to pass through them.

Polymer: A large molecule made up of many small, identical units bonded together; plastics, rubber, cellulose, and starch are examples of polymers.

Positron: The antiparticle of the electron, with the same mass but the opposite electric charge.

Postulate: Today, synonymous with "axiom"; originally, postulates were empirically self-evident relationships.

Potential difference: Work done in moving a unit charge from one point to another point in an electric field.

Potential energy: The energy that an object has as a result of its position in a field of force of some kind (such as gravitational or electromagnetic).

Precession: The motion of a rotating rigid object in which the spin axis changes directions because of the presence of an applied torque.

Precision: The accuracy with which a given measurement or calculation is made, regardless of whether the measured datum is a true and accurate representation.

Pressure: The force per unit area exerted by the system on its surroundings.

Prime number: A positive integer that has exactly two divisors, namely, the number 1 and itself; examples are 1, 2, 3, 5, 7, 11.

Principle of equivalence: The rule that, in a limited region of space-time, the effects of the acceleration of a given frame of reference are not distinguishable from those of a gravitational field.

Principle of relativity: A statement about how different observers agree on the laws of physics.

Probability of an event: The likelihood that an event will occur when more than one outcome is possible; the likelihood that a tossed coin will come up "heads," for example, is one in two, or ½.

Projective geometry: A branch of geometry in which invariant properties of figures under projections or transformations are studied.

Protein: A biological polymer in which the links are amino acids.

Proton: A subatomic particle with a charge of $+1.60219 \times 10^{-19}$ coulomb and a mass of 1.67265×10^{-27} kilogram; the nucleus of the hydrogen atom.

Protostar: An embryonic stage of a star that is in the process of formation.

Pulsar: A rapidly spinning neutron star that emits radio pulses.

Quadratic equation: An equation that can be put in the form $ax^2 + bx + c = 0$, where a, b, and c are real numbers.

Quadratic formula: The roots of a quadratic equation $ax^2 + bx + c = 0$ are $x = (-b \pm \sqrt{b^2 - 4ac})/2a$.

Quantitative analysis: The process of determining the amount of a compound or element present in a sample.

Quantized: Having only certain discrete values, such as the energy of an atom.

Quantum chromodynamics (QCD): The theory of the strong interaction that has colored gluons and quarks as fundamental components; hadrons are the colorless bound states of the quarks confined to a bubble or "bag."

Quantum efficiency: The percentage of incident radiation that is detected or converted to a useful purpose by a device.

Quantum electrodynamics (QED): The quantum theory of the electromagnetic field and its interactions with charged particles.

Quantum mechanics: A branch of physics that details the workings of the basic constituents of matter and is characterized by the use of mathematical probabilities.

Quantum numbers: A set of numbers representing certain fundamental properties of a system and that define its overall state.

Quark: The fundamental constituent of the hadrons; these particles have a fractional electrical charge.

Quasar: A quasistellar radio source, an intense emitter of electromagnetic radiation; quasars may be the oldest and most distant objects in the universe.

Radical: Refers to any atom or molecule containing an unpaired electron; radicals are very reactive because of the unpaired electron.

Radio astronomy: A subdiscipline within astronomy that observes the universe at radio wavelengths.

Radioactive isotope: An unstable isotope of an element that is capable of decaying by emitting some form of radiation; usually converts to a more stable isotope through this decay or a series of these decays.

Raman effect: An increase or decrease in the wavelength of light related to quantum interactions of photons with molecules of a transparent substance.

Random numbers: A sequence of numbers that cannot be predictably computed using a formula; the next number in the sequence cannot be determined from previous ones.

Rational number: Any number expressible as the quotient of two integers, a/b, where b is not zero.

Real part: The real number x in $z = x + iy$, is called the real part of complex number z; if the real part is zero, the number is called imaginary.

Red giant: A large, cool star of high luminosity plotted to the upper right of the main sequence on the Hertzsprung-Russell diagram.

Redox reaction: Any chemical reaction in which one reactant is oxidized (loses valence electrons) and another one is reduced (gains valence electrons).

Redshift: The increase in wavelength of light from stars moving away from Earth, causing the lines of their spectra to shift toward the red (also known as the Doppler shift).

Reduction reaction: The gain of one or more electrons from an atom, molecule, or ion; in organic reactions, it is the gain of hydrogen or the loss of oxygen.

Reflector telescope: A telescope that works principally by using curved mirrors to focus light.

Refraction: The bending of light rays as they pass from a medium of one refractive index to a medium with a different refractive index.

Refractive index: The ratio of the speed of light in a vacuum to its speed in a particular material.

Relativistic: Refers to moving at high speeds that are comparable to the speed of light.

Resistance: The hindrance to the flow of electricity through a material; it depends upon the material, its size, and its shape.

Resistor: A device used in a circuit to create resistance.

Resolution: Spatially, the ability to distinguish clearly, according to some criterion, two or more closely separated objects; temporally, the discrimination of two or more nearly simultaneous events.

Resolving power: The capacity to show two nearby objects as separate.

Resonance: The enhanced response of any vibrating system to an external stimulus that has the same driving frequency as the natural frequency of the system.

Rest mass: The mass of an object when measured at rest with respect to an inertial reference frame.

Roche limit: The closest orbital distance a satellite that is held together by gravity can approach a planet or other body before the tidal forces of the larger body break it apart.

Salts: Historically, the compounds formed by reaction of an acid with a base in water; generically, any simple ionic compound, excluding the strong bases.

Scalar quantity: A quantity that has magnitude but not a direction, such as the nondimensional measurements of time or temperature.

Schrödinger equation: The fundamental equation of nonrelativistic quantum mechanics; its solution gives the wave function for a quantum system.

Second law of thermodynamics: One form of energy cannot be completely transformed into useful work; the energy conversion is always less than 100 percent efficient.

Seiche: A standing wave in an enclosed water body that does not progress forward but continues as a pendulum-like oscillation after the originating force has ceased.

Selection rules: A statement of how the quantum numbers may change as a quantum state makes a transition to a different quantum state.

Semiconductor: A material, usually a periodic group IVA element (for example, silicon, germanium) or an equivalent composite, that conducts energy at a level in between metal conductors and nonmetal resistors.

Seyfert galaxy: A spiral galaxy with an active core that is similar to a quasi-stellar object.

Shear stress: Tangential stress produced by a force acting along a surface of a solid body (rather than normal to it); shear stress is produced by parallel pairs of oppositely directed forces.

Shell models: Models which describe the basic properties of an atomic nucleus by ordering the protons and neutrons inside the nucleus or electrons in an atom in their most stable configurations.

Shock wave: A zone of compression and heating of matter traveling faster than the speed of sound in the material.

Signal-to-noise ratio (SNR): The ratio of the desired amplitude or energy from a signal to that from unwanted noise.

Snell's law: An equation that relates seismic or light velocity to angle of approach of a wave to a boundary; this relationship allows one to predict the angle of refraction, or bending, of a wave across a boundary.

Solar flares: Explosive outbursts of ionized gas from the sun.

Solar wind: A completely ionized plasma composed mainly of hydrogen that originates in the sun and streams outward at a velocity of about 400 kilometers per second.

Solubility: The maximum amount of a substance (solute) that will dissolve in a medium (solvent).

Solute: A substance dissolved in solution.

Solvent: That which dissolves or can dissolve another substance.

Sonar: An acronym for sound navigation ranging; an underwater sensing technology, originally designed for military purposes, utilizing sound pulses in water.

Sonic boom: A loud, transient explosive sound caused by a shock wave preceding an object traveling at supersonic speeds.

Special theory of relativity: Albert Einstein's theory unifying the concepts of space and time; as a consequence, matter and energy become equivalent according to the relation $E = mc^2$.

Specific heat: The amount of energy required to raise the temperature of a fixed amount of a substance by 1 degree Celsius, generally expressed in joules per mole per degree Celsius.

Spectral classes: An arrangement of stars using the letters O, B, A, F, G, K, and M, from hot blue stars to dim red ones: the sun is a G star, with a surface temperature of about 5,000 Kelvins.

Spectroscopy: The technique of producing spectra and analyzing their constituent wavelengths to determine quantities such as the chemical composition, density, and temperature of the emitting object.

Spectrum: The entire range of electromagnetic radiation from long-wavelength radio waves to short-wavelength gamma rays; also, a limited range of wavelengths in which an instrument separates the component elements.

Specular reflection: A coherent, mirrorlike change in the direction of light or electromagnetic rays when they strike a smooth surface.

Spin: A fundamental quantum property; although it has an obvious classical analog, it can be truly understood only in terms of relativistic quantum mechanics.

Sputtering: The bombardment of solid surfaces by high-energy particles, such that atoms and molecular fragments are eroded from the surface and the surface chemistry is altered.

Starch: A carbohydrate, or polysaccharide; a common polymer of glucose monomers that is generated within plant cells as a reserve of energy.

Statics: The study of the state of rest of a system—that is, static equilibrium.

Stefan-Boltzmann law: A relationship between the temperature of a "black body" and the rate at which it radiates energy.

Stimulated emission: The process responsible for emission in lasers in which all the excited molecules emit radiation of identical energy in the same direction.

Straggling: The statistical variation of the path length for monoenergetic particles in going through a material.

Strain: The change in form produced by external forces acting on a nonrigid, solid body; strains may lengthen (tensile strain), shorten (compressive strain), or change the angle between adjacent faces (shear strain) of a body.

Strangeness: A quantum mechanical property of baryons and mesons which is conserved in strong nuclear interactions but not in weak interactions.

Strong interaction: The force that binds the protons and neutrons in nuclei; on a more fundamental level, the forces between quarks.

Sublimation: A phase change in which particles of matter shift from the solid phase to the vapor phase without passing through a liquid phase.

Superconductor: A material that has zero resistance to a flow of current, which occurs when certain materials are cooled to very low temperatures, such as a few degrees above absolute zero; superconducting wires can carry current without any loss or through heating, which occurs in normal conductors.

Supercooled liquid: A liquid that has a temperature below its melting point but remains a liquid.

Superfluid: A fluid that flows without friction, as exemplified by liquid helium.

Superheated liquid: A liquid which is maintained at a temperature and pressure which are higher than the pressure for that liquid's boiling point at that temperature and for which the pressure is suddenly decreased.

Supernova: The final phase of a massive star's life, in which it explodes, thereby releasing most of its mass and energy, followed by collapse into a neutron star or black hole.

Supersaturated solution: An unstable solution in which there is more solute dissolved in the solvent than it can hold at a given temperature.

Supersonic: Having a velocity greater than the speed of sound in a given medium.

Superstring theory: A theoretical proposal that defines elementary particles as tiny strings in ten dimensions.

Supersymmetry: A theory that extends the symmetry relationships of the standard model to the Planck scale.

Surface tension: The property of a liquid to adhere to capillary surfaces.

Susceptibility: Also called permeability; an intrinsic property of magnetic material that indicates the ease with which a material is magnetized in a magnetic field.

Symmetry: A mathematical or geometrical rule that describes the relationship among points in an object or pattern.

T-Tauri stage: A temporary stage of instability in the early life of a main sequence star when it experiences great mass loss and intense solar wind.

Tauon: A charged lepton with a mass about thirty-five hundred times the electron mass, which decays into a tau-neutrino plus either a muon and mu-neutrino or an electron and electron neutrino.

Taylor series: A representation of a function by an infinite sum whose coefficients involve the derivatives of the function evaluated at a given point.

Temperature: A measure of the magnitude of thermal energy contained in a quantity of matter.

Tesla: The unit of measurement of magnetic fields; one tesla equals one newton (coulomb times meter per second).

Theorem: A property or relationship that can be logically derived from an axiom and/or a previously established theorem.

Thermal energy: The energy of molecular motion possessed by objects at temperatures above absolute zero; any material containing heat is composed of units that move or vibrate randomly.

Thermodynamics: The branch of science dealing with heat, work, and the thermal properties of matter.

Tidal forces: The gravitational attraction exerted between two bodies that may potentially distort their shape and cause internal deformation and heating.

Time dilation: The slowing down of a moving clock relative to a clock which is stationary with respect to an observer.

Torque: A force applied to an object so as to cause a rotation.

Trajectory: The path followed by a moving particle.

Transducer: Any device or instrument that converts one type of energy into another.

Transmission electron microscope (TEM): A magnifying tool that uses a beam of electrons to penetrate thin specimens; an image is formed by the electrons that go through the object.

Transverse wave: A wave in which the vibrational direction of the medium is at right angles to the direction of wave motion; also called shear wave.

Triple point: A unique temperature where a substance's solid, liquid, and vapor phases all exist together in equilibrium.

Turbulence: A thermodynamic term used to describe chaotic deviations from the normal, smooth periodic behavior of a process.

Ultraviolet light: Electromagnetic radiation in the wavelength range from 4×10^{-7} meter to 2×10^{-8} meter; occurs between visible light and X rays.

Uncertainty principle: It is impossible, in principle, to measure precisely certain pairs of properties, such as position and momentum, simultaneously.

Unit cell: The smallest three-dimensional geometric unit of a crystal that can create the complete crystal structure through repetition.

Vacancy: An irregularity in a crystal lattice that occurs when a site normally occupied by an atom or ion is unoccupied.

Valence electrons: The relatively loosely held electrons associated with an atom that participate in the chemical bonding process; two valence electrons form each bond in a molecule.

Van Allen radiation belts: Belts of electrons and protons trapped within the earth's magnetic field.

Van der Waals force: A force between molecules arising from instantaneous electrical polarization of molecules.

Vapor pressure: The pressure exerted by the vapors of a volatile substance at equilibrium and at a specific temperature, usually room temperature.

Vector: Geometrically, a mathematical quantity having magnitude and direction; can be abstracted to include physical or mathematical spaces of more than three dimensions.

Velocity: The speed and direction of a moving object.

Virtual image: An image that can be seen only by looking directly into a lens or into the rays that are reflected by a mirror.

Viscosity: The measure of a fluid's resistance to shearing forces, that is, those forces that deform it, measured in kilograms per meters times seconds.

Volatile: Undergoing phase changes (such as vaporization) at relatively low temperatures.

Voltage: A measure of electrical potential (potential energy per unit charge) across a circuit element that is usually measured in units of volts; associated with the electrical forces that drive currents through an electrical circuit.

Volume: The amount of space occupied by a substance.

Wave function: A mathematical function representing the state of the system; the square of the amplitude of this function gives the probability of finding the system in a particular state. The square of the amplitude of a particle's wave function gives the probability of finding a particle in a particular state.

Wavelength: The distance between successive crests or compressions of a wave, whether a wave of water or a wave of electromagnetic radiation.

Weak interaction: One of the four fundamental forces, which is responsible for radioactive beta decay of nuclei.

White dwarf: A low-mass star at the end of its life; typically, it has exhausted all of its thermonuclear fuel and contracted to a size roughly the same as that of Earth.

Work: A form of energy transfer generated by a force acting through a finite displacement.

World line: A curve in space-time representing all events in the history of a particle; world lines of particles with mass are timelike.

X ray: A short-wavelength, high-energy photon of electromagnetic radiation used in radiology because of its high tissue-penetrating power.

Zero point energy: The energy retained by a substance even at the lowest temperatures; the existence of such an energy was predicted by quantum theory.

Zeroth law of thermodynamics: The observation that two bodies, each in thermal equilibrium with a third body, are in thermal equilibrium with each other.

MAGILL'S SURVEY OF SCIENCE

ALPHABETICAL LIST

CATEGORY LIST

MAGILL'S SURVEY OF SCIENCE

INDEX

Page ranges appearing in boldface type indicate that an entire article devoted to the topic appears on those pages; a single page number in bold denotes definition of a term in the Glossary.